Ethnobotany in the New Europe

Environmental Anthropology and Ethnobiology

General Editor: **Roy Ellen**, FBA
Professor of Anthropology, University of Kent at Canterbury
Interest in environmental anthropology has grown steadily in recent years, reflecting national and international concern about the environment and developing research priorities. This major new international series, which continues a series first published by Harwood and Routledge, is a vehicle for publishing up-to-date monographs and edited works on particular issues, themes, places or peoples which focus on the interrelationship between society, culture and environment.

Volume 1
The Logic of Environmentalism: Anthropology, Ecology and Postcoloniality
Vassos Argyrou

Volume 2
Conversations on the Beach: Fishermen's Knowledge, Metaphor and Environmental Change in South India
Götz Hoeppe

Volume 3
Green Encounters: Shaping and Contesting Environmentalism in Rural Costa Rica
Luis A. Vivanco

Volume 4
Local Science vs Global Science: Approaches to Indigenous Knowledge in International Development
Edited by Paul Sillitoe

Volume 5
Sustainability and Communities of Place
Carl A. Maida

Volume 6
Modern Crises and Traditional Strategies: Local Ecological Knowledge in Island Southeast Asia
Edited by Roy Ellen

Volume 7
Traveling Cultures and Plants: The Ethnobiology and Ethnophamacy of Migrations
Edited by Andrea Pieroni and Ina Vandebroek

Volume 8
Fishers and Scientists in Modern Turkey: The Management of Natural Resources, Knowledge and Identity on the Eastern Black Sea Coast
Ståle Knudsen

Volume 9
Landscape Ethnocology: Concepts of Biotic and Physical Space
Leslie Main Johnson and Eugene S. Hunn

Volume 10
Landscape, Process and Power: Re-evaluating Traditional Environmental Knowledge
Edited by Serena Heckler

Volume 11
Mobility and Migration in Indigenous Amazonia: Contemporary Ethnoecological Perspectives
Edited by Miguel N. Alexiades

Volume 12
Unveiling the Whale: Discourses on Whales and Whaling
Arne Kalland

Volume 13
Virtualism, Governance and Practice: Vision and Execution in Environmental Conservation
Edited by James G. Carrier and Paige West

Volume 14
Ethnobotany in the New Europe: People, Health and Wild Plant Resources
Edited by Manuel Pardo-de-Santayana, Andrea Pieroni and Rajindra K. Puri

Volume 15
Urban Pollution: Cultural Meanings, Social Practices
Edited by Eveline Dürr and Rivke Jaffe

Volume 16
Weathering the World: Recovery in the Wake of the Tsunami in a Tamil Fishing Village
Frida Hastrup

Volume 17
Ethnobotany in the New Europe: People, Health and Wild Plant Resources
Edited by Manuel Pardo-de-Santayana, Andrea Pieroni and Rajindra K. Puri

Volume 18
Things Fall Apart? The Political Ecology of Forest Governance in Southern Nigeria
Pauline von Hellermann

Ethnobotany in the New Europe

People, Health and Wild Plant Resources

Edited by
Manuel Pardo-de-Santayana,
Andrea Pieroni and Rajindra K. Puri

berghahn
NEW YORK • OXFORD
www.berghahnbooks.com

First published in 2010 by

Berghahn Books
www.berghahnbooks.com

First paperback edition published in 2013

Library of Congress Cataloging-in-Publication Data

Ethnobotany in the new Europe: people, health and wild plant resources/ edited by Manuel Pardo-de-Santayana, Andrea Pieroni and Rajindra K. Puri.
pages cm. — (Environmental anthropology and ethnobiology ; volume 14)
ISBN 978-1-84545-456-2 (hardback) — ISBN 978-1-78238-124-2 (paperback) — ISBN 978-1-78238-125-9 (ebook)
1. Human-plant relationships—Europe. 2.Ethnobotany—Europe. 3. Medicinal plants—Europe. 4. Plants and civilization—Europe. 5. Plants, Medicinal—Europe. 6. Plants, Edible—Europe.
QK46.5.H85 E84 2010
581.6/3094

2010006436

British Library Cataloguing in Publication Data

A catalogue record for this book is available from the British Library

Printed in the United States on acid-free paper.

ISBN: 978-1-78238-124-2 paperback ISBN: 978-1-78238-125-9 retail ebook

Contents

List of Figures

List of Tables

List of Appendices

CHAPTER 1

The Ethnobotany of Europe, Past and Present

MANUEL PARDO-DE-SANTAYANA, ANDREA PIERONI AND RAJINDRA K. PURI

This book reports on an old and venerable discipline, the study of European wild food plants and herbal medicines, invigorated by a new generation of researchers pursuing modern ethnobotanical studies in new contexts. It offers new insights into the past and contemporary uses of wild plant resources, which despite decades in decline still play an important role for many rural communities. Recently, some of these wild plants and the practices associated with them have received renewed attention as symbols of local identities – or forms of intangible cultural heritage[1] – perceived to be under threat or as new resources for local economic growth. However, the future of these traditions is uncertain, as some are not practised any more and for others the resources themselves are under pressure due to continuing expansion and intensification of human environments. An important theme to emerge from these studies is the need for new theoretical and practical approaches that link the revaluation of plant-based cultural heritage with the conservation and use of biocultural diversity.

This book bridges biological and social science disciplines such as medicine, food science, human ecology, environmental science, history, anthropology and linguistics, and is intended to benchmark the development of the subject, for scientists and scholars active in the field, for those who make and implement policy, and generally for all those with an interest in biocultural diversity issues. Being at the interface of these various disciplinary perspectives, the researchers have made use of a variety of methods for obtaining information. Most of the data were provided by personal interviews and observations, but folk songs, historical texts, ethnographies and literature were also surveyed and analysed.

The authors and studies presented here reflect work being conducted in many European regions, including Portugal, Albania, Norway and Malta, and provide an overview of current ongoing field studies in Europe. Highlighting the rich diversity of cultural traditions still found here, the findings demonstrate both the common European heritage of folk knowledge on wild and cultivated plants and the diversity of local knowledge found across Europe's many areas. These studies tell the story of the ongoing evolution of human–plant relations in Europe, one of the most bioculturally dynamic places on the planet.

This dynamism derives in part from a long history of interaction among Europe's forty-five countries, city-states and principalities, which contain a quarter of the world's population living on less than 7 per cent of its land, but speaking 239 languages (Gordon 2005). Language groups are further subdivided into regional dialects, and the unique embedding of local cultural heritage and specific ways of perception and management of natural resources have generated myriad 'senses of place' (what in France is called '*terroir*';[2] see Bérard et al. 2005). Europe's incredible diversity is in part due to the geographically fragmented nature of the continent – separated by high mountain ranges and seas, and with only rivers to unite particular regions – and the multiple historical trajectories of tribes, kingdoms, empires and nation-states that have been battling for control of regions, or indeed the whole continent, for thousands of years (Diamond 1998; Llobera 2004; Stacul, Moutsou and Kopnina 2005). Such geographical, linguistic and historical richness has led to a multitude of ecological conditions, agroecosystems, cultures and ethnobotanical traditions.

Europe is therefore considered a crossroads of civilization, where human migrations and displacements have played a major role not only today but also historically, and these exchanges of people have led to a constant exchange of ideas, customs and knowledge (Rietbergen 1998; Stacul, Moutsou and Kopnina 2005). These old population movements are reflected in many ethnic, linguistic and religious minorities which still survive today with their own characteristic knowledge systems, of which ethnobiological studies have investigated only a very few, such as the descendents of Greeks living in Calabria, Italy, investigated by Sabine Nebel and Michael Heinrich in chapter 8; the Albanian descendents/ Arbëreshë of Lucania, also of southern Italy (Pieroni et al. 2002); and the old descendents of Romanian-speaking populations living in the Croatian northern part of Istria (Pieroni et al. 2003).

Ethnobotanical Studies in Europe: Past and Present

The history of the study of useful European plants dates back to ancient Greek times. One of the earliest works is *De Materia Medica*, published in AD 77 by the Greek surgeon Pedanius Dioscorides of Anazarbus, in which he compiled information about the use of six hundred plants in the Mediterranean. Later, from Medieval and Renaissance periods to the nineteenth century, scholars and explorers continued collecting and describing the indigenous uses of plants worldwide. For instance, the Swedish botanist Linnaeus, the founder of modern scientific botany, also published books such as *Flora Lapponica*, where he included not only plants of Lapland but also their local uses (Linnaeus 1737). Later, modern botanical and medical science itself was built on studies of Medieval Europeans' use of the food plants and medicinal herbs that graced the tables of both nobles and peasants (Atran 1990).

Since the nineteenth century, folklore studies in Central and Northern Europe have occasionally focused on traditional uses of plants (e.g., Marzell 1938; Butura 1979; Pettersson, Svanberg and Tunón 2001; De Cleene and Lejeune 2003; Allen and Hatfield 2004; Tunón, Pettersson and Iwarsson 2005) or the ethnolinguistics of useful plants (Marzell 1943; Borza 1968; Sejdiu 1984; Sella 1992; the last two referring to comprehensive works conducted in Kosovo and Albania, and North-western Italy, respectively).

While the development of ethnobiology and ethnobotany as interdisciplinary subjects is relatively recent in Europe, modern ethnobotanical studies focused on European territories have been growing very quickly, especially in southern European countries such as Italy and Spain. Moreover, the discipline is now turning its attention to long neglected regions such as the Balkans (Pieroni in chapter 2; Redzic 2006) and the East, including Poland, Lithuania, Romania and Bulgaria (de Boer in chapter 5; Bernáth 1999; Kathe, Honnef and Heym 2003; Ploetz and Orr 2004; Šeškauskaitė and Gliwa 2006; Łuczaj and Szymański 2007).

Many researchers in this book have linked the present use of plants to their historical roots, usually by studying the continuity of popular plant names and uses in archival material and literature, but also more recently through historical linguistic analysis of popular names (Pardo-de-Santayana, Blanco and Morales 2005; Nebel, Pieroni and Heinrich 2006). For instance, Torbjørn Alm and Marianne Iversen's study of the history of the use of *Rhododendron tomentosum* Harmaja by Sami in Norway found continuity in vernacular names and medicinal uses from the early eighteenth to the twentieth centuries, with only a loss in use as a salt substitute (see chapter 13). The study of cognates to local plant names give us clues to the historical relationship between cultures, while the meaning of many plant names reveals their local uses and perceptions (Pardo-de-Santayana 2008). Sabine Nebel's comparison of names for edible greens among *Grecanico* speakers in Calabria (Italy) and Ancient and Modern Greek literature shows remarkable continuity of language and traditions. For example, *Portulaca oleracea* L. (purslane) is called *andrácla* in Gallicianò and *andrakla* in Greece. The uses of many of these wild plants are, in effect, living relics of ancient Greek culture (see chapter 8). Manuel Pardo-de-Santayana and Ramón Morales also use an historical-linguistic approach to link the Spanish use of plants known as *manzanilla* (chamomile) in drinking infusions back through the ages to Moorish practices in the twelfth and thirteenth centuries, and even further back to Dioscorides in ancient Greece (see chapter 14). Daiva Šeškauskaitė and Bernd Gliwa present a rare glimpse into Lithuanian ethnobotanical classification by tracing and indeed unravelling the origins of cognate local names for sycamore maple (*Acer pseudoplatanus* L.), plane tree (*Platanus* spp.), black poplar (*Populus nigra* L.), guelder rose (*Viburnum opulus* L.) and sacred wreaths made from harvested rye. They demonstrate the value, and dangers, of using folk texts and ethnographic data, such as songs, riddles and children's verses, as ethnobotanical evidence for reconstructing the etymology and symbolic history of botanical nomenclature (see chapter 12). Timothy Tabone found that the Maltese shock/fright–jaundice syndrome seems to have resulted from syncretism of the South Italian *mal d'arco* and the Spanish *susto*, probably a legacy of the centuries when these territories were under Spanish control (see chapter 4).

Some researchers focus on the contemporary uses of wild plants, not just because of their continuity with past practices or re-emergence in new markets, but also because of their important dietary functions. In general, wild greens are nutritious due to their high content of minerals and vitamins (Ansari et al. 2005; Pardo-de-Santayana et al. 2007). Maria Barão and Alexandra Soveral Dias (chapter 9) show that the consumption of common golden thistle (*Scolymus hispanicus* L.) among poor farmers in Alentejo, Portugal, has a long history, also stretching back to ancient Roman times, and has now become popular among tourists. Underlying the use of this particular thistle, though, is the fact that it manages to maintain its high nutritional value regardless of the quality of the soils in which it grows. Local farmers have recognized this uniqueness and thus ignore all other thistles that grow in the area.

Other social aspects such as gender relations are also of special interest across Europe. Although deep knowledge of wild greens is said to be characteristic of women in many countries (Howard 2003), to gather and prepare thistles in Portugal is a man's work (Barão and Soveral Dias, chapter 9). A very unusual example is provided by Andrea Pieroni about women who become men in the Albanian Alps: in this archaic form of transgenderism there is convergence of the ethnobiological knowledge of 'typical' men, concerning fodder and ethnoveterinary plants, wild fruits, and the ethnobotanical knowledge of women, concerning weedy food and medicinal plants (see chapter 2).

One of the main goals of these newer ethnobotanical studies has been to document the dynamics of traditional knowledge about plants primarily gathered by rural communities. This is a key part of European biocultural heritage, which due to migration from rural areas and many deep social, economic and cultural changes since the last world war, in the West, and the break up of the Soviet Bloc, in the East, has suffered significant erosion. In fact, most young people today prefer the new ways of life, and their lack of interest in traditional plant use has led to a loss of this rich heritage (Pardo-de-Santayana and Gómez Pellón 2003; Pieroni 2003; Vallès, Bonet and Agelet 2004).

The Dynamism of the European Ethnobotanical Heritage

Europe's folk botany has always been dynamic and changing. Consider, for example, all the new plants and plant products introduced by explorers, traders and colonizers during the 'Age of Discovery and Mercantile Capitalism' (Crosby 1972). Many of these, such as the tomato, the capsicum, the potato and beans, have since achieved a kind of culinary keystone status for the cuisines of Europe, and at a more general level have come to symbolize these cultures (Fernández Pérez and González Tascón 1990). In spite of such monumental changes, many communities continued to hold on to old recipes and traditions, while others adopted and enculturated these exotic plants and remedies into their diets and pharmacopoeias in new and creative ways (Teti 1995; Nabhan 2004). Now, in the twenty-first century, in the age of the European Union (EU) and globalization, European folk botany is once again dynamically responding to changing economic, political and cultural contexts.

Widespread socioeconomic changes – modernization, industrialization, mechanization of agriculture – beginning in post-Second World War reconstruction across Europe, and following the dissolution of the Soviet Bloc in the east, have led to radical transformations in the lifestyle of rural societies (Abrahams 1996), which often relied on knowledge of plants to secure many of their basic needs (Gómez Pellón 2004). Accompanying this shift, from a rural, agriculturally-based, subsistence economy to a market-oriented one, has been a rapid erosion of ethnobotanical knowledge (Pardo-de-Santayana and Gómez Pellón 2003), and practices which many of the authors have described and endeavoured to explain for their particular field sites.

Some of this erosion is due to the simple fact that there are fewer farmers. Across Europe, pensions, tourism income and EU or member-state subsidies have become the main sources of income for rural regions

(Pérez Díaz 1996–2003; López Pérez 2003; Psaltopoulos, Balamou and Thomson 2006). This has led to less dependence on wild plants for food and medicine, and also less direct contact with nature, so many of the species are not gathered any more, or at best only seldom. In fact, several of the plant-use traditions described in this book are no longer practised, or persist only in the memory of the elderly. Those who do still collect wild plants often have less time to do so and thus cannot range as far as their parents or grandparents might have in the past. Furthermore, many of the species once collected are now difficult to find due to modifications of habitat, such as watercress in Spain (*Rorippa nasturtium-aquaticum* (L.) Hayek, syn: *Nasturtium officinale* R. Br. in W.T Aiton). Anja Christanell and colleagues discovered in Austria that species too labour intensive to process or difficult to find are usually rejected (see chapter 3). Exacerbating the problem is a concomitant rejection of communal social institutions that once bound local communities together and insured transmission of traditional botanical knowledge (Gómez Pellón 2004).

Changes in culture – shared beliefs, values and meanings of plants and plant traditions – are also responsible for changes in gathering practices, as when wild edible plants come to be considered as symbols of poverty or backwardness, often because of their importance during times of food scarcity (González Turmo 1997). Many authors demonstrate how modernization downgrades and devalues wild resources, especially among the youth who are very conscious of fitting into the new, modern Europe (see Christanell et al., chapter 3; Tardío, chapter 10; Carvalho and Morales, chapter 7).

Despite all these changes, continuity in plant use across Europe can sometimes be startling. Globalization may be making Europe smaller, in terms of faster communication and reduced travel times, and the EU may be attempting to unify and streamline economic and political systems (see Stacul, Moutsou and Kopnina 2005), particularly those of Eastern European and former communist countries, but that does not imply a necessary homogenization of culture (Llobera 2004; Vaishar and Greer-Wootten 2006), or, in this case, plant use. In fact, EU policy supports decentralized 'regionalism' within nations, through the EU Common Agricultural Policy (CAP) and Rural Development schemes (Nogués 2004; EU 2006a, b). In the face of increased global competition for tourists and other markets, regional identities, characterized by regional foods, music, artefacts and products, are seen by some politicians and businessmen as critical marketing tools for local economic growth (Tellstrom, Gustafsson and Mossberg 2005).

Although most chapters describe declining gathering practices of food and medicinal plants, some of these practices are not only not disappearing: they are becoming more popular. This is often a result of the new regionalism and the accompanying tourism that demands local authenticity in food, wine, architecture and even landscape. Across Europe's many small markets, numerous local, plant-based products are appearing. For instance, there are infusions, such as rock tea (*Jasonia glutinosa* (L.) DC.) in Spain, elderflower wine (*Sambucus nigra* L.) in Central European countries such as Germany, Austria and Slovenia (A. Pieroni, pers. observ.), and gourmet liqueurs and marmalades made from wild fruits such as elderberries (*Sambucus nigra*), blackberries (*Rubus* spp.), blackthorn berries (*Prunus spinosa* L.) or wild apples (*Malus sylvestris* (L.) Mill.) (e.g., Bonet and Vallès 2002; Pardo-de-Santayana, Blanco and Morales 2005; Pardo-de-Santayana, Tardío and Morales 2005; Pieroni et al. 2005). In restaurants and cafés one can find salads and other dishes made from commonly gathered wild greens such as wild chicory (*Cichorium intybus* L.), wild asparagus (*Asparagus acutifolius* L. and other *Asparagus* species), bladder campion (*Silene vulgaris* (Moench) Garcke) and wall rocket (*Diplotaxis muralis* (L.) DC.) (Picchi and Pieroni 2005; Tardío, Pardo-de-Santayana and Morales 2006). The common golden thistle (*Scolymus hispanicus*) is now in great demand in Spain and Portugal (see Barão and Dias, chapter 9; Tardío, chapter 10; Carvalho and Morales, chapter 7.)

Economic diversification has been a good strategy for mountainous and remote areas where it is too risky to be specialized in only one resource (e.g., Andersson and Ngazi 1998). A combination of some cash income from activities such as farming, selling by-products like cheese and jam, or providing beds for tourists and other off-farm labour, and subsidies, has become a successful strategy across Europe today (Van Lier 2000).

On the other hand, political and other crises may have the effect of increasing dependence on wild foods. This appears to be especially true in the post-communist countries, which can be said to have been in transition and in some cases in crisis since the end of the Cold War (Ekström et al. 2003). Elsewhere, the war in the former Yugoslavian countries, or the collapse of the state health system in Albania or Bulgaria (see chapters 2 and 5), for instance, have pushed people to use many of their wild resources that had been previously abandoned. Not only does the use of wild plants prevail, but also the ideas, concepts and beliefs about illness and remedies that underpin these uses are maintained. This is the case with the Doctrine of Signatures, still prevalent in the Albanian population surveyed by Pieroni and his research team (see chapter 2).

There are also cultural reasons for increased attention to some wild plant resources. Along with an emphasis on developing regional economies, or perhaps because of it, regional identities have also grown in strength and importance. Since local and regional identities are always in part composed of natural symbols, it is not surprising that wild plant products and the shared knowledge and values surrounding them would also attract more attention as regional identities began to be asserted (Wu 2003). In Scotland, clootie trees, once worshipped, have become tourist attractions precisely because they are emblematic (see chapter 11), and in Austria the blessing of flowers in Catholic celebrations has also been popularized for tourists (see chapter 3).

Identity markers often arise to maintain social boundaries (Barth 1969; Cohen 2000), and so wild plants and their uses also can persist when they symbolically distinguish two competing or entangled peoples or regions. For instance, Alm and Iverson describe how Norwegians consider *Rhododendrum tomentosum* a pleasant, scented plant and call it with the borrowed name *rosmarin*, equating it with the herb rosemary, while Sami people say that it smells very bad. The authors suggest that this distinction helps to highlight ethnic differences and maintains the plant tradition as well (see chapter 13). Thus the maintenance or revival of plant use, the reassertion of regional and cultural identities, and even the renewed interest in ethnobotany, have emerged together in many European countries.

Many of the studies in this book clearly demonstrate that the reasons for still gathering wild plants are rarely ever entirely economic. People have emotional reasons, such as a love of nature, a desire to conserve an old tradition or a way of remembering their parents; social reasons, such as the obligation to give gifts, share or barter products with friends and relatives; gastronomical reasons, such as the enjoyment of homemade delicacies; or health reasons, gained from the supposed healthiness of wild and self-cultivated plants and a preference for self-medication. Finally, earning a small amount of money on the side can be a motivator to maintain wild plant use (see Christanell et al., chapter 3), while Christine Wildhaber reports that cultivating organic vegetables in allotment gardens can be a way to save money (see chapter 16).

Studies on European homegardens have also shown the importance of environmental, health, educational, emotional and recreational reasons for taking care of a garden. An activity that was traditionally for obtaining food now has multiple functions and is thus receiving more attention from European researchers (Vogl, Vogl-Lukasser and Puri 2004; Buckingham 2005; British Homegardens Project 2008; Wildhaber, chapter 16).

European Ethnobotany in the Future

The authors and research reported in this book only begin to scratch the surface of what is happening in Europe today, in terms of the variation of changing human–environment relationships that involve the use of wild and medicinal plants, and the techniques and methods being developed by researchers to document and explain these new relationships. Much more research needs to be conducted to cover the vast array of experiments in living being carried out on hundreds of farms, among small communities and even in urban neighbourhoods across the continent. Research may reveal patterns of change and new innovations, and even serve as conduits to link all these variable areas.

One of the key areas for ethnobiological research today is the interaction between autochthonous populations and newcomers, often from other parts of Europe but increasingly from other continents. Researchers are interested to know what happens to the traditional knowledge and practices of migrants when they settle in new ecological and cultural contexts. Do migrating people bring their plants with them, and does that go some way to alleviating the stress of moving or the unfamiliarity of a new home? Do they have to rely on new plants to maintain old traditions, and if so how do they choose these new plants? Or do they create trade networks and establish new markets to provide traditional plant resources? Are traditions hybridized or just lost over time? Simultaneously, migrant groups living in Europe face varying difficulties in maintaining and transmitting their traditional practices to new generations and this raises very relevant issues for public health and nutrition policies. Answering these questions will go a long way toward better understanding the dynamics of ethnobotanical knowledge systems as well as the importance of the environment for migrants more generally (e.g., Pieroni and Vandebroek 2007; Pieroni et al., chapter 6).

Finally, in the context of Europe's dynamic past and present, the sustainable use of plant resources into the future is a common interest of many of the authors. Since many of these are wild plants, their conservation and sustainable use is problematic. With land being squeezed for expanding cities, housing, roads and pasture, where are wild plants going to survive? Who controls wild plants? Can public policies regulate these resources?

In Bulgaria, Hugo de Boer reports that quota systems for regulating medicinal plant collection by professional harvesters have shown promising initial results. However, since many of the species gathered are locally abundant and easily accessible, harvesters are often unaware of the risk of overharvesting the more rare species. Identifying local specialists as key informants has been shown to be valuable for detecting local declines in the more uncommon medicinal plants (see chapter 5).

It turns out that many of the wild plants studied are in fact found in managed areas, and their status as wild is now questioned by some (Van den Eynden 2004). We must take into consideration that there is a gradient between plants that grow wild and those that are cultivated. Some wild plants may be tolerated in gardens or fields, lightly promoted through weeding out competitors, managed more heavily through pruning, or finally transplanted into better conditions. In fact, the role of homegardens in increasing biodiversity, including agrobiodiversity, needs to be more seriously considered, both in terms of the potential benefits to the farmer and the effect it has on regional levels of biodiversity and ecosystem services that benefit the wider public (Eyzaguirre and Linares 2001; British Homegardens Project 2008: see Wildhaber, chapter 16). This is the case of trees or shrubs, which grow in homegardens or on village lands, and thorny bushes, which typically grow in hedgerows that mark boundaries. Finally, other taxa are grown in meadows and receive special inputs such as natural fertilizers (see McCune, chapter 15). Thus wild plants appear to be found in a variety of habitats, some more anthropogenic than others.

The status of wild versus cultivated is critical, because tenure is often closely tied to management, and concepts of ownership have been changing across Europe for more than a decade now, especially in the Eastern countries (Abrahams 1996; Ortega Valcárcel 2004). Trees, for instance, in many regions have an owner, except for those growing far away from villages or cattle-grazing grounds. These single or communal owners are responsible for the planting, protection, grafting, pruning and exploitation of their wood (San Miguel 2004). Recognizing tenure, individual or communal, and thus responsibility for plants in both legal and social contexts, will be critical in promoting conservation and development initiatives in Europe in the coming years.

Jenny L. McCune's chapter focuses on the interest of using ethnobiological tools in the study of grassland management by livestock farmers and its relation with the conservation status of these environments. She suggests that they have deep, site-specific knowledge of grassland flora and animal fodder species that can greatly assist in conservation efforts of state agencies (see chapter 15). There are other cases where overharvesting may in fact be problematic for certain species, such as *sahlep* in Albania (Pieroni, chapter 2), *Artemisia granatensis* Boiss. in Spain (Pardo-de-Santayana and Morales, chapter 14), medicinal plants in Bulgaria (de Boer, chapter 5), or mushrooms in general (Christanell et al., chapter 3). On the other hand, people sometimes cultivate or transplant wild herbs that are scarce or threatened into homegardens to avoid over-exploitation.

Across Europe, the related fields of economic botany, ethnobotany, ethnopharmacology, food anthropology, agriculture and organic farming are emerging as important and overlapping endeavours with unique resources at their disposal: old botanic gardens and plant collections, even older archives, new centres for research and public awareness (such as the Eden Project), and a variety of academic institutions with growing interdisciplinary and often international programmes.[3]

We offer here an initial glimpse into an exciting and growing field of European ethnobotany, and a call for scientists and students to join us in unravelling a small part of this grand experiment that is Europe in the twenty-first century, this dynamic diversity so characteristic of Europe's past, present and probable future.

Notes

1. The UNESCO Convention for the Safeguarding of Intangible Cultural Heritage, signed at the 32nd Session of UNESCO in Paris on 17 October 2003, includes knowledge and practices concerning nature and the universe, ethnobiology and ethnosciences (Pieroni, Price and Vandebroek 2005). It is a crucial turning point for recognizing all orally transmitted traditional knowledge (TK) systems as an integral part of the worldwide cultural heritage that has to be protected and sustained.
2. *Terroir* is a French term, originally referring to the special characteristics of food production within a given, unique, biocultural locality. *Terroir* is considered the sum of the effects that the local environment and the immaterial heritage of the local culture has on the production, processing/ technology and manufacture of a specific food product.
3. There are programmes at Canterbury, Kent (Department of Anthropology), Vienna (Institute of Organic Farming, BOKU), Wageningen (Department of Social Sciences), Madrid (Universidad Autónoma de Madrid), Bra, Italy (International University of Gastronomic Sciences), and Uppsala (Department of Evolutionary Biology), to name but a few.

References

Abrahams, R. (ed.). 1996. *After Socialism: Land Reform and Social Change in Eastern Europe.* Oxford: Berghahn Books.

Allen, D.E. and G. Hatfield. 2004. *Medicinal Plants in Folk Traditions: An Ethnobotany of Britain & Ireland.* Portlans/Cambridge: Timber Press.

Andersson, J. and Z. Ngazi. 1998. 'Coastal Communities' Production Choices, Risk Diversification, and Subsistence Behavior: Responses in Periods of Transition – A Case Study From Tanzania', *Ambio* 27(8): 686–693.

Ansari, N.M., L. Houlihan, B. Hussain and A. Pieroni. 2005. 'Antioxidant Activity of Five Vegetables Traditionally Consumed by South-Asian Migrants in Bradford, Yorkshire, UK', *Phytotherapy Research* 19: 907–911.

Atran, S. 1990. *Cognitive Foundations of Natural History: Towards an Anthropology of Science.* Cambridge: Cambridge University Press.

Barth, F. 1969. *Ethnic Groups and Boundaries: The Social Organization of Cultural Difference.* Long Grove, Ill.: Waveland Press.

Bérard, L., M. Cegarra, M. Djama, S. Louafi, P. Marcheney, B. Rousell and F. Verdeaux. 2005. *Local Ecological Knowledge and Practice: an Original Approach in France.* Paris: IDDRI.

Bernáth, J. 1999. 'Biological and Economical Aspects of Utilization and Exploitation of Wild Growing Medicinal Plants in Middle- and South Europe', *Acta Horticulturae* (ISHS) 500, 31–42.

Bonet, M.A. and J. Vallès. 2002. 'Use of Non-crop Food Vascular Plants in Montseny Biosphere Reserve (Catalonia, Iberian Peninsula)', *International Journal of Food Sciences and Nutrition* 53: 225–248.

Borza, A. (ed.). 1968. *Dicţionar etnobotanic cumprizînd denumirile populare româneşti şi în alte limbe ale plantelor din România*. Bucharest: Academiei Republicii Socialiste România.

British Homegardens Project 2008. *The Ethnobotany of British Homegardens: Diversity, Knowledge and Exchange*. Retrieved 24 October 2008 from http://www.kent.ac.uk/anthropology/department/research/environmental/homegardens.html

Buckingham, S. 2005. 'Women (Re)construct the Plot: the Regen(d)eration of Urban Food Growing', *Area* 37(2): 171–179.

Butură, V. 1979. *Enciclopedie de etnobotanică românească*. Bucharest: Editura Ştiinţifică şi Enciclopedică.

Cohen, A.P. (ed.). 2000. *Signifying Identifies: Anthropological Perspectives on Boundaries and Contested Values*. London: Routledge.

Crosby, A.W. 1972. *The Columbian Exchange: Biological and Cultural Consequences of 1492*. Westport, Conn.: Greenwood Press.

De Cleene, M. and M.C. Lejeune. 2003. *Compendium of Symbolic and Ritual Plants in Europe*. Ghent: Man & Culture Publishers.

Diamond, J. 1998. *Guns, Germs and Steel: The Fates of Human Societies*. New York: W.W. Norton & Company.

Ekström, K.M., M.P. Ekström, M. Potapova and H. Shanahan. 2003. 'Changes in Food Provision in Russian Households Experiencing Perestroika', *International Journal of Consumer Studies* 27(4): 194–301.

EU (European Union) 2006a. *Community Strategic Guidelines for Rural Development. Council Decision of 20 February 2006*. EUROPA: SCADPlus. Retrieved 25 July 2007 from http:/europa.eu/scadplus/leg/en/lvb/160042.htm

——— 2006b. *General Provisions on the European Rural Development Fund, the European Social Fund, and the Cohesion Fund (2007–2013). Council Regulation (EC) No. 1083/2006 of 11 July 2006*. EUROPA: SCADPlus. Retrieved 15 July 2007 from http://europa.eu/scadplus/leg/en/lvb/g24231.htm

Eyzaguirre, P.B. and O.F. Linares. 2001. 'A New Approach to the Study and Promotion of Home Gardens', *People and Plants Handbook* 7: 30–33. WWF-UNESCO-RBG Kew.

Fernández Pérez, J. and I. González Tascón (eds). 1990. *La agricultura viajera: Cultivos y manufacturas de las plantas industriales y alimentarias en España y en la América Virreinal*. Madrid: Real Jardín Botánico, CSIC.

Gómez Pellón, E. 2004. 'A Rural World in Change: on Cultural Modernisation and New Colonisation', in S. Nogués (ed.), *The Future of Rural Areas*. Santander: Universidad de Cantabria, pp. 301–326.

González Turmo, I. 1997. *Comida de rico, comida de pobre: Los hábitos alimenticios en el Occidente andaluz (Siglo XX)*. Sevilla: Universidad de Sevilla.

Gordon, R.G., Jr. (ed.). 2005. *Ethnologue: Languages of the World*, 15th edn. Dallas, Tex.: SIL International. Online version at http://www.ethnologue.com

Howard, P.L. 2003. *Women and Plants: Case Studies on Gender Relations in Biodiversity Management and Conservation*. London: Zed Press.

Kathe,W., S. Honnef and A. Heym. 2003. *Medicinal and Aromatic Plants in Albania, Bosnia-Herzegovina, Bulgaria, Croatia and Romania*. Report by WWF Deutschland and TRAFFIC Europe-Germany. Bonn: German Federal Agency for Nature Conservation (BfN).

Linnaeus, C. 1737. *Flora Lapponica*. Amsterdam: Salomo Schouten.

Llobera, J. 2004. *Foundations of National Identity: from Catalonia to Europe*. Oxford: Berghahn Books.

López Pérez, P.M. 2003. 'La reforma del subsidio agrario: Una aproximación a la realidad rural extremeña', *Revista de Estudios Extremeños* 59(3): 1291–1322.

Łuczaj, Ł. and W. Szymański. 2007. 'Wild Vascular Plants Gathered for Consumption in the Polish Countryside: a Review', *Journal of Ethnobiology and Ethnomedicine* 3: 17.

Marzell, H. 1938. *Geschichte und Volkskunde der deutschen Heilpflanzen*. Stuttgart. Reprinted by Reichl, St. Goar, 2002.

——— 1943. *Wörterbuch der deutschen Pflanzennamen*. Volumes 1–5. Leipzig: Verlag von S. Hirzel.

Nabhan, G.P. 2004. *Why Some Like it Hot: Food, Genes, and Cultural Diversity*. Washington/Covelo/London: Island Press/Shearwater Books.

Nebel, S., A. Pieroni and M. Heinrich. 2006. '*Ta chòrta*: Wild Edible Greens Used in the Graecanic Area in Calabria, Southern Italy', *Appetite* 47(3): 333–342.

Nogués, S. (ed.). 2004. *The Future of Rural Areas*. Santander: Universidad de Cantabria.

Ortega Valcárcel, J. 2004. 'Rural Transition in Spain: The Country in Urbanised Society', in S. Nogués (ed.), *The Future of Rural Areas*. Santander: Universidad de Cantabria, pp. 89–114.

Pardo-de-Santayana, M. 2008. *Estudios etnobotánicos en Campoo (Cantabria)*. Madrid: Consejo Superior de Investigaciones Científicas.

———, E. Blanco and R. Morales. 2005. 'Plants Known as *té* in Spain: An Ethno-pharmaco-botanical Review', *Journal of Ethnopharmacology* 98(1–2): 1–19.

——— and E. Gómez Pellón. 2003. 'Etnobotánica: aprovechamiento tradicional de plantas y patrimonio cultural', *Anales del Jardín Botánico de Madrid* 60(1): 171–182.

———, J. Tardío, E. Blanco, A.M. Carvalho, J.J. Lastra, E. San Miguel and R. Morales. 2007. 'Traditional Knowledge of Wild Edible Plants Used in the Northwest of the Iberian Peninsula (Spain and Portugal): A Comparative Study', *Journal of Ethnobiology and Ethnomedicine* 3: 27.

———, J. Tardío and R. Morales. 2005. 'The Gathering and Consumption of Wild Edible Plants in Campoo (Cantabria, Spain)', *International Journal of Food Sciences and Nutrition* 56 (7): 529–542.

Pérez Díaz, A. 1996–2003. 'La política agraria común y la reconversión del campo extremeño', *Norba: Revista de Historia* 16(2): 685–699.

Pettersson, B., I. Svanberg and H. Tunón (eds). 2001. *Människan och naturen: Etnobiologi i Sverige, vol. 1.* Stockholm: Wahlström & Widstrand.

Picchi, G. and A. Pieroni. 2005. *Le Erbe: Atlante dei prodotti tipici.* Rome: INSOR, RAI, AGRA.

Pieroni, A. 2003. 'Wild Food Plants and Arbëresh Women in Lucania, Southern Italy', in P.L. Howard (ed.), *Women & Plants: Case Studies on Gender Relations in Biodiversity Management & Conservation.* London: Zed Press, pp. 66–82.

———, M.E. Giusti, H. Münz, C. Lenzarini, G. Turkovic and A. Turkovic. 2003. 'Ethnobotanical Knowledge of the Istro-Romanians of Žejane in Croatia', *Fitoterapia* 74: 710–719.

———, S. Nebel, C. Quave, H. Münz and M. Heinrich. 2002. 'Ethnopharmacology of Liakra, Traditional Weedy Vegetables of the Arbëreshë of the Vulture Area in Southern Italy', *Journal of Ethnopharmacology* 81: 165–185.

———, S. Nebel, R.F. Santoro and M. Heinrich. 2005. 'Food for Two Seasons: Culinary Uses of Non-cultivated Local Vegetables and Mushrooms in a South Italian Village', *International Journal of Food Sciences and Nutrition* 56(4): 245–272.

———, L. Price and I. Vandebroek. 2005. 'Welcome to the Journal of Ethnobiology and Ethnomedicine', *Journal of Ethnobiology and Ethnomedicine* 1: 1.

——— and I. Vandebroek (eds). 2007. *Travelling Cultures, Plants and Medicines: the Ethnobiology and Ethnopharmacy of Migrations.* Oxford: Berghahn Press.

Ploetz, K. and B. Orr. 2004. 'Wild Herb Use in Bulgaria', *Economic Botany* 58: 231–241.

Psaltopoulos, D., E. Balamou and K.J. Thomson. 2006. 'Rural-urban Impacts of CAP Measures in Greece: an Inter-regional SAM Approach', *Journal of Agricultural Economics* 57(3): 441–458.

Redzic, S. 2006. 'Wild Edible Plants and Their Traditional Use in the Human Nutrition in Bosnia-Herzegovina', *Ecology of Food and Nutrition* 45, 189–232.

Rietbergen, P. 1998. *Europe: A Cultural History.* London: Routledge.

San Miguel, E. 2004. *Etnobotánica de Piloña (Asturias): Cultura y saber popular sobre las plantas en un concejo del Centro-Oriente Asturiano*, Ph.D. dissertation. Madrid: Department of Biology, Universidad Autónoma de Madrid.

Sejdiu, S. 1984. *Fjalorth Etnobotanik i Shqipes.* Prishtinë: Rilindja.

Sella, A. 1992. *Flora Popolare Biellese.* Alessandria: Edizioni dell'Orso.

Šeškauskaitė, D. and B. Gliwa. 2006. 'Some Lithuanian Ethnobotanical Taxa: a Linguistic View on Thorn Apple and Related Plants', *Journal of Ethnobiology and Ethnomedicine* 2: 13.

Stacul, J., C. Moutsou and H. Kopnina (eds). 2005. *Crossing European Boundaries: Beyond Conventional Geographical Categories.* Oxford: Berghahn Books.

Tardío, J., M. Pardo-de-Santayana and R. Morales. 2006. 'Ethnobotanical Review of Wild Edible Plants in Spain', *Botanical Journal of the Linnean Society* 152: 27–72.

Tellstrom, R., I.B. Gustafsson and L. Mossberg. 2005. 'Local Food Cultures in the Swedish Rural Economy', *Sociologia Ruralis* 45(4): 346–359.

Teti, V. 1995. *Il Peperonicino: un Americano nel Mediterraneo*. Vibo Valentia: Monteleone Editore.

Tunón, H., B. Pettersson and M. Iwarsson (eds). 2005. *Människan och floran: Etnobiologi i Sverige, vol. 2*. Stockholm: Wahlström & Widstrand.

Vaishar, A. and B. Greer-Wootten. 2006. 'Sustainable Development in Morovia', in Z. Bochniarz and G.B. Cohen (eds), *The Environment and Sustainable Development in the New Central Europe*. Oxford: Berghahn Books, pp. 218–231.

Vallès, J., M.À. Bonet and A. Agelet. 2004. 'Ethnobotany of *Sambucus nigra* L. in Catalonia (Iberian Peninsula): the Integral Exploitation of a Natural Resource in Mountain Regions', *Economic Botany* 58(3): 456–469.

Van den Eynden, V. 2004. *Use and Management of Edible Non-crop Plants in Southern Ecuador*, Ph.D. dissertation. Gent: University of Gent.

Van Lier, H.N. 2000. 'Land Use Planning and Land Consolidation in the Future in Europe', *Zeitschrift fur Kulturtechnik und Landentwicklung* 41(3): 138–143.

Vogl, C.R., B. Vogl-Lukasser and R.K. Puri. 2004. 'Tools and Methods for Data Collection in Ethnobotanical Studies of Home Gardens', *Field Methods* 16(3): 285–306.

Wu, X. 2003. '"Turning Waste into Things of Value": Marketing Fern, Kudzu, and *Osmunda* in Enshi Prefecture, China', *Journal of Developing Societies* 19(4): 433–457.

CHAPTER 2

People and Plants in Lëpushë

Traditional Medicine, Local Foods and Post-communism in a Northern Albanian Village

ANDREA PIERONI

Introduction

This is the story of a village and its surrounding alpine settlements in a very remote, mountainous area in upper Kelmend, northern Albania. This story tells of the people who have remained in this place since the fall of the communist regime in 1991, either because it was so special to them that they did not want to leave, or because they could not find the means to migrate, either legally or illegally, to the United States or Western Europe, as most of the other inhabitants of northern Albania have done. It also tells of how the villagers have managed to provide their own healthcare as they continue living in the mountains, sustained by their cows, pigs, potatoes, cabbages and corn, while facing the dramatic collapse of their national economy and their institutionalized health and transport systems. This is the brief story of their traditional medicine (TM) and local foods, of their interactions with their natural environment where they gather wild plants for food and medicines, and of the gender relations prescribed by these traditional practices. It also tells of the dramatic changes this village has had to face since the end of the communist era. This story, like every story, expresses a personal view: it describes what I myself have observed and reflected on, and is therefore undoubtedly biased by my own subjectivity.

Albania

Albania is located in the Balkan Peninsula, and bordered by Montenegro and Kosova in the north and northeast, the former Yugoslav Republic of Macedonia in the east and Greece in the south. To the west are the Adriatic and Ionian seas. The country covers an area of 28,750 km^2 and apart from its flat coastline, is primarily mountainous. Albania's population is younger than that of other European countries, with one third of its 3.1 million inhabitants under the age of fifteen years, and 40 per cent younger than eighteen years (Nuri 2002). A high proportion of Albanians live in rural areas: 58 per cent in 2001 according to the Albanian National Institute of Statistics, INSTAT (quoted in World Bank 2003). However, since restrictions on the freedom of movement were lifted in the 1990s, there has been an unprecedented level of internal migration from rural to urban areas. The population of the capital, Tirana, for example, increased from 250,000 inhabitants in 1990 to an estimated one million in 2005. Many people have also left the country to migrate abroad – between 600,000 and 800,000 depending on various estimates (King and Vullnetari 2003). It appears that this emigration process is not yet slowing down, and large numbers of Albanians are still leaving, both legally and illegally.

About 97 per cent of the Albanian population is ethnic Albanian, 1.9 per cent are Greek, while other groups are represented in small numbers. Islam is the religion of 70 per cent of the population, while 20 per cent are Orthodox Christian, with 10 per cent Roman Catholic. These figures reflect the religion of origin, since at least up until now religion has not generally been a crucial identifying element in Albanian society. Nevertheless, with the return of religious freedom, many mosques and churches that were closed in 1967 have now been reopened, and I have the impression that the issue of religion is becoming increasingly central in building cultural identity, at least in the mountainous, rural areas in the north.

Historical Background

The ancestors of the Albanians, the Illyrians, preserved their own language and culture despite the establishment of Greek colonies in the seventh century BC, and subsequent centuries of Roman rule. Illyria became part of the Byzantine Empire in the division of the Roman Empire of AD 395. Migrating Slavic and Germanic groups invaded the region throughout the fifth and sixth centuries, while various neighbours contested control of the region. In 1344 the country was annexed by Serbia, which in turn was occupied by the Turks in 1389. A national Albanian hero, Skenderbeg, led the resistance opposing the Ottomans; nevertheless in 1479 Albania was finally incorporated into the Ottoman Empire, in which it remained a very poor, rural province for several centuries. Many Albanians at that time migrated to Southern Italy, where they still live today (approximately 100,000 persons) and they define themselves and are known as Arbëreshë. Nowadays these Arbëreshë have been largely assimilated into South Italian culture, even if a few of them still speak the Albanian language (for a medical anthropological field study, which lasted three years, conducted on the Arbëreshë of northern Lucania, see Pieroni and Quave 2005, 2006).

Albania achieved independence from the Ottoman Empire in 1912, while Kosova, which made up nearly half of Albania at the time, was transferred to Serbia at the 1913 peace conference. Albania was overrun by successive armies in the First World War and became a kingdom only in 1928, under King Zog I. Italy occupied the country during the Second World War, and King Zog fled to the United Kingdom.

The Albanian Communist Party, founded by Enver Hoxha, led the resistance against first the Italians and then the Germans. In January 1946, the People's Republic of Albania was proclaimed, and Hoxha became its president, remaining in power until his death in 1985. Albania initially followed Soviet-style economic policies, but in 1960/1961 it broke off diplomatic relations with the Soviet Union and aligned itself with China. In 1978 Albania isolated itself, politically and socially, from the rest of the world. After the collapse of communism in eastern Europe in the early 1990s, the Albanian population staged demonstrations, finally forcing the government to agree to allow opposition parties.

The governing Labour Party, led by Ramiz Alia, won the election in March 1991 after promising to privatize state land. The Democratic Party led by Sali Berisha won the election in March 1992. March 1997 saw the collapse of several pyramid savings schemes, in which perhaps two thirds of the population had invested money, with an estimated loss of $1,000 million (US). People blamed the government for complicity in the schemes, and widespread violence followed. The country's economic growth rate of the previous four years was reversed, inflation and unemployment rose, and economic recovery was severely curtailed. In the wake of this scandal, Berisha was forced to call new parliamentary elections in June 1997, and he resigned the presidency when the Socialist Party won. Fatos Nano became prime minister, and except for a short period from 1998 to 1999 when Pandeli Majko was appointed prime minister, Nano governed until July 2005, when Berisha was re-elected. The democratization of Albania is a work in progress, however, and if the past elections are any indication, the political processes still do not fully meet international standards (OneWorld 2005).

Economy

Albania is one of the poorest countries in Europe. According to the most reliable estimates, the gross national product (GNP) per capita in 1999 was

$930 (US). Gross domestic product (GDP) per capita, once adjusted for purchasing power parity (PPP), was estimated at $2,892 (US) (Nuri 2002). While the economy is slowly recovering after the financial crisis and the tragic events of 1997–1999, the per capita GDP remains low for the region, as revealed by a comparison with the 1999 figures for the former Yugoslav Republic of Macedonia ($4,590 (US)) and Bulgaria ($5,070 (US)).

Agriculture and forestry are the main sources of employment and income in Albania. However, in recent years the construction, transport and service sectors have been growing. Remittances from emigrants form another important part of the economy, since 28 per cent of Albanian families receive them, and they represent 13 per cent of the total income among Albanian households (World Bank 2003). The registered unemployment rate increased from 12.3 per cent in 1996 to 18.3 per cent in 1999, but these figures underestimate the true situation. According to the World Bank, approximately 20 per cent of the population lives in very poor conditions, with 10 per cent on less than $2 (US) per day (World Bank 2003).

The Albanian writer Fatos Lubonja is a prominent voice of moral authority in his country, having spent seventeen years in communist prisons, and having a family history of intellectual resistance. He recently had this to say:

> Only someone who does not know Albania well can think that Albania is a country in transition that is approaching a liberal democracy and the free market. Albania requires a yearly budget of 2 billions of dollars. Of these, 350 million are derived from the legal economy. Another 350 come from the trafficking of drugs and human beings; 175 million comes from prostitution, and 600 from the remittances of Albanians who work in foreign countries. The rest of Albania's income is from foreign loans and foreign aid. If these are the actual figures, can you reasonably think that the trafficking of human beings will come to an end? Of course not. It is an integral part of the economy. (Bazzocchi 2004 [Translation: A. Pieroni]).

Public Health

The healthcare sector in Albania is publicly owned, and the Ministry of Health (MOH) is the major provider of services, while the private sector is limited to the distribution and commercialization of pharmaceuticals and to dental-care services. Nevertheless, the last decade has seen the establishment of a few private outpatient practices and diagnostic centres in urban areas, especially Tirana. A health insurance scheme is offered by the Health Insurance Institute (HII), covering general practitioner (GP) services and subsidizing approximately 350 pharmaceuticals. According to the 'Law on Health Insurance', it is obligatory for all economically active individuals in Albania, whether they are employees, employers, self-employed, or unpaid family workers, to contribute to the scheme and obtain a 'license', while the state pays for children, students, retired people, disabled people, unemployed people, pregnant women, and citizens under compulsory military service. However, data from 2002 show that only about 39 per cent of the total population and only 17 per cent of people living in the mountain regions report having health insurance.

Primary healthcare (PHC) is provided by healthcare centres and clinics in rural areas. In addition, PHC in urban areas is also provided by large polyclinics offering specialized outpatient care. Secondary and tertiary care is offered by public hospitals and clinics. Albania has one of the lowest ratios of hospital beds per inhabitants in all of Europe. Its number of beds per 100,000 inhabitants is less than half the average of the EU countries. Despite its low per capita income level and limited health services, life expectancy in Albania is seventy-two years for men and seventy-eight for women, one of the highest life expectancies in the area, and higher than the average of the EU and all Europe. A vegetable-based diet and a healthy lifestyle, especially in the south, have been cited as possible reasons for these positive indicators (Gjonça and Bobak 1997).

Infant mortality is particularly high in rural areas, where most of the poorest people live. Circulatory diseases are still the most important cause of mortality in Albania. Fewer are dying of infectious, respiratory or gastric diseases, but deaths due to injuries and tumours have increased markedly.

About 30 per cent of Albanians buy nonprescription drugs when ill. While there are no income differences among people buying these pharmaceuticals, there are large regional differences. As many as 43 per cent of Albanians living in the mountainous areas in the north and northeastern parts of the country are likely to buy nonprescription drugs when ill, regardless of their age (Nuri 2002). Recent scientific surveys (Hotchkiss et al. 2005; Vian et al. 2006) have clearly underlined another more serious barrier to healthcare services, particularly among the poor, in the form of unofficial payments to healthcare providers for services that are supposed to be provided at no charge to the patient. Factors promoting these informal payments include perceived low salaries of health staff, a belief that good health is worth paying for, the desire to get better service, the fear of being denied treatment, and the tradition of giving gifts to express gratitude. However, there is ample evidence that a large proportion of the nonformal payments are not voluntary, but are either requested or expected (Lewis 2000).

Lëpushë, Upper Kelmend, Northern Albania

On 14 July 2004, I arrived in the village of Lëpushë, which is located in the Northern Albanian Alps, and belongs to the upper part of the northern commune of Kelmend, in the District of Malësia e Madhe (Figure 2.1).

For many years during the nineteenth century, life in the mountains of the north was managed by networks of patrilineal clans, known as *fis* (Doja 1999). The Kelmendi were originally a northern Catholic tribe whose origins Johann von Hahn (1854: 183–184) has located in a mountainous area on the present-day border between Albania and Montenegro. The patriarch of the tribe, Clement, settled in Bestána, a place in the lower Kelmend area, which von Hahn visited on his travels in the middle of the nineteenth century and described as presenting 'the remains of an old church, a few houses, and vines that had reverted to a wild state' (von Hahn 1854: 184). Clement had seven sons, who became the founders of various *fis*. Selcë was founded by Kola, the oldest son; Vukël was founded by Vuko; Nikç was founded by Nika; and Vuthaj, which is not far from present-day Gusinje on the other side of the mountains in Montenegro, was founded by the sons Balla and Unthai. We do not know the names

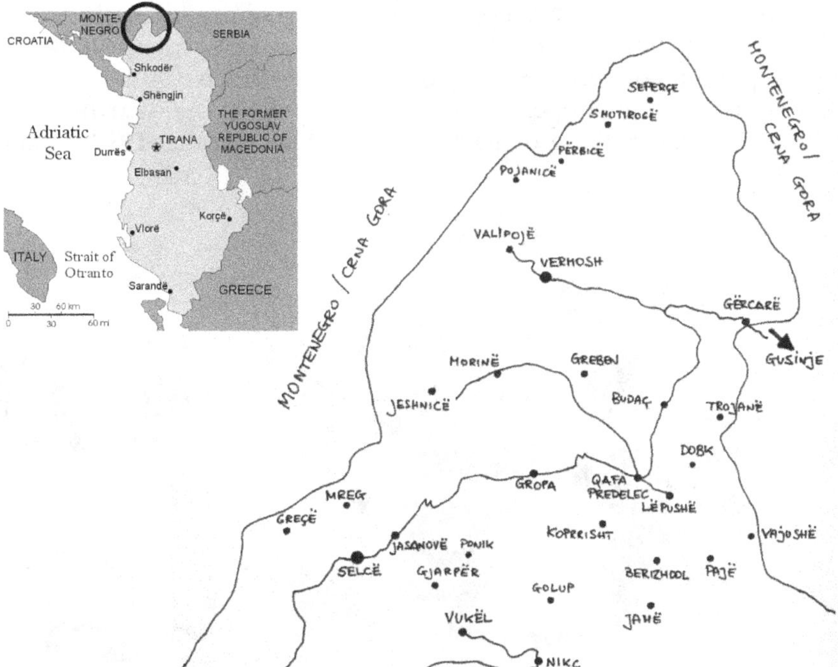

Figure 2.1. Albania and the Kelmend commune (map on the left from www.national-symbol.com)

of the remaining two sons who are believed to have founded Martinovic, from where further descendents founded Bukova in the Dukagjini area (to the east of the present-day Kelmend commune) and Lapo in Kosova (von Hahn 1854: 184; Baxhaku and Kaser 1996: 110–111).

The Kelmendi tribe retained its Catholicism, which it strongly defended against the Ottoman occupation, with only a few families converting to Islam in Nikç. Catholicism was also secretly practised during the communist era from 1944 to 1991.

The people in Kelmend speak Albanian Gheg, which is also spoken in Kosova. The name of the village, Lëpushë, means in the local dialect 'butterbur' (*Petasites hybridus* (L.) P. Gaertn., B. Mey. & Scherb.), a herbaceous plant (also known as 'bog rhubarb') which grows in large amounts during the summer in the small glacial valley where the village is located (Figure 2.2).

The area around Lëpushë is mountainous, with an average summer temperature of 16°C. In winter there are frequent, large snowfalls, and the average temperature is –3°C (Progni 2002). The landscape is characterized by alpine pastures and beech forests. At 1260 metres above sea level, Lëpushë is the highest village in Albania, and it is also one of the most isolated in the entire Albanian Alps. Situated in a small valley less than one kilometre from the Montenegro border, it consists of about twenty-five households, with an overall population of approximately one hundred. There are official plans of the Albanian Government to enlarge a nearby national park in northern Albania, so that it takes in some of the northern and eastern parts of upper Kelmend including the area surrounding Lëpushë. However, these plans have not yet been realized (Hoda and Zotaj 2004).

Figure 2.2. Panoramic view of the glacial valley of Lëpushë, in a photograph and in a painting by Ilir Grishaj

The first inhabitants of Lëpushë settled in the present area at the beginning of the twentieth century, having moved further north into the mountains from the lower villages of Vukël and Nikç (Progni 2002: 251). Lower Kelmend is a very inhospitable area, characterized by the two valleys of the Cem rivers (Cemi i Vuklit and Cemi i Selcës), and completely covered by stones, where to gain even a small piece of arable land has surely been an extremely hard task. As the French Consul in Shkodër, Hyacinthe Hecquard, wrote in 1853: 'At first glance it seems impossible that such a sad country, so poor in arable lands, could be inhabited by humans' (Hecquard 1853: 176 [Translation: A. Pieroni]).

That is why people from Vukël, Nikç, and Tamara have generally looked for green pastures in the higher mountains for grazing animals and for cultivating small fields of potatoes and rye during the summertime. In fact, Lëpushë was originally a settlement for shepherds, who lived there only during the summer months; only later on did people begin to remain there during the winter too.

The first precise description of the valley of Lëpushë was made by Franz Baron von Nopcsa, the most important German-speaking scholar in the history of Albanian studies (Nopcsa 1909, 1925, 1927). At the beginning of September 1907, while on one of his many expeditions to northern Albania, he visited Lëpushë when it was apparently uninhabited, and wrote:

> Everywhere there are broad flat valleys with beech forests, and occasionally a few sparse fir trees and clearings; we could believe that we had been transported into the Viennese forest, were it not for the looming top of the Trojanë mountain (Nopcsa 1910: 88 [Translation: A. Pieroni]).

Lëpushë was also not mentioned in a census carried out in Kelmend between 1916 and 1918 by occupying Austro-Hungarian troops (Seiner 1922: 29–30).

Interestingly, almost everyone in Lëpushë could tell me the story of how the village was built by shepherds who came from Vukël many years before, but nobody could tell me when the settlement became permanent, even though the village's history must go back less than a hundred years. A similar phenomenon was observed in the lower Kelmend village of Kalcë by Krasztev (2001), who conducted fieldwork there in 1997. He explained it as 'amnesia' produced by the communist era.

Post-communism in Northern Albania

After 1991, the process of democratization in Albania initiated a series of major changes that brought about the rapid privatization of lands and herds to even the most rural and mountainous areas.[1]

In Lëpushë, my informants described to me how the former landowners took back their properties, and how the elderly men of the village played a central role in this process because they remembered who owned the parcels of land and where the boundaries had been before the communist era. From 1991, a more traditional form of political organization developed alongside the institutionalized local authorities in which the male heads of the households (*zoti i shtëpies*) gathered together in village assemblies. These assemblies still exist today in Lëpushë, where the men of the villages meet generally once a year. Every few years they also meet to elect someone, usually the community's most respected male elder, to be the village spokesman. There has been considerable speculation in the social science literature about the complexities Albanians have had to face in returning to their 'roots' following the fall of communism. One of the main concerns has been the resurgence of ancient blood feuds and their tragic consequences; for example, at times young boys are unable to attend school because of the danger of reprisals. This has become a crucial issue, not only in a few rural and mountainous areas of the north, but also in new urban arenas, where a new criminal class has attempted to appropriate arguments and issues from past 'traditions' to create new conflicts and anarchy (Schwandner-Sievers 1999; Krasztev 2001; Saltmarshe 2001). Other villages in Kelmend have been heavily affected by the return of blood feuds, such as Selcë and its surrounding smaller villages, traditionally the centre of blood feuds, where much of the male population has been decimated over the years (Seiner 1922: 29).[2]

While Lëpushë has remained untouched by blood feuds, the village has been much more affected by the collapse of the local economy immediately after 1991. With the elimination of the collectives and the reduction of jobs in the public sector, public transport was cancelled, regular maintenance of the roads ceased, and unemployment became a widespread phenomenon in the north. In the last fifteen years, the majority of the region's younger inhabitants have left, either to find work in Shkodër or Tirana, or they have migrated to Italy, Greece, the United Kingdom and, especially, the United States.

In fact, since 1991 more people have left the village than have stayed. The majority of those remaining in the mountains are elderly people who do not want to leave, or members of the 'middle-generation' who have young children and who generally have not been successful in their attempts to leave the country illegally.

Even more dramatic for the local people in upper Kelmend was the political upheaval of 1997, when Albania experienced a few months of quite complete anarchy. At that time, all forms of communication infrastructure in Kelmend were destroyed, including telephone cables, and as of 2004 they still had not been repaired. The upper part of Kelmend has been largely cut off from the world ever since, and for several months of the year it is

Figure 2.3. Potato fields in the higher part of the valley on the outskirts of Lëpushë

completely inaccessible. The Montenegrin mobile phone network works irregularly in the mountains of Kelmend, although there is a project being undertaken by a multinational telecommunications company to establish a network in northern Albania that will cover the entire Kelmend.

The economy of Lëpushë today is largely subsistence oriented. Each household owns a few cows and pigs, and perhaps a few sheep. The villagers cultivate their own potatoes, corn and cabbages for the production of sauerkraut, and a few vegetables such as onions and garlic in their homegardens. Fields of potatoes are also cultivated in the shallow topsoil of the glacial valleys higher up the mountains (Figure 2.3) and in the alpine pastures where the hamlets of shepherds are located.

A few households in upper Kelmend are able to earn some cash by selling cheese and potatoes in Shkodër, and more interestingly, by the uncontrolled exploitation of timber (Figure 2.4) and dried *sahlep* (bulbs of wild orchids, Orchidaceae), which also find their way to Shkodër. The timber purchased in Shkodër is sold in Tirana, Italy or the Balkans, while the dried *sahlep* usually ends up in Turkey.

After 1991, the border between Albania and Montenegro/Yugoslavia lost its strategic importance, and the people of Lëpushë finally began to have free access to the alpine pastures located along the border. This was an area that had remained off-limits for many decades, having been tightly

Figure 2.4. Timber ready to be transported to Shkodër

controlled by the military border police of the communist regime. In fact, it was not until 2002 that the last military forces left Lëpushë, giving the villagers the opportunity to 'colonize' parcels of virgin land higher up the mountains, which are now generally used for grazing animals and cultivating potatoes and rye.

Nevertheless, as mentioned earlier, the major source of revenue comes from the remittances from emigrants to the EU and U.S.A., who send money back to the members of their extended family who have remained 'at home', generally via the Western Union office in Shkodër. Other authors have emphasised how remittances are perceived in Albania as being many families' best if not only means of escaping poverty (De Soto et al. 2002: 39; King and Vullnetari 2003: 48; Uruçi and Gedeshi 2003). On the other hand, if return migration is an increasingly relevant phenomenon in other parts of Albania, it is not an issue in upper Kelmend.

Provision of Healthcare in Lëpushë

> Well, it is a good thing that you are beginning this study on our traditional medicinal plants here. We need to use these herbs, since GPs and hospitals are too far away for us now…

The head of the Kelmend commune made this comment in July 2004 after I had explained the research that I had begun a few days before. People in upper Kelmend nowadays feel they have been abandoned by the institutionalized health system, because since the fall of the communist regime, they have lost all their doctors and nurses. The head of the household (*zoti i shtëpies*), and my host, explained:

> You know, we actually still should have one nurse here in Lëpushë, but her current salary is so low that during the entire summer she lives in her hut at 2,000 metres above sea level, hours away from the village, in order to let her few cows and sheep graze the pastures there.

Thus proper medical care is practically nonexistent at the moment in Lëpushë. The nearest health centre and doctor is in Tamara, which is more than an hour-and-a-half's drive away by four-wheel drive vehicle. Also, from November until April communication and travel between Lëpushë and Tamara are often blocked for many weeks by snow. According to my informants in Lëpushë, the health centre of Tamara is equipped with very few clinical facilities and has only a few pharmaceuticals, many of which are already out of date. During the summer, a general practitioner (GP) from Shkodër makes irregular visits to a few families, if they are in desperate need and request her help; these consultations are paid for in cash. The case of a patient living in Lëpushë with very advanced multiple sclerosis springs to mind: this man has been looked after solely by his wife and daughters, having not received a single medical visit for more than three years. Tragically, in the spring of 2004 a young girl aged fourteen years died of appendicitis in Lëpushë because no medical assistance was available there and because of a heavy snowfall it was impossible for her to reach the nearest hospital in Shkodër.

Healthcare in Lëpushë means self-medication, provided within the households. Not one of my interviewees has ever been enrolled in the national insurance system. When major troubles occur, these people have to go to the hospital in Shkodër or even to Tirana, which costs them a month's wages. When they visit Shkodër, they generally stock up on prescription pharmaceuticals and over-the-counter (OTC) products, which they store at home in case they are needed. Another crucial part of their healthcare is represented by the use of TMs, delivered generally by the women in the house, or very occasionally by the last remaining 'herbal healer', Simon, aged 64 years, who lives in the nearby village of Gropa (Figure 2.5). According to local people in Lëpushë, the use of TMs increased after the health system collapsed, as many of them went back to the many traditional remedies, whose detailed knowledge is generally retained by the oldest members of the community.

Figure 2.5. Simon, the last herbal healer of Gropa

Gathering Bime Mjeksore in Lëpushë

Up until this study, no ethnobotanical research had been carried out in Albania; the only existing reference was a literature survey on botanical folk names in diverse areas of the Albanian-speaking Balkans (Sejdiu 1984). The only ethnographic account published in the West during the last fifty years was a survey of local biographies in Albania, which included a chapter on Kelmend, and impressive photographic documentation (Sheer and Senechal 1997).

The aim of this field study, conducted during the summers of 2004 and 2005, was to document TM and traditional knowledge (TK) related to the use of local medicinal and wild food plants in Lëpushë and the surrounding shepherds' hamlets (*stanë*) of Berizhdol, Koprrisht, Trojanë, Dobk, Vajushë and Pajë (Figure 2.6), which are inhabited only during the summer.

The settlement of Berizhdol, which is located close to the border with the former Yugoslavia, was in existence during the communist era. However, the other *stanë* have been established since early in the 1990s, when the privatization of land and herds was completed. These settlements are inhabited only in summer and mainly by shepherds from Lëpushë and especially from Vukël.

The methodological framework chosen for this study was that generally used in ethnobotany (Alexiades and Sheldon 1996; Cotton 1996), cognitive

Figure 2.6. A few huts (*stanë*) in the hamlet of Pajë

anthropology (Berlin 1992; D'Andrade 1995), ethnography (O'Reilly 2004) and ethnopsychiatry (Nadig 2000). The fieldwork was carried out by using in-depth participant observation and unstructured and semistructured interviews and discussions, with individuals and groups, among approximately ninety persons in Gheg Albanian (with the help of a simultaneous interpreter) and Italian languages (a large portion of the inhabitants of northern Albania speak Italian, generally learned by watching Italian television programmes). Prior informed consent (PIC) was obtained verbally, while ethical guidelines adopted by the American Anthropological Association (AAA) and by the International Society of Ethnobiology (ISE) were rigorously followed. Identification of the plant taxa was conducted in the field following the standard botanical reference *Albanian Flora* (Paparisto and Qosja 1988–2000). Voucher specimens were deposited at the Pharmacognosy Herbarium of the University of Bradford.

Bime mjeksore (medicinal plants) are gathered in the summer pastures by the local people for their domestic needs. They are dried and used in the households during the winter for minor health troubles. We recorded approximately 70 botanical taxa and 160 folk pharmaceutical preparations, mostly derived from plants, but also from animal products and minerals (Pieroni et al. 2005). Table 2.1 shows the most commonly gathered, wild medicinal plants in upper Kelmend together with their folk medicinal uses.

Table 2.1. The most commonly gathered plants for medicine in Lëpushë

Botanical taxa, families and voucher specimen codes	Albanian folk names recorded in Lëpushë	Parts used	Administration	Folk medicinal uses
Asplenium trichomanes L. (Aspleniaceae) LEP-ASP	*Fier guri*	Aerial parts, dried	Decoction	To treat kidney stones
Chelidonium majus L. (Papaveraceae) LEP-CHE	*Bar saralleku* / *Bar verrçi*	Aerial parts, fresh	Decoction, drunk with sugar in small portions (half a coffee cup), eventually increasing the dose over time	To treat hepatitis (*saralleku*)
Gentiana lutea L. (Gentianaceae) LEP-GEN	*Kshanza*	Roots	Macerated in wild plum distillate (*raki*) or in cold water for 1–2 days	To prevent heart disease
Hypericum maculatum Crantz (Hypericaceae) LEP-HYP	*Balsam* / *Caj verdhë* / *Bar pezmet* / *Caj bjeshke*	Aerial parts, dried	Decoction	To treat digestive troubles and diarrhoea (also used as a veterinary preparation, especially for sheep); to treat stomach ache; as a tranquillizer; drunk every morning as a diuretic; to treat flu, sore throat, coughs and bronchitis; as an antihelmintic (used as a veterinary preparation for calves)
			Oleolite	To treat burns

Table 2.1. *continued*

Botanical taxa, families and voucher specimen codes	Albanian folk names recorded in Lëpushë	Parts used	Administration	Folk medicinal uses
Lilium martagon L. (Liliaceae) LEP-LIL	*Bar tamthi*	Tubers, dried	Decoction	To treat all liver diseases (also used as a veterinary preparation)
Origanum vulgare L. (Lamiaceae) LEP-ORI	*Caj* *Caj malhit*	Aerial parts (including flowers)	Infusion	Originally used to treat coughs; nowadays drunk as a recreational beverage, and also as diuretic and digestive
Phyllitis scolopendrium (L.) Newman. (Aspleniaceae) LEP-SCO	*Bar mushknisë*	Leaves, fresh	Decoction	To treat all respiratory afflictions and lung disorders
Plantago major L. (Plantaginaceae) LEP-PLA	*Dejča*	Leaves, fresh	Applied topically	Haemostatic, anti-bacterial and suppurative
		Inflorescence, dried	Decoction	Diuretic
Rosa canina L. (Rosaceae) LEP-ROS	*Kaça*	Pseudo-fruits	Decoction	To improve digestion and treat flu; anti-rheumatic; used to prevent various illnesses
			Decoction, applied externally	To heal eczema
Sedum telephium L. (Crassulaceae) LEP-SED	*Babanik*	Leaves, fresh	After the epicuticular portion is removed, the lower part of the leaf blade is applied to the skin	Suppurative

Table 2.1. *continued*

Botanical taxa, families and voucher specimen codes	Albanian folk names recorded in Lëpushë	Parts used	Administration	Folk medicinal uses
Sempervivum tectorum L. (Crassulaceae) LEP-SEM	*Houseleek*	Juice from the leaves	Applied in the ear	To treat earache
Tussilago farfara L. (Asteraceae) LEP-TUS	*Thundërmushka / Bar mushkë / Xhur mushk*	Leaves, dried	Decoction	To treat coughs
			Applied externally for about 12 hours	To treat wounds; anti-rheumatic
		Leaves, fresh	Applied externally for about 12 hours	Anti-rheumatic
Vaccinium myrtillus L. (Ericaceae) LEP-VAC	*Boronica*	Fruits, dried	Decoction; cooked in syrup and jams; macerated in wild plum distillate (*raki*)	To treat intestinal troubles; anti-diarrheic; 'to strengthen the stomach' and the eyes (especially in children); for 'cleansing the blood'
		Fruits, fresh	Jam	To treat abdominal pains

When the informants talked about *bime* (plants) and especially *bime mjeksore* (medicinal plants), they immediately divided them into the following two categories:

1. Plants that were gathered during the communist era by the local collectives, and then transported to the main northern Albanian centre of Shköder.
2. Plants that have been and are still gathered for use in the household.

During the communist era, many medicinal and toxic plants were gathered by individuals who sold them to local pharmacies. These included yarrow (*Achillea millefolium* L.), club moss (*Lycopodium clavatum* L.), burdock (*Arctium lappa* L.), wild strawberry leaves (*Fragaria vesca* L.), autumn crocus (*Colchicum autumnale* L.) and elderberry (*Sambucus nigra* L.), but the trade ceased after 1991. Only wild strawberries are still collected for domestic use and also for producing homemade jams, which have been sold since 2004 in Shköder.

Wild sage (*Salvia officinalis* L., *sherbel* or *medër*) is still gathered in *lower* Kelmend and in many areas in Albania (though not in upper Kelmend, in areas north of Selcë, where sage does not grow) and exported to western Europe and U.S.A. in huge amounts. It is estimated that one thousand tons of the dried herb were exported in 2001 (Kathe, Honnef and Heym 2003: 79).

In Upper Kelmend, most of the *bime mjeksore* are gathered for personal consumption and are not destined to be traded. Only in the summer of

Figure 2.7. Woman showing *caj* (*Origanum vulgare*)

2005 did the people in upper Kelmend begin to gather *caj* or *caj malhit* (wild oregano, *Origanum vulgare*, Figure 2.7) and *balsam* or *caj verdhë* or *caj bjeshke* (the local St John's wort, *Hypericum maculatum*) for commercial purposes, due to the initiative of the local Italian Franciscan priest. These dried herbs are now sold to the priest, who in turn sells them and other food products from the mountains in a small shop in Shkodër owned by the Catholic Church.

During the summer, *sahlep* (wild orchid bulbs, *Orchis* spp.) are also gathered in the alpine pastures, dried and sold in Shkodër (Figure 2.8). Many wild orchids are of course included in the IUCN Red List of threatened species. However, there is debate about whether those who gather or sell *sahlep* and those who buy them should be criminalized for doing so (see Kasparek and Grimm 1999, and response by Ertuğ 2000 and Pieroni 2000).

Gathering *sahlep* is at present one of the most profitable activities undertaken by the poorest shepherd families of the mountains, where a household's monthly income is often less than fifty euros. One kilogram of dried *sahlep* tubers can be sold for between sixty and seventy euros. Other authors estimate that at least one hundred tons of dried *sahlep* tubers are gathered yearly in Albania (Kathe, Honnef and Heym 2003), probably mainly for the Turkish market. In Turkey, ground *sahlep* is frequently used to prepare the homonymous national beverage, which is drunk hot, especially in the winter. The illegal gathering of *sahlep* and the uncontrolled exploitation of timber are the most critical conservation issues in upper Kelmend today.

I have observed that a few species of wild herbs (e.g., *Lilium martagon, Sedum telephium*) are sometimes transplanted into the villagers' homegardens, which are located close to the house, so that families can readily harvest these herbs when they are needed.

Figure 2.8. Gathering and drying *sahlep* in the Northern Albanian Alps

An important category of wild medicinal plants is represented by the folk generic *caj* (Figure 2.9). The term *caj* has its linguistic origin in the Chinese *tschai,* which roughly translates to the English word *tea. Caj* in standard Albanian means both the leaves of the tea plant, *Camellia sinensis,* and the hot beverage obtained from them. In northern Albania, however, locals use the term *caj* to indicate all wild, herbaceous plants or herbs that are dried and used in infusions, generally with a lot of sugar added, and drunk as both a social beverage and as a medicine. The prototype of *caj* in northern Albania is the simple *caj* or *caj malhit* ('mountain tea', *Origanum vulgare*). Its aerial parts are gathered during the flowering season in July and August, dried and used to prepare an infusion, which is served frequently in the morning and also in the evening. Dried *caj* also represents the most common 'travelling plant' that Albanian migrants take with them when they move abroad; hence it has strong cultural importance.

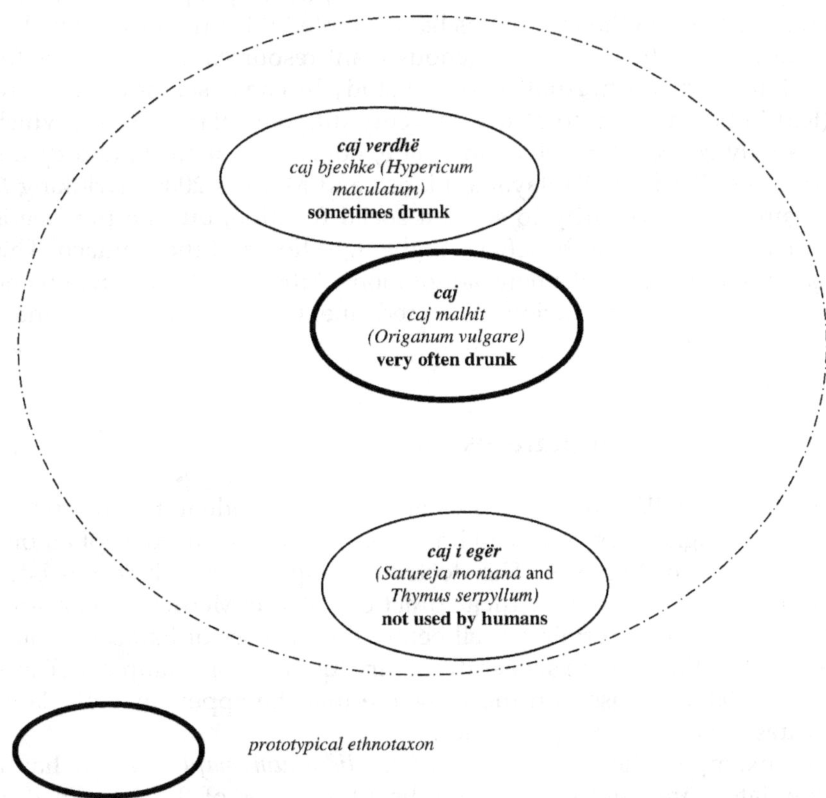

Figure 2.9. Representation of the ethnotaxonomy of the folk generic *caj*

Hypericum maculatum (alpine St John's wort, *caj verdhë* [yellow *caj*], or *caj bjeshke* [alpine pastures' *caj*]) is also gathered and used in upper Kelmend, but much less often than the true *caj*. *Caj verdhë* is perceived primarily as an herb to be used in decoctions for healing purposes. It is very rarely consumed as a social beverage. A St John's wort species was also mentioned by Edith Durham as being externally used in the Shala Valley, east of the Kelmend commune, to treat wounds and cuts (Durham 1909: 115). This use is very common in the medical folklore of Europe.

Caj e egër is the term local people generally give to aromatic plants that are morphologically similar to the 'true' *caj* (*O. vulgare*), but are not very useful. For example, a few informants in Lëpushë consider *Thymus serpyllum* L. (*caj e egër*) to be inferior to *O. vulgare* because the aerial parts of *T. serpyllum* are considered too small and immature for making *caj* (Figure 2.9).

The habit of drinking *caj*, as well as that of drinking coffee, were introduced during the long period of Ottoman domination in Albania (fifteenth to twentieth centuries), and are regarded as essentially an urban pastime. It is interesting to note that 'peripheral' people living in the countryside and in the mountains have adopted this urban tradition, but that they utilize their own indigenous plant resources as substitutes for the original urban tea. Another recent study has analysed how the word *té* (tea) in Spain refers to at least seventy different plant species, which are usually collected in the countryside, boiled dry or fresh, and drunk after meals (Pardo-de-Santayana, Blanco and Morales 2005). Drinking *té* in Spain seems primarily to serve a social function, but the practice is perceived also as being beneficial to the digestion and the stomach. This might be another case of a new 'adaptation' of the older tradition of using local wild plants in decoctions as a medicine, which could be why these teas are so popular.

The Doctrine of Signatures

From the data collected during my fieldwork, it is evident that the archaic 'doctrine of signatures' is still an important ethnomedical concept among the Albanians of Lëpushë. The doctrine of signatures, which has been widespread in the folk medical practices of the Mediterranean area since ancient times, certainly well before the treatises of Paracelsius and Giambattista Porta of the sixteenth century (quoted, for example, in Dafni and Lev 2002), is based on the principle that the appearance of plants indicates their medicinal properties.

For example, the aerial parts of *Chelidonium majus*, which has a yellow latex, are used to treat jaundice; the leaves of the fern *Phyllitis scolopendrium*, which is called locally 'the herb of the lungs', are thought to be useful in treating all respiratory and lung afflictions since the shape of lower leaf lamina resembles a human lung. Likewise, the bulbs of *Lilium martagon*, known locally as the 'herb of the liver', are used to treat liver diseases, since the colour of the flowers resembles that of the liver (Figure 2.10); and the mineral jasper, which is blood red in colour, is ground into milk and given to sheep to drink because it is believed this mineral allows blood to converge in the animals' heads, thereby indirectly healing any ailments. Finally, the membrane from a hen's stomach is used to treat human kidney stones, because there are often small stones present in the muscular stomach of poultry, which have been ingested by the hen to assist in grinding its food. All of these remedies are reportedly still commonly used in Lëpushë.

Local Foods

People in Lëpushë state that the variety of food they can now afford to produce for their own consumption is a dream compared with the meagre daily diet and the hunger they experienced during the days of communism. All dairy produce at that time was taken by the cooperatives, and families had only potatoes, bread – sometimes covered with a very small amount of grated cheese – and onions to eat as staples. The issue of hunger is often raised in the narratives of the people of Lëpushë, and all of them have described the food situation they are experiencing at present as 'far better than it was before'.

Today's daily diet of a typical household in Lëpushë is based on many dairy products, potatoes, a few vegetables (especially onions), corn for

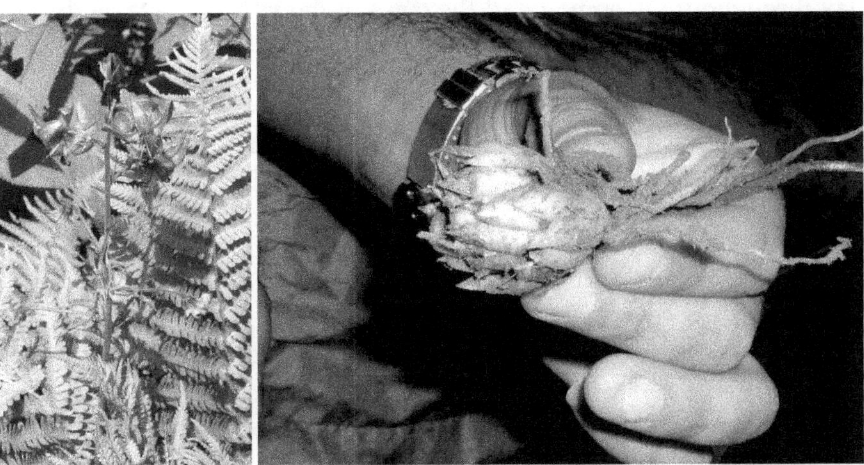

Figure 2.10. *Lilium martagon*

Figure 2.11. Corn bread and *byrek*, filled with *nena*

baking the traditional corn bread (Figure 2.11), and, especially in winter, sauerkraut and pork. The good health of their two to four cows and the production of milk is a critical factor for each family in the village. Every day throughout the summer all the cows of the village are taken by the children and teenagers to graze in the higher alpine pastures. Every evening they are brought back and milked just as darkness falls. During the winter the cows are kept in the village and fed with hay. It is during haymaking that the village clearly shows its complex social networks, as it is not unusual to see male members of diverse families helping one another to make hay, to let it dry, or to bring it in, if rain looks imminent (Figure 2.12).

Haymaking in the summer is hard, physical work, but it is also one of the most important tasks for each household. Without enough hay for the cows families could find themselves in a desperate situation because without their cows 'a family can die of hunger'. This point was made repeatedly by my informants. The belief in the evil eye – that an envious gaze can cause harm, often unintentionally, to those being watched – is quite widespread in Lëpushë. Locals think the evil eye can badly affect a cow's health so that the animal becomes weak and will not produce any

Figure 2.12. Collecting and storing hay

more milk. It is therefore common to see cows grazing in the pastures wearing talismans that protect against the evil eye. These are made out of a piece of red cloth that contains one garlic clove, coffee beans and grains of salt (Figure 2.13). When an animal gets very sick and the talismans appear not to be working any more, a few old women in the village are called in to treat the illness, generally by making the animal drink hot

Figure 2.13. Cow wearing amulet against the evil eye

water mixed with ashes, or by performing ritual healing by reciting special incantations.

Also crucial to the domestic economy of each family are pigs. Each family owns one to three pigs, which range freely in the village. Although it surprises many foreign travellers, it is actually not uncommon to see pigs running everywhere in the villages of northern Albania and their surrounding areas. The pigs are butchered in winter and their meat is generally cured and stored and used throughout the year. Fodder beet (*pangjari*) and rye (*theker*) are expressly cultivated as pig fodder, while the leaves of beech trees (*Fagus sylvatica* L., *ahu*), nettle (*Urtica* spp., *hithër*) and alpine dock (*Rumex alpinus* L., *nena elpiet*) are gathered from the wild. All these plants are boiled and then given to the pigs. For 'strengthening' the pigs and as a galactagogue for cows, locals frequently gather aerial parts of the cow parsnip (*Heracleum sphondylium* L., *barovina*), while wild thyme (*Thymus serpyllum*, *caj e egër*) is thought to give the milk a very nice flavour if it is fed to sheep and cows.

Local dairy products are the basis of the diet and include: milk (*tomël*), yoghurt (*kos*), cheese (*diathë*), butter (*burrofresko*), *tëlynë* (a sort of clarified butter which is prepared by heating butter for an hour and then filtering it), cream (*mazë*), *jardun* (obtained by heating sheep milk with salt for approximately one hour), curd cheese (*gjizë*), whey (*hirra* or *qumësht*), colostrum (*koloster*) – when the cows have given birth — and *mishavin*. *Mishavin* is a type of creamy cheese that is prepared by fermenting grated cheese and salt for at least two months in a receptacle sealed on the top but with a few holes in the bottom through which the liquid is continuously expelled. The result is a creamy, dense and fragrant product, of which large quantities are consumed during the cold winter months as a 'strengthening' food.

Locals use their cheese to prepare two of their most traditional dishes: *diathe zje*, a kind of porridge prepared by heating cheese with *burrofresko*, salt and flour; and *kaçimak*, which involves mixing boiled potatoes with the same ingredients as *diathe zje*. From colostrum, it is customary to prepare a sweet dish called *perpeq*, which is made with eggs, honey or sugar, and soda.

Throughout Albania families usually produce large amounts of their own alcohol. In Lëpushë, this *raki* is made by distilling fermented small, red plums. The drink appears to serve as a social lubricant, and is thus consumed whenever people congregate or socialize. It would be considered shameful for a *zoti i shtëpies* (male head of the family) to have guests and not be able to offer them *raki*. A family of five members are said to easily consume as much as 250 litres of *raki* in a year.

During my field studies I observed the increased consumption of other forms of alcohol besides *raki*, and tobacco, among males of upper Kelmend. Up until 2004 there were only two bars in nearby Qafa Predelec, which sold beer, spirits and cigarettes. In the summer of 2005 I counted five bars. The consumption of beer is new to the area and is perceived as a modern and sophisticated pastime. The price of a bottle of beer is very high by Albanian standards (roughly one euro), and that is maybe why a few informants came to me with the suggestion that 'a good idea for the future development in our areas would be to establish a small brewery so that we could be self-sufficient'.

It is interesting to note how the issue of 'being self-sufficient' – in other words, being independent from any external influences – is raised repeatedly in northern Albania. Being self-sufficient is perceived by the locals as their 'destiny' and a virtue. Others have emphasised how the issue of isolation has been important in building the Albanian identity (Krasztev 2002; Schwandner-Sievers 2002), but a recent ethnoarchaeological and ethnohistorical fieldwork has shown instead that population changes in the Shala Valley (Northern Albanian Alps) were influenced by several external forces 'despite the seeming isolation of the northern tribes' (Lee and Galaty 2007).

Among the most gathered wild plants, it is important to mention a few herbs and fruit such as strawberries, raspberries and bilberries, which are usually eaten raw or in the form of syrups or jam.

More than half of all recorded wild-food plants in upper Kelmend are considered by the locals to be medicinal foods – foods that are deliberately ingested in order to obtain a specific medicinal effect – or what I and my colleague Cassandra Quave recently defined as 'folk functional foods', which are foods that are ingested because they are reputed to be 'healthy' (Pieroni and Quave 2006). These perceptions once again confirm that in the provision of domestic healthcare the difference between food and medicine is unclear, and that food plants are often ingested because they are perceived to prevent or even cure certain diseases (Pieroni and Price 2006).

Bara e Egër *as Food*

'Wild herbs' (*bara e egër*) that are used for food are gathered during the late spring and summer only. Among the most culturally important taxa are wild garlic (*lertha, Allium triquetrum* L.), which is eaten raw or boiled in soups, nettle (*hither, Urtica dioica* L.) and *nena* (edible *Chenopodium* and *Rumex* spp.) leaves, which are boiled and used as a filling for *byrek* (see Figure 2.11).

People in northern Albania use the term *nena* to indicate various wild plants, whose leaves are generally consumed boiled, and used as a stuffing together with *tëlynë*, cream or cheese when preparing the typical Albanian pie called *byrek*, which is well know also in other Balkan countries and especially in Turkey.

Chenopodium bonus-henricus L., Good King Henry, is the prototype of this group of taxa. It is called simply *nena*, or *nena e butë*, which literally means 'true' *nena*, or *nena e bjeshkes* ('alpine pasture *nena*'). The other members of this group (*Chenopodium album* L. (fat hen), and *Rumex alpinus* (alpine dock)) are classified in the folk taxonomy of upper Kelmend with folk specifics *nena e egër* ('wild' *nena*) and *nena elpiet*. The former (*C. album*) is gathered and cooked in the same manner as the true *nena* (*C. bonus-henricus*), but only if true *nena* is unavailable as fat hen generally is not considered to be edible. The latter (*R. alpinus*) is gathered and used as fodder for pigs, who apparently consider it to be a very tasty vegetable (Fig 2.14).[3]

Gender Roles in Wild-plant Gathering in Upper Kelmend

As in all of the most traditional areas of northern Albania, in upper Kelmend the men are the only ones allowed to walk far from their houses. Women generally must remain at home, where it is customary for them to dedicate their time to managing the homegarden, cooking and other domestic tasks. They can help men in the fields during the labour-intensive

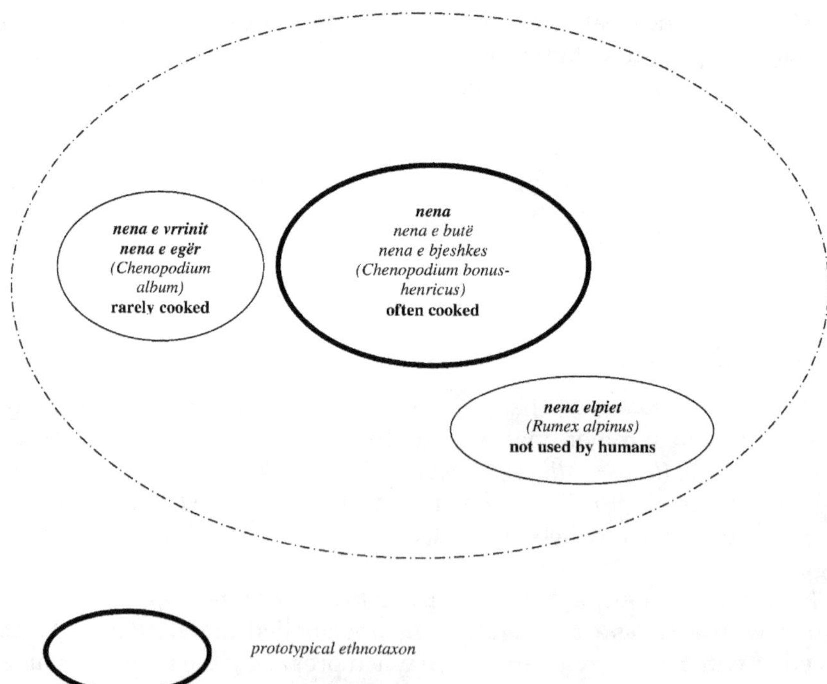

Figure 2.14. Representation of the ethnotaxonomy of the folk generic *nena*

process of making hay in the summer, but only in the presence of the more senior males of the family, for example their fathers-in-law or husbands.

According to customary laws (*Kanun*) in northern Albania, which have always been orally transmitted, but were compiled in written form for the first time by a Kosovar Franciscan in the 1920s (Gjeçov 1989: 238), 'a woman is a sack, made to endure'. Littlewood and Young (2005: 77) have summarized the meaning of gender roles in the *Kanun* in this way: 'women have to perform an essentially child-bearing role as the property of their husbands'. According to the *Kanun*, women cannot be the target of a blood feud, and neither can they inherit any property nor refuse their arranged marriage.

In practice, gender roles are similarly well defined and strictly enforced. Women are only allowed to occupy the domestic sphere. Even amongst younger couples, women have still in a few cases to ask their husbands' permission to go out of the house during the day, even if they just want to buy a few things in the nearby Qafa Predelec, where there is a tiny food store selling a few imported vegetables (tomatoes, sweet peppers), soaps, and a few minor textile items. Moreover, women are generally not allowed to own cash. If they want to buy something, they have to ask their men to give them some money.

Women are the main gatherers of cultivated plants, which they grow in their homegardens or collect from wild areas close to home, particularly nettles and Good King Henry leaves for preparing *byrek*. Children and men are the main gatherers of the medicinal plants, *bime mjeksore*, that grow in the fields and areas far from the house.

Plant Knowledge of Women Who Become Men

I was greatly surprised when the head of the household in which I was staying suggested that I visit the summer settlements in the alpine pastures, which were many hours' walk from the village, with his sister for a guide. His sister, G, is forty years old, cuts her hair short like a man, dresses like a man, and is known by the male name of D or *djalë* ('lad'). The case of G/D is one of the last remaining examples of what in the literature is referred to as northern Albanian 'sworn virgins': women who become men (see Durham 1909: 80 for an early reference and more recent mentions in Grémaux 1993; Young 2000; Littlewood 2002; Littlewood and Young 2005).

During the days I spent in upper Kelmend, I developed a deep friendship with G/D after she/he took me to meet the shepherds and plant gatherers in the very isolated alpine hamlets. I was amazed to observe how G/D's knowledge of the environment encompassed both the homegardens close to the houses, which is generally the domain of the women, and the alpine pastures located very far from the village, which is generally a male domain. G/D learnt about the homegardens during her/his childhood and especially when she/he was a teenager, but now she/he has taken on a male role in gathering plants. Hence she/he would never gather nettles or Good King Henry leaves, although she/he knows how to recognize and process them. As she/he explained: 'I do not like to cook and do not like any of the things women do at home: washing cloths, ironing, and sewing'.

As Littlewood stated in his account of this phenomenon, 'the occurrence of this third gender in northern Albania does not influence at all the strictly binary gender system of northern Albania, nor does it contribute to a less subdominant position of women' (Littlewood 2002). It is instead a product of this extremely binary system, where gender roles are very strict, and the dominant code is characterized by a strong 'hetero-normative' masculinity (Haller 2001). 'Sworn virgins' are then seen as supporters of patriarchy (Young 2006). As G/D told me once, while we were walking across the mountains:

> I did it [changed gender role] after my father died ten years ago. I wanted to be free; now I can walk for hours in the mountains and nobody can touch me; I am free now; I can do what I want and nobody can tell me anything; I can even have my money and buy what I want.

Erosion of TK and Prospects of Eco-sustainable Development in Upper Kelmend

The end of the communist era has allowed the people of Lëpushë to regain their land and their herds, to manage them according to what they consider to be the 'traditional' way, and to 'rediscover' their traditional medicine and local foods. On the other hand, prosperity and new opportunities abroad have also led to the outmigration of the younger members of the village, thereby reducing the likelihood that this TK is transmitted to future generations.

During my last visit, in August 2005, I met a few members of the extended families of Lëpushë who had emigrated a few years earlier and were now back in their village for a short summer holiday. It was common to hear them say: 'How do we cook that, mum? I do not remember it'; or 'I have forgotten how to do this stuff. We live so well in Italy, we don't need it there'. One told me, 'I've been helping my mum and dad store hay for the winter, but I'm really looking forward to going back to Italy soon'.

So while the few remaining children and young people in Lëpushë will retain the villagers' knowledge of how to recognize and gather wild plants, it is doubtful that this TK will be transmitted in the decades to come. Much will depend on how people think about their future and what form it will take in upper Kelmend. The development of ecotourism as well as the implementation of the Balkan Peace Park Project (BPPP

2005) could have an immense impact on northern Albania and upper Kelmend. The Balkan Peace Park Project was conceived and is being led by a world expert on Albania, Antonia Young, and her husband, the peace studies scholar Nigel Young. The Youngs could have a great influence on the daily lives and future prospects of the people of Lëpushë, especially if these communities can be convinced to participate in planning and building the park. Once realized, the park could offer an employment alternative for the young and dynamic members of the community, who at present believe their future lies in escaping from the mountains and finding adventure and opportunity by migrating to New York, Bari or Florence. In time, they may decide it is more attractive to remain in their village and let the world come to them. There could be a huge potential in the Albanian mountains for trekkers and agrotourism (Pieroni 2007); for the sustainable gathering and organic cultivation of high-quality plant products, including medicinal herbs; and for improved production of speciality dairy products, which could be sold to niche markets in Tirana and, perhaps, throughout western Europe. Another key issue is interregional cooperation with bordering Montenegro. Gusinje is less than thirty minutes' drive from Lëpushë by four-wheel drive vehicle, so the barriers to be overcome are more psychological than logistical or linguistic. The BPPP could bring about true reconciliation among the peoples of the region, by improving their understanding of each other, and by providing opportunities for regular exchanges between the two sides of the Trojanë Mountain, especially among the younger generations.

To make all this feasible, the present, appalling state of the roads and telecommunications in upper Kelmend will have to be improved. Perhaps this could be done through foreign aid, but if we analyse international programmes currently working in the area, the scenario is extremely discouraging. The recent United Nation Environmental Programme (UNEP) initiative 'Enhancing Trans-boundary Biodiversity Management in South Eastern Europe' (UNEP 2006), for example, seems to suggest more restrictive laws for improving biodiversity conservation policies in the area. In the UNEP document terms like 'traditional knowledge' or 'cultural heritage', as well as references to 'emic' views of management of the environment, are completely omitted; moreover, a genuine participatory approach is missing and one of the assumptions seems to be that local people in the area simply overexploit the environment.

The history and the people of Lëpushë show us instead how this particular natural environment might be dynamically managed by taking into account their complex and sophisticated local knowledge systems.

Political and social changes since 1991 have allowed the people in the mountains of northern Albania to 'return' to their roots. This has had a remarkable influence on their perceptions of themselves, their interactions with their natural environment and the way they deal with the extreme difficulties they have to face every day just to survive. With their natural dynamism and vitality, northern Albanians may be able to work together with external stakeholders to ensure the continuation of their deeply embedded relationship with nature, so that they and their descendents will be able one day to enjoy a secure and viable future in their marvellous mountains.

Acknowledgements

Special thanks are due to the entire community of Lëpushë, and all the fantastic people who generously agreed to share their TK with me.

Special thanks are due also to the European Commission, who funded the research consortium RUBIA (coordinated by A. Pieroni, #ICA3-2002-10023; www.rubiaproject.net), who in turn funded the author's stay in northern Albania in 2004 and 2005; to the SCH Group (Anke Niehof) at the Department of Social Sciences of Wageningen University (The Netherlands), who funded the 2005 research travel to Albania; and last but not least, to Antonia Young, University of Bradford, and Joan Vallès, University of Barcelona (Catalonia, Spain), for their encouraging comments on a previous draft of this chapter.

Notes

1. A unique, audiovisual, ethnographic document of what actually happened in 1991 when communally owned land and animals were distributed among all the households has been filmed in the village of Rrogam in the Northern Albanian Alps by David Watson and by the unforgettable (and well remembered) Norwegian anthropologist and Albanian specialist Berit Backer (Watson and Backer 1991), who shortly after having produced this video was killed in Oslo by one of her Albanian friends, for whom she had dedicated her entire life.
2. The census of northern Albania, 1916–1918, showed that of all villages in Kelmend, Selcë had 30 per cent more women than men (Seiner 1922: 29).
3. The use of the adjective *e egër* (wild) to define plants that are similar to prototypical ones, but not as useful because they are inedible for humans, has been observed in other contexts; for example, among the Arbëreshë Albanians in southern Italy (Maddalon and Belluscio 1996). As Maddalon and Belluscio write in their very impressive ethnolinguistic analysis of Arbëreshë folk plant names in southern Italy, lexemes produced by adding suffixes do not necessarily represent 'folk specifics' as described by Berlin (1992), since they can indicate taxa belonging to diverse folk generics. The function of the suffix in these cases could be to express functional concepts, for example, edible or inedible for humans (Figure 2.14).

References

Alexiades, N.M. and J.W. Sheldon (eds). 1996. *Selected Guidelines for Ethnobotanical Research: a Field Manual*. New York: New York Botanical Garden.

Baxhaku, F. and K. Kaser 1996. *Die Stammesgesellschaften Nordalbaniens: Berichte und Forschungen österreichischer Konsuln und Gelehrter (1861–1917)*. Vienna: Böhlau Verlag.

Bazzocchi, C. 2004. 'Morti nel canale di Otranto: intervista a Fatos Lubonja'. Rome: *Osservatorio sui Balcani*, 14 January 2004. Retrieved 10 January 2006 from http://www.osservatoriobalcani.org/article/articleview/2732/1/41

Berlin, B. 1992. *Ethnobiological Classification: Principles of Categorization of Plants and Animals in Traditional Societies*. Princeton: Princeton University Press.

BPPP, 2005. *Balkan Peace Park Project*. Retrieved 6 January 2006 from http://www.balkanspeacepark.org

Cotton, C.M. 1996. *Ethnobotany: Principles and Applications*. Chichester: Wiley.

Dafni, A. and E. Lev. 2002. 'The Doctrine of Signature in Present-day Israel', *Economic Botany* 56: 328–334.

D'Andrade, R. 1995. *The Development of Cognitive Anthropology*. Cambridge: Cambridge University Press.

De Soto, H., P. Gordon, I. Gedeshi and Z. Sinoimeri. 2002. *Poverty in Albania: a Qualitative Assessment. Technical Paper 250*. Washington DC: World Bank.

Doja, A. 1999. 'Morphologie traditionnelle de la societe albanaise', *Social Anthropology* 7: 37–55.

Durham, E. 1909[2000]. *High Albania*. London: Phoenix Press.

Ertuğ, F. 2000. 'Reply to Kasparek and Grimm's Article (Orchid Trade for Salep)', Letter, *Economic Botany* 54: 421–422.

Gjeçov, S. 1989[1933]. *Kanuni i Lekë Dukagjinit*. Translated by L. Fox. New York: Gjonlekaj.

Gjonça, A. and M. Bobak. 1997. 'Albanian Paradox, Another Example of Protective Effect of Mediterranean Lifestyle?', *The Lancet* 350: 1815–1817.

Grémaux, R. 1993. 'Woman Becomes Man in the Balkans', in G. Herdt (ed.), *Third Sex, Third Gender: Beyond Sexual Dimorphism in Culture and History*. New York: Zone Books, pp. 241–281.

Haller, D. 2001. 'Die Entdeckung des Selbstverständlichen: Heteronormativität im Blick', *kea. Zeitschrift für Kulturwissenschaften* 14: 1–28.

Hecquard, H. 1853. *Histoire et Description de la Haute Albanie ou Guégarie*. Paris: Arthus Bertrand.

Hoda, P. and A. Zotaj. 2004. *Data on Albanian Biodiversity and Mapping*. Power Point Presentation. Retrieved 28 November 2004 from http://lynx.uio.no/lynx/ballynxco/04_wildlife-management/4_4_biodiversity/Pdfs/Hoda_&_Zotaj_-_Albanian_Biodiversity.pdf

Hotchkiss, D.R., P.L. Hutchinson, A. Malaj and A.A. Berruti. 2005. 'Out-of-pocket Payments and Utilization of Healthcare Services in Albania: Evidence from Three Districts', *Health Policy* 75: 18–39.

Kasparek, M. and U. Grimm. 1999. 'European Trade of Turkish Sahlep with Special Reference to Germany', *Economic Botany* 53: 396–406.

Kathe, W., S. Honnef and A. Heym. 2003. *Medicinal and Aromatic Plants in Albania, Bosnia-Herzegovina, Bulgaria, Croatia and Romania.* Bonn: WWF Germany/German Federal Agency for Nature Conservation (BfN).

King, R. and J. Vullnetari. 2003. *Migration and Development in Albania. Working Paper C5.* Brighton: Development Research Centre on Migration, Globalisation and Poverty.

Krasztev, P. 2001. 'Back to the Torn-out Roots: Reflections on Vendetta in Contemporary Albania', *Slovo* 13, no pages reported. Retrieved 6 January 2006 from http://www.kodolanyi.hu/szabadpart/szam5/tnt/back_to_the_torn_out_roots.htm

——— 2002. 'The Price of Amnesia: Interpretation of Vendetta in Albania', *Identities: Journal for Politics, Gender and Culture* 1, page not reported. Retrieved 8 January 2006 from http://www.realitymacedonia.org.mk/web/specials/amnesia

Lee, W.E. and M. Galaty. 2007. 'Warfare, Politics, and Rural Population Movements: Analyzing Houses, Neighborhoods, and Abandonment in the Shala Valley of Northern Albania, 1450–2006'. Paper delivered at the 108th Annual Meeting of the Archaeological Institute of America, San Diego, CA, January 4-7, 2007.

Lewis, M. 2000. *Who Is Paying for Healthcare in Eastern Europe and Central Asia?* Washington DC: World Bank.

Littlewood, R. 2002. 'Three into Two: the Third Sex in Northern Albania', *Anthropology & Medicine* 9: 37–49.

——— and A. Young. 2005. 'The Third Sex in Albania: an Ethnographic Note', in A. Shaw and S. Ardener (eds), *Changing Sex and Bending Gender.* Oxford: Berghahn, pp. 74–84.

Maddalon, M. and G. Belluscio. 1996. 'Proposte preliminari per l'analisi del lessico fitonimico Arbëresh in una prospettiva semantico-cognitiva', *Quaderni del Dipartimento di Linguistica dell'Università della Calabria* 6: 68–95.

Nadig, M. 2000. 'Interkulturalität im Prozess: Ethnopsychoanalyse und Feldforschung als methodischer und theoretischer Übergangsraum', in H. Lahme-Gronostaj and M. Leuzinger-Bohleber (eds), *Identität und Differenz: zur Psychoanalyse des Geschlechterverhältnisses in der Spätmoderne.* Opladen: Westdeutscher Verlag, pp. 87–102.

Nopcsa, F.B. 1909. 'Beiträge zur Vorgeschichte und Ethnologie Nordalbaniens', *Wissenschaftliche Mitteilungen aus Bosnien und Herzegowina* 11: 82–90.

——— 1910. *Aus Šala und Klementi.* Sarajevo: Daniel A. Kajon Verlag.

——— 1925. *Albanien. Bauten, Trachten und Geräte Nordalbaniens.* Berlin/Leipzig: Verlag von Walter de Gruyter & Co.

——— 1927. 'Ergänzungen zu Meinen Buche über die Bauten, Trachten und Geräte Nordalbaniens', *Zeitschrift für Ethnologie* 59: 279–281.

Nuri, B. 2002. *Healthcare Systems in Transition: Albania.* Copenhagen: European Observatory on Healthcare System/WHO.

OneWorld. 2005. 'Albanian Elections Weakened by Insufficient Political Will; System Open to Abuse', *OneWorld Southeast Europe.* Retrieved 30 December 2005 from http://see.oneworld.net/article/view/114669/1

O'Reilly, K. 2004. *Ethnographic Methods.* Abingdon: Routledge.

Paparisto, K. and X. Qosja (eds). 1988–2000. *Flora e Shqipërisë/Flore de l'Albanie.* Volumes 1–4. Tirana: Akademia e Shkecave e Republikes se Shqipërisë.

Pardo-de-Santayana, M., E. Blanco and R. Morales. 2005. 'Plants Known as "*Té*" (Tea) in Spain: An Ethno-pharmaco-botanical Review', *Journal of Ethnopharmacology* 98: 1–19.

Pieroni, A. 2000. 'Reader Ponders Trade in Sahlep (Ground Orchid Bulbs) ', Letter, *Economic Botany* 54: 138.

——— 2007. *People and Mountains in Kelmend: Documentation of a Field Study on Traditional Knowledge in the Northern Albanian Alps.* Wageningen: RUBIA Project/WUR.

———., B. Dibra, G. Grishaj, I. Grishaj and S.G. Maçai. 2005. 'Traditional Phythotherapy of the Albanians of Lëpushë, Northern Albanian Alps', *Fitoterapia* 76: 379–399.

——— and L.L. Price (eds). 2006. *Eating and Healing: Traditional Food as Medicine.* Binghamton, NY: Haworth Press.

——— and C.L. Quave. 2005. 'Traditional Pharmacopoeias and Medicines among Albanians and Italians in Southern Italy: A Comparison', *Journal of Ethnopharmacology* 101: 258–270.

——— and C.L. Quave. 2006. 'Functional Foods or Food-medicines? On the Consumption of Wild Plants among Albanians and Southern Italians in Lucania', in A. Pieroni and L.L. Price (eds), *Eating and Healing: Traditional Food as Medicine.* Binghamton, NY: Haworth Press, pp. 101–129.

Progni, K. 2002. *Malësia e Kelmendit.* Shkodër: Camaj-Pipa.

Saltmarshe, D. 2001. *Identity in a Post-communist Balkan State.* Aldershot: Ashgate.

Schwandner-Sievers, S. 1999. 'Humiliation and Reconciliation in Northern Albania: The Logic of Feuding in Symbolic and Diachronic Perspectives', in G. Elwert, S. Feuchtwang and D. Neubert (eds), *Dynamics of Violence: Process of Escalation and De-escalation in Violent Group Conflicts.* Berlin: Duncker and Humblot, pp. 133–152.

——— 2002. *Albanian Identities: Myth and History.* Bloomington, Indiana: Indiana University Press.

Seiner, F. 1922. *Ergebnisse der Volkszählung in Albanien in dem von den Österr.-ungar. Truppen 1916–1918 Besetzten Gebiete.* Vienna: Hölder-Pichler-Tempsky, Verlag der Akademie der Wissenschaften in Wien.

Sejdiu, S. 1984. *Fjalorth Etnobotanik i Shqipes*. Prishtinë: Rilindja.
Sheer, S. and M. Senechal. 1997. *Long Life to Your Children! A Portrait of High Albania*. Amherst, Mass.: The University of Massachusetts Press.
UNEP. 2006. *Enhancing Trans-boundary Biodiversity Management in South Eastern Europe*. Vienna: United Nations Environment Programme.
Uruçi, E. and I. Gedeshi. 2003. *Remittances Management in Albania: Working Paper 5/2003*. Rome: CESPI.
Vian, T., K. Grybosk, Z. Sinoimeri and R. Hall. 2006. 'Informal Payments in Government Health Facilities in Albania: Results of a Qualitative Study', *Social Science & Medicine*, 62: 877–887.
von Hahn, J.G. 1854. *Albanesische Studien*. Jena: Verlag von Friedrich Mauke.
Watson, D. and B. Backer. 1991. *Disappearing World: The Albanians of Rrogam*. VHS/PAL. London: Royal Anthropological Institute (RAI).
World Bank. 2003. *Albania Poverty Assessment: Report No. 26213-AL*. Washington DC: World Bank.
Young, A. 2000. *Women Who Become Men: Albanian Sworn Virgins*. Oxford: Berg Press.
——— 2006. '"Sworn Virgins" as Supporters of Patriarchy', *Albanian Journal of Politics* 2: 7–25.

CHAPTER 3

The Cultural Significance of Wild-gathered Plant Species in Kartitsch (Eastern Tyrol, Austria) and the Influence of Socioeconomic Changes on Local Gathering Practices

ANJA CHRISTANELL, BRIGITTE VOGL-LUKASSER, CHRISTIAN R. VOGL AND MARIANNE GÜTLER

Introduction

Gathering of wild plant species is a typical and important activity in rural communities of the European Alps, so not surprisingly there are hundreds of books that focus on plant species to be gathered and their potential uses. Among these, books of alternative recipes featuring wild plant species and their products are especially popular, and there are seminars and courses offered in a variety of activities associated with harvesting and preparing wild plants. Despite all this attention, the gathering activities of *farmers* in European Alpine regions are rarely researched by scientists.

In this chapter we report on one of a series of studies being conducted by Austrian ethnobotanists to investigate the traditional practices of alpine farmers (Christanell 2003; Vogl and Vogl-Lukasser 2003). We investigated the gathering of wild plants by farmers in Kartitsch, a village in Eastern Tyrol, Austria. In addition to documenting ethnobotanical knowledge of wild plants, we focused our fieldwork on the cultural significance of wild plants in the daily life of the farmers; in particular, their role in religious practices. We also investigated whether recent socioeconomic changes had affected local gathering practices.

Wild Plant Gathering in Europe

Gathering of wild plants is an integral part of livelihood strategies throughout the world (Plenderleith 1999; Cunningham 2001). The notion that wild plant gathering is equally present in Europe may be less widespread, yet evidence dates back to the Bronze Age (French Alps, Bouby and Billaud 2005) and modern-day gathering has been investigated in at least twelve countries, mainly in the Mediterranean region.[1]

In comprehensive studies, plants falling along the food–medicine continuum (Etkin 1994) occupy a major part of the species lists (e.g., Forbes 1976; Brüschweiler 1999; Ertuğ 2000) and a large number of specific studies focus on edible and medicinal plants (e.g., Bonet and Vallès 2002). Similar to the *quelites* of the Tarahumara, Mexico (Bye 1981), specific terms for wild food plants also exist in the Greek Argolid (wild *horta*; Forbes 1976) and with the Arbëreshë of southern Italy (*liakra*; Pieroni et al. 2002). Gathering has been found to serve subsistence purposes and be a means of overcoming periods of scarcity in Greece (Forbes 1976), Spain (Tardío, Pascual and Morales 2002; Pardo-de-Santayana, Tardío and Morales 2005) and Italy (Pieroni et al. 2005a). In Anatolia, wild plant gathering is more related to taste and perceived nutrition than to poverty alleviation and is an important social activity (Ertuğ 2000, 2003). Austrian women claim that gathering is meaningful and brings joy (Gruber 2005), while a Spanish study suggests that it produces highly appreciated edible plants (Tardío, Pascual and Morales 2002). In Greece, Italy and Turkey, women are predominantly responsible for plant gathering, with men typically contributing speciality plants (Forbes 1976; Pieroni 1999; Ertuğ 2000, 2003; Pieroni 2003). Often conducted in groups, gathering strengthens Anatolian women's social relations (Ertuğ 2003), and through the associated custom of gift-giving extends southern Italian Arbëresh women's social space (Pieroni 2003). Gathered wild plants form part of meals and processions on religious holidays in Italy (Pieroni 2003) and are used as incense, amulets and protective charms in Switzerland (Brüschweiler 1999), Anatolia (Ertuğ 2000) and Italy (Pieroni 2003).

Wild plant gathering is an active, living custom in Europe today, although it has experienced change along with people's socioeconomic situation. The shift from agriculture to wage labour has reduced gathering opportunities in many countries (e.g., Tardío, Pascual and Morales 2002). In Italy, people no longer pass by sites to gather plants on their way home from work (Pieroni 2003), and remote areas have been abandoned (Pieroni 1999). In Spain, Italy, Croatia and Greece, fewer noncultivated vegetables are consumed than in previous decades (Pardo-de-Santayana, Tardío and Morales 2005; Pieroni et al. 2005a; Pieroni et al. 2003; Forbes 1976). In Northern Spain and Southern Italy this has been found to be partly due to the fact that wild plant consumption is perceived negatively, as a symbol of poverty (Pardo-de-Santayana, Tardío and Morales 2005; Pieroni et al. 2005a). Attempts have been made to semidomesticate a few species whose gathering is time-consuming in Italy, Anatolia and Greece (Forbes 1976; Ertuğ 2003; Pieroni 2003; Pieroni et al. 2005b). Of late, a high number of popular-science field guides to edible plants and mushrooms have been published in the German language area (e.g., Hanf 1998) and culinary interest in wild plants has increased (e.g., Trum and Lotter 1998; Till 2001).

The research described in this chapter reinforces this emergent and general picture of wild plant use in Europe, but by detailing how wild plants are used and how related practices have undergone change in a contemporary community in the Austrian Alps, we also emphasise the importance of regional and local variation in people–plant relations across Europe.

The Eastern Tyrol

Our research area is located in the Alpine region of Eastern Tyrol. Geographically, Eastern Tyrol is separated from Northern Tyrol. Instead it borders the regions of Salzburg, Carinthia (*Kärnten*) and the Italian Autonomous Province of Southern Tyrol (*Südtirol*) (Figure 3.1). This alpine landscape is characterized by spruce forests up to 1,700 m above sea level and alpine pastures up to 2,500 m (Gärtner 2001). Annual precipitation in the region is 826–1,354 mm and the mean annual temperature is 2.8–6.9 °C, though local conditions can vary depending on exposure to the sun and altitude. This broad range of natural conditions within a small area has lead to a highly diverse pattern of human–environment relationships (Staller 2001).

The name 'Eastern Tyrol' (*Osttirol*) was first mentioned in the year 1837 and includes, as it has done since the end of the First World War, the district of Lienz, which is part of the Austrian province of Tyrol (Figure 3.1). The district contains thirty-three municipalities and 50,678 inhabitants (Ingruber 2001).

Despite the increasing secularization of the last decades, there are still many religious beliefs and customs practised that have their origin in the Roman Catholic religion. Social events on Catholic holidays are still very popular and celebrated by the local population. Associations like the *Musikkapellen* (brass bands playing folk music) and folk and church choirs are a lively part of the Eastern Tyrolean culture (Köck 2001; Strasser 2001). The local dialects of Eastern Tyrol belong to the Southern-Bavarian language area and share many terms and expressions with the neighbouring valley of Südtiroler Pustertal and Upper Carinthia (Hatzer 2001).

In Eastern Tyrol, 2,313 farms are managed by families. An additional 445 farms are managed by associations of varying legal status. Of the

Figure 3.1. The study area: the village of Kartitsch in Eastern Tyrol, Austria (retrieved May 2006 from http://www.mygeo.info; modified)

total of 2,758 farms, 20.3 percent of them are run on a full-time basis; the rest combine farming with off-farm labour. Eighty-three farms are not accessible by car; eleven farms are accessible only by foot (Brugger 2001).

The historical form of agriculture in this region can be described as *mountain cereal grazing* (Netting 1981), where arable farming (cereal cultivation, field vegetables, fibre crops etc.) and dairy cattle were the main components of the subsistence system until the 1970s. Large parts of today's meadowlands were once tilled up to 1,700 m altitude.

Farming systems in Eastern Tyrol have been in a process of change in the last few decades. Cultivation of cereals (e.g., rye, wheat), fibre crops (e.g., flax, hemp) and field vegetables (e.g., pea, broad bean, turnip) has been declining during the last three decades, due to unfavourable economic circumstances and their need for high inputs of labour. The economy is dominated by meadowland in lower zones, where hay is produced for winter fodder, and by pastureland in the higher alpine zones, where cattle remain throughout the summer. Nowadays the majority of mountain farms in Eastern Tyrol are based on cattle breeding, milk production and timber harvesting for cash income. Some farmers offer beds for tourists and/or process milk, meat and other products from the farm. Farmers have also diversified their own food production by adding activities such as keeping sheep, goats, pigs, hens or bees, and/or growing fruit, herbs or vegetables (e.g., potatoes). Farming is combined with different kinds of off-farm labour and federal subsidies play an important role as income (Vogl-Lukasser 1999).

Research in Kartitsch

We selected the municipality of Kartitsch because previous studies (e.g., Vogl-Lukasser 1999) showed that gathering is still a custom practised by farmers there. Kartitsch is a village of 59 km^2 at 1,353 m altitude with 896 inhabitants. The village is situated between the Dolomites to the north and the Carnic Alps to the south (Statistik Austria 2001). Today, the economic panorama of Kartitsch is still dominated by agriculture, forestry and tourism. Many farmers work part time in off-farm labour and the majority of the inhabitants in nonagricultural business work in neighbouring villages (Goller and Kofler 2001). Of the 129 farms in total, 20 percent are operated privately on a full-time basis, 36 percent on a part-time basis, and 44 percent are owned by corporate bodies (Statistik Austria 2001).

The purpose of the study was to work with farmers in Kartitsch who were knowledgeable about gathering plant species. Kartitsch farmers known to the authors were asked if they themselves gathered plant species and if they were willing to participate in data collection. They were then asked if they could recommend additional people who had knowledge of gathered plant species and were known to be active gatherers. The result was a list of forty-three potential respondents. From the forty-three names we excluded

1. Those who did not live in Kartitsch (twelve names).
2. Those who refused to participate in an interview (eleven names).

The remaining twenty participants (fifteen women, five men) lived in seventeen households of the village of Kartitsch. The age of the informants varied between thiry-four and seventy-nine years, and most of them had been born and had grown up in the village.

Most of the interviews required more than one visit, sometimes conducted with the participation of other family members or neighbours. During these interviews, respondents spoke in their own dialect; there was no need for translators because of the authors' ability to speak and understand the local idiom.

Interviews were conducted between July and September 2002. Not all respondents participated in all data collection events, due to time constraints on their part or lack of willingness. During the course of the interviews most male respondents decided not to participate any more, because – as they said – 'their wives knew more about the topics introduced by the interviewer'.

Methods used for data collection included (Bernard 2002):

- Free lists (i.e., answering spontaneously to a request, mentioning all items that come to the mind of a respondent when thinking about a given topic or domain) about gathered plant taxa (n=16; 12 female, 4 male).
- Semistructured interviews; first on motivations for gathering plants, with attention to religious practices/beliefs related to gathered plant species (n=16; 12 female, 4 male), and second, on the most quoted religious practice only (n=14; 13 female, 1 male).
- Structured interviews, first on sociodemographic data of the respondents (n=20; 15 female, 5 male). and second, on the uses of the most quoted plant species in the free lists (n= 15; 14 female, 1 male).
- Pile sorts, to elicit the local classification system for gathered plant species (n=14; 13 female, 1 male). Pile sorts were done with pressed herbaria specimens and the first author's photos, fixed on numbered cards and protected with a transparent film. Informants were asked to name the plants on the cards in their local dialect and to sort them into as many piles of similar items as they wanted, unconstrained by any particular criteria for sorting. The comments of the respondents regarding their choice of criteria for sorting were recorded.
- Informal interviews with older people, which investigated changes in the gathering of plants in the study area.

In addition, nonparticipant and participant observation was carried out during gathering, processing and cooking of gathered plant species, as well as during other agricultural activities, such as haying. Transect walks, with two very knowledgeable persons, elicited in-depth information on their choices of places and plants for gathering. Plant voucher specimens were taken of the most frequently mentioned plant species for use in pile sorts.

With the consent of the respondents, interviews were recorded on tape whenever possible and were partly transcribed afterwards. Pictures taken during interviews and transect walks were returned to the farmers and results were shown to the local community in a public presentation. The final report of this study, written in German, was handed over to the village library of Kartitsch.

Voucher specimens, photographs and audio recordings of each interview were deposited at the University of Natural Resources and Applied Life Sciences in Vienna (BOKU). Collected data was stored in an Microsoft® Access database. Free lists and Pile sorts were analysed according to frequency, average rank, salience and consensus (Weller and Romney 1988; Bernard 2002) with ANTHROPAC 4 (Borgatti 1996a).[2] The network data was visualized with VISONE.

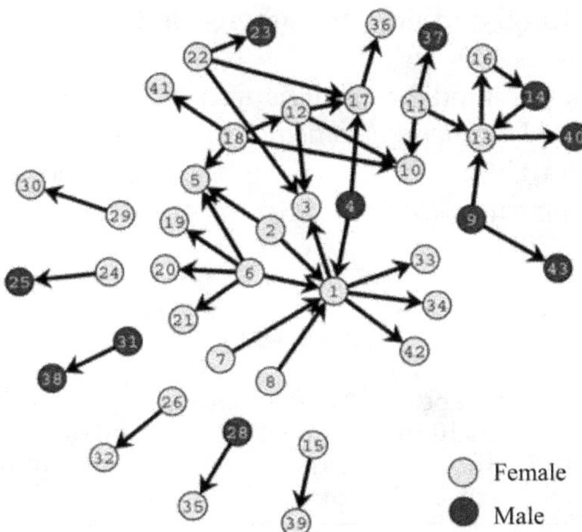

Figure 3.2. The gendered distribution of knowledge on wild plants in the village of Kartitsch: network of recommendations of inhabitants of the village who were asked to recommend people that have a huge knowledge on wild plants and that are still gathering (n=31)

Women as Local Experts on Gathering

The findings of the study show that the majority of the respondents recommended people of their own village as experts on this topic. The recommended people are mostly women (n=31), while men (n=11) are positioned only peripherally in the network, that is, being recommended rarely and recommending few other people (Figure 3.2). The female informants mainly recommended other women as experts and male informants recommended more women than men. The densely connected 'knots' in Figure 3.2 are exclusively women. Also, confirmed by participant observation, we can say that women have extensive exchange relations for their knowledge of wild plants (e.g., exchanging recipes on how to prepare wild plants for local dishes). Nearly all women listed as experts by more than one informant also had an important social function in the village in the past or still have this social status in the present (e.g., several women were heads of the Women Farmers Association in the past).

The Location of Gathered Plants

The sixteen respondents who free listed gathered plants mentioned 393 items in total. Twenty-one generic or higher-level classes were listed without relation to a specific plant species. The remaining 372 items referred to 106 plant species. The listed plant species can be classified according to their location in the environment, in this case using the categories of the authors:

1. Plant species, fungi and lichens gathered in the wild (72 species; Table 3.1).
2. Plant species cultivated in gardens and arable plots (29 species, e.g., *Mentha x piperita* L., *Calendula officinalis* L., *Althaea officinalis* L.).[3]
3. Plant species bought in local shops or chemists (5 species, e.g., *Pimpinella anisum* L., *Coriandrum sativum* L., *Acorus calamus* L.).

Table 3.1. Free-listed plant species, fungi and lichens gathered in Kartitsch (Eastern Tyrol, Austria), including frequency of respondents, percentage (total n=16), average rank, Smith's Salience Index and percentage of the actually gathered species in the year 2002; sorted by frequency

Plant species	Frequency	Percentage (n=16)	Average rank	Smith's Salient Index	Actually gathered in 2002: percentage (n=16)
Vaccinium myrtillus L.	15	94	7.667	0.554	75
Arnica montana L.	13	81	2.385	0.753	69
Vaccinium vitis-idaea L.	12	75	9.417	0.391	69
Picea abies (L.) H. Karst	11	69	9.909	0.384	44
Plantago lanceolata L.	11	69	9.455	0.403	31
Rubus idaeus L.	11	69	10.000	0.347	56
Urtica sp.	11	69	10.273	0.390	56
Hypericum sp.	10	63	6.300	0.466	38
Sambucus nigra L.	10	63	10.500	0.278	50
Achillea millefolium L.	9	56	7.111	0.447	56
Alchemilla sp.	8	50	8.750	0.347	31
Boletus edulis Bull.: Fr.	7	44	11.714	0.181	31
Cantharellus cibarius (Fr.: Fr.) Fr.	7	44	8.571	0.231	31
Cetraria islandica (L.) Ach.	7	44	9.429	0.242	25
Taraxacum sp.	7	44	12.857	0.218	25
Thymus sp.	7	44	10.714	0.275	25
Allium schoenoprasum subsp. *alpinum* (DC.) Čelak.	6	38	10.833	0.168	31
Betula sp.	6	38	18.000	0.137	31
Equisetum sp.	6	38	10.667	0.205	13
Fragaria vesca L.	6	38	14.500	0.165	19
Tussilago farfara L.	6	38	15.833	0.173	13
Rhododendron sp.	5	31	25.200	0.032	25
Sambucus racemosa L.	5	31	18.000	0.049	0
Carum carvi L.	4	25	14.250	0.120	6
Gentiana sp.	4	25	2.500	0.202	6
Juniperus communis L.	4	25	14.500	0.143	6
Plantago major L.	4	25	14.500	0.133	13
Salix sp.	4	25	17.000	0.086	25
Larix decidua Mill.	3	19	13.667	0.071	13

Table 3.1. *continued*

Plant species	Frequency	Percentage (n=16)	Average rank	Smith's Salient Index	Actually gathered in 2002: percentage (n=16)
Peucedanum ostruthium (L.) W. Koch	3	19	20.333	0.062	19
Pinus cembra L.	3	19	16.333	0.079	13
Rosa sp.	3	19	17.667	0.038	6
Veronica sp.	3	19	18.333	0.075	0
Achillea clavenae L.	2	13	11.000	0.065	6
Artemisia sp.	2	13	24.000	0.039	0
Leucanthemum sp.	2	13	3.000	0.113	0
Macrolepiota procera (Scop.: Fr.) Singer	2	13	11.500	0.064	6
Nasturtium sp.	2	13	23.500	0.034	6
Nigritella nigra L.	2	13	19.000	0.014	0
Rumex sp.	2	13	21.500	0.040	0
Sorbus aucuparia L.	2	13	14.000	0.067	13
Valeriana celtica L.	2	13	11.000	0.032	0
Viola sp.	2	13	15.500	0.067	6
Amanita muscaria (L.: Fr.) Hook.	1	6	18.000	0.027	0
Berberis vulgaris L.	1	6	25.000	0.013	0
Campanula latifolia L.	1	6	1.000	0.063	6
Campanula sp.	1	6	6.000	0.054	0
Carlina acaulis L.	1	6	27.000	0.010	6
Dianthus barbatus L.	1	6	29.000	0.006	6
Euphrasia sp.	1	6	11.000	0.043	6
Gentiana lutea L.	1	6	8.000	0.043	6
Gentiana punctata L.	1	6	9.000	0.047	6
Humulus lupulus L.	1	6	31.000	0.002	6
Lamium sp.	1	6	30.000	0.011	0
Leontopodium alpinum Cass.	1	6	22.000	0.003	0
Mentha longifolia (L.) Huds.	1	6	21.000	0.023	6
Myosotis sp.	1	6	5.000	0.055	0
Pinus mugo Turra	1	6	20.000	0.024	6
Pinus sylvestris L.	1	6	18.000	0.028	0
Primula sp.	1	6	3.000	0.059	0
Ramaria aurea (Schaeff.) Quél.	1	6	32.000	0.007	0
Ribes nigrum L.	1	6	5.000	0.050	6
Rubus fruticosus agg.	1	6	20.000	0.025	0
Ribes uva-crispa L.	1	6	10.000	0.044	0
Russula vesca Fr.	1	6	14.000	0.039	0
Salvia pratensis L.	1	6	10.000	0.006	0
Stellaria media (L.) Vill.	1	6	7.000	0.045	6
Tilia sp.	1	6	5.000	0.055	0
Usnea sp.	1	6	21.000	0.022	0
Vaccinium uliginosum L.	1	6	27.000	0.008	0
Valeriana sp.	1	6	23.000	0.020	0
Verbascum sp.	1	6	14.000	0.026	6

For the purposes of this study, further analysis used only those plant species gathered in the wild (72 species, including fungi and lichens). The species of the categories 2 and 3 (above) were excluded after checking with every respondent the site where these listed species were gathered. Plant species that are cultivated in local gardens or bought in local stores or pharmacies are mainly used for medicinal purposes or as supplements in daily nutrition (e.g., spices or beverages). Most of the plant species of the categories 2 and 3 belong to the group locally called *kräuter, kreita* or *kreitlan* (herbs).

The Importance of Free-listed Wild-gathered Species

Vaccinium myrtillus was the most frequently quoted wild plant species, listed by 94 per cent of respondents. Other frequently quoted species are *Arnica montana* (81 per cent), *Vaccinium vitis-idaea* (75 per cent), *Urtica* sp. (69 per cent), *Rubus idaeus* (69 per cent), *Picea abies* (69 per cent) and *Plantago lanceolata* (69 per cent) (Table 3.1).

If sorted by Smith's Salience Index, a measure of cognitive importance based on frequency and average rank, *Arnica montana, Hypericum* sp. and *Achillea millefolium* appear higher in the list than if sorted by frequency, suggesting they are more likely to be mentioned earlier in the free list (i.e., of higher rank). Nevertheless, the most frequently quoted (≥ 50 per cent) species are also those with the highest Smith's Salience (see Table 3.1). This result suggests that informants generally agreed on which plants are typical of the domain 'wild plants that are gathered'. These results do *not* imply that the most salient plants are also the most frequently harvested or used. Specific questions directed to actual use of the plants would be required to provide that information.

A consensus analysis of the free lists supports this finding. ANTHROPAC's consensus-analysis program creates a hypothetical model of what the culturally correct answer would be to the free list question, 'What wild plants that can be gathered are there?' In this case, the model generated includes ten plant species – the same ten species that were listed most frequently by respondents (see Table 3.1). The pseudoreliability of this model is calculated at 0.92, which suggests that variation within the group is small enough for the model to represent the group consensus. Each informant's free list is compared to the model to see how typical, or close to the consensus answer, their responses were. The average of all these comparisons, known as the 'average informant agreement', was about 66 per cent (range: 33–87%), which again shows some variation among the group, but overall supports the one-culture assumption of the model (Borgatti 1996b).

Local Categories of Most-quoted Wild Plant Species

The pile-sorting exercise was done with the eleven most frequently mentioned plant species (≥ 50 per cent; Table 3.1). Each informant created piles of the species s/he thought were most similar, and was then asked to explain the reasoning behind these groupings. The aggregated data of all the pile sorts was analysed using ANTHROPAC and can be displayed in several ways. Figure 3.3 is a tree diagram showing which plants were more likely to be placed in the same pile. The shorter the distance between any two plants the more similar they are assumed to be. From this diagram and our interview data, we suggest that respondents group these wild plants into four categories. These are:

1. *Beeren, beern* (berries): *Vaccinium myrtillus, Vaccinium vitis-idaea, Sambucus nigra* and *Rubus idaeus.*
2. *Kräuter, kreita, kreitlan* (herbs): *Arnica montana* and *Hypericum* sp.
3. *Kräuter, kreita, kreitlan* (herbs): *Achillea millefolium, Urtica* sp., *Alchemilla* sp. and *Plantago lanceolata.*
4. *Baam, baamda* (trees): *Picea abies.*

Respondents used the same label (*kräuter, kreita, kreitlan*) for groups 2 and 3, and so these may be considered subcategories within a covering group of 'herbs'. Respondents distinguished the two groups based on their means of preparation as either a tea or an alcohol/oil macerate, respectively. Local

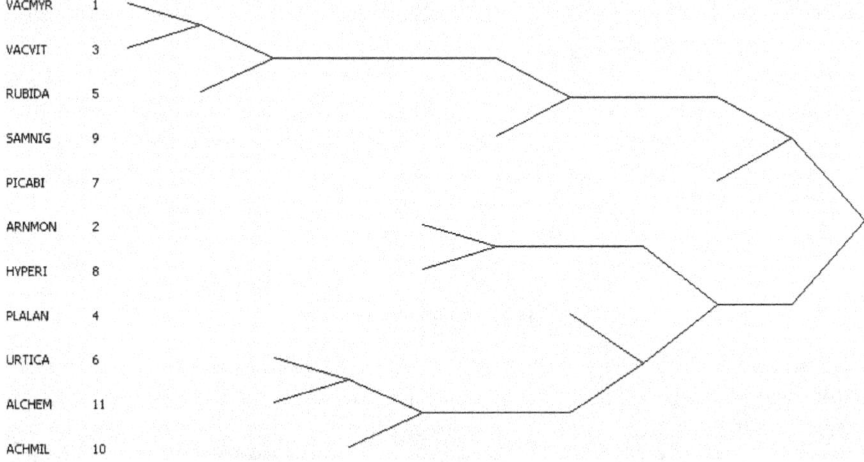

Figure 3.3. Tree diagram of the similarities between *Johnson's Hierarchical Clustering* for the eleven most frequently mentioned wild-gathered plant species according to *Pilesorts.* n=14 (1 *Vaccinium myrtillus,* 2 *Arnica montana,* 3 *Vaccinium vitis-idaea,* 4 *Plantago lanceolata,* 5 *Rubus idaeus,* 6 *Urtica* sp., 7 *Picea abies,* 8 *Hypericum* sp., 9 *Sambucus nigra,* 10 *Achillea millefolium,* 11 *Alchemilla* sp.)

terms used for the respective names (as mentioned above) differ only in pronunciation.

In fact, the piles and the sorting criteria given show that most respondents sorted the species according to a combination of the plant parts that they collected (fruits, flowers and leaves) and the typical processing methods:

1. Fruits: eaten fresh or processed as juice or jam.
2. Flowers: macerated in alcohol or oil for medicinal use.
3. Leaves and flowers: dried for medicinal tea.
4. Several plant parts and uses.

The respondents separated *Picea abies* into its own category because of the diversity of plant parts that can be used, its many uses (food, medicine, decoration, construction, firewood), and its life form and size (Table 3.2).

Reported Uses of Most-quoted Wild Plant Species

Usually respondents reported more than one use for each plant species. The eleven most frequently free-listed species are primarily used for human medical treatment and nutrition. Other uses mentioned by the informants included animal medical treatment, ornamentation, symbolic purposes, pest management, fodder for animals and fuelwood. The most frequently collected plant parts are fruits, leaves and flowers (see Table 3.2).

The most frequently applied remedies are teas, made from leaves or flowers, for internal treatment, and distillation of *Arnica montana* for external treatment. Fruit jam is the most popular food product derived from wild plant species.

The use of some plant species has undergone change, as reported for *Urtica* sp. (stinging nettles). Some of the older respondents mentioned that in their childhood stinging nettles were fed to pigs and hens or used for haircare or as a condiment to a traditional dish – *Schlipfkropfn* (a type of stuffed pasta). The modern use of the dried leaves as blood-cleansing tea has been introduced via popular magazines and books.

Knowledge versus Actual Behaviour

During the study period, the respondents gathered 69 per cent of free-listed species. Ten of the eleven most frequently free-listed plant species were also the most frequently gathered plant species, the exception being *Plantago lanceolata*, which was eclipsed by *Boletus edulis* in 2002 (see Table 3.1). The three most frequently mentioned species were also the three most frequently collected. Thus, the findings support the idea that plant salience is for the most part utilitarian; that is, the most salient plants are

Table 3.2. Categories of uses, part(s) used and number of quotations per use for the eleven most frequently free-listed wild-gathered plant species in Kartitsch (Eastern Tyrol, Austria) (n=15). Numbers comprise actual as well as past uses

		Categories of use and sum of number of quotations per category of uses					
Species	Part(s) used	Human medicinal treatment	Animal medicinal treatment	Human nutrition	Decoration	Symbolic meaning	Other uses
Vaccinium myrtillus L.	Berries, leaves, aerial parts	12	1	43		1	
Arnica montana L.	Flowers, aerial parts, ligules	31	3			2	1
Vaccinium vitis-idaea L.	Berries, leaves, aerial parts	16		21	7	4	
Plantago lanceolata L.	Leaves, aerial parts					2	
Picea abies (L.) H. Karst	Young shoots, twigs, needles, stem, cones, resin, bark	19		6	9	7	29
Rubus idaeus L.	Berries, leaves	3		31		2	
Urtica sp.	Leaves, aerial parts, roots, whole plant including roots	13		11			16
Sambucus nigra L.	Berries, flowers	11		26	1		2
Hypericum sp.	Leaves, flowers, aerial parts	23				1	1
Achillea millefolium L.	Leaves, flowers, aerial parts	17			5	6	3
Alchemilla sp.	Leaves	16	1		1	1	

those that are used more often. On the other hand, there were a few wild plants mentioned in the free list that were rarely or not at all gathered by respondents in 2002. The reasons for this are:

- Modifications of the habitat caused a decrease in abundance (e.g., *Carum carvi, Nasturtium* sp.).
- Too labour-intensive processing (e.g., *Rosa canina*).
- No need to gather every year because stocks of past years still available (e.g., *Juniperus communis*).
- Related customs and uses not common any more (e.g., *Achillea clavenae*). The difference between free-listing and gathering is especially high for some plant species used typically for medicinal purposes (e.g., *Cetraria islandica, Thymus* sp., *Equisetum* sp., *Tussilago farfara*).

Only a small number of free-listed wild plant species are not gathered at all (see Table 3.1). The reasons for not gathering are, according to the respondents:

- Plant species not growing wild in Kartitsch (e.g., *Tilia* sp.).
- Plant species and purpose only known from literature; but purpose not usual in Kartitsch (e.g., *Salvia pratensis, Valeriana* sp., *Lamium* sp.).
- Plant species that are endangered have been protected by law and may not be gathered any more (e.g., *Leontopodium alpinum, Nigritella nigra, Valeriana celtica*).
- Plant species or fungi still grow in Kartitsch, but are difficult to find (e.g., *Campanula latifolia, Ramaria aurea, Russula vesca*).

Motivation for Gathering

Thirty-one per cent of the respondents mentioned emotional motives (e.g., loving nature, enjoying self-produced delicacies) and 26 per cent gave health motives (e.g., self-medication with medicinal plants, healthiness of wild plants) as reasons for gathering wild plants:

> [I gather] simply as a hobby and out of love for nature and for plants and because it all is life. And because it seems to me that one can talk with them, with the plants, and that they almost give answers.[4]

> I enjoy it – to be in nature. A little stems from grandmother, a little from mother, and I enjoy it myself when I walk and everything is in bloom. And that I've got something if someone has a stomach ache or something else, to then have something in store.[5]

The quality of homemade products derived from wild plants (said to be 'pure' or 'natural') and the contribution of wild plants to daily subsistence ('no need to buy anything during the whole winter') were mentioned by 15 per cent and 10 per cent of the respondents, respectively. Some respondents want to perpetuate the gathering tradition of their parents and grandparents (5 per cent). Saving money (5 per cent) or selling products derived from wild plants at local markets (3 per cent) are less important to the respondents (Table 3.3).

It is very common that women prepare fruit jams, syrups and digestive drinks (schnapps and liqueurs) from gathered wild fruits (*Rubus idaeus, Vaccinium myrtillus, Vaccinium vitis-idaea*) for barter and as gifts. Nevertheless, respondents do not explicitly mention their contribution to the local, informal, nonmonetary food economy when asked for their motivation.

In interviews with older respondents, information about current aspects of gathering was combined with stories about gathering in the respondents' personal history (starting from 1930, approximately). Thus, changes in the motives for gathering between earlier times and nowadays were emphasised. For one older woman, who knows much about medicinal plants, the gathering of wild medicinal plants is strongly related to memories of her grandfather. In her childhood, she frequently joined him during gathering or was sent alone to gather the plants he needed. Her grandfather was well known for healing sick persons with medicinal plants in the village at a time before formally educated doctors started to practise there. Also, other informants mentioned that consultations with doctors and purchase of remedies in pharmacies was difficult in the village because of a lack of health insurance; the long distance and lack of transport to the doctors; there being few available doctors;and scant financial resources of farmers to pay doctors and buy remedies.

In addition to the dependence of people on gathering for medicinal needs, respondents mentioned the importance of gathering for subsistence (for food and animal fodder) and as a means to generate income. Older informants remembered, for example, the harvesting of huge amounts of

Table 3.3. Motives for gathering wild plant species in Kartitsch (Eastern Tyrol, Austria) (n=16). Multiple answers were possible

Motive	Frequency	Percentage of sum of topics mentioned
Joy	12	31
Health and medicinal purposes	10	26
Food quality	6	15
Subsistence	4	10
Maintain traditions	2	5
Saving money	2	5
Commercialization	1	3
No answer	2	5
Total	39	100

wild fruits like *Vaccinium myrtillus* and *Vaccinium vitis-idaea* for markets and the village's grocery. The district pharmacy was provided with medicinal herbs (e.g., *Arnica montana*) gathered in the region. This was an important economic activity during the Second World War and in post-war times for poor people of the village.

The Social Act of Gathering

Some respondents prefer gathering on their own: they enjoy the gathering as an individual contemplative, relaxing activity, leaving behind all the work at the farm. Others enjoy gathering in nature with friends and relatives. One older woman considered it very important to be accompanied by her granddaughter, teaching her the uses of the plants on the roadside.

Independent of these individual preferences, the social organization of gathering depends on what is gathered, whether wild greens and herbs or berries and fungi; because they differ in length of harvest season, labour requirements, distance to growing site, pattern of distribution and amount of plant material needed.

Wild greens or herbs (*Urtica* sp., *Achillea millefolium*) are usually (except for on Assumption Day) collected around or near the house or the farm, at the edges of meadows and forests. This gathering is rarely planned and therefore, in most cases, a spontaneous activity. It 'happens' during a walk or in conjunction with other agricultural activities, such as haying or looking after the cows in the pasture.

> When you go to the valleys to the animals it is an automatic reflex to take something away with you .... I just go – every time when I go to the valley I take something away with me, whatever I see on the way, I take it with me.[6]

> And really one would just gather when one walked in the place by chance and passed by, then one would just take something, so one would not go on purpose.[7]

The number of people that participate in this gathering process depends on the situation. This way of gathering was in former times predominant, because of frequent hikes of people on foot as part of their farming activities.

Contrary to the gathering of wild herbs and greens, gathering of wild berries (*Vaccinium myrtillus, Vaccinium vitis-idaea and Rubus idaeus*) is often combined with a day trip into the mountains with the whole family or a group of friends. It is a good occasion to take a day off, to chat and to have a picnic surrounded by nature. Places where wild berries grow are well known by local people (though favourite places are sometimes kept secret) and often far from the farms. The gathering of wild berries can be more time-intensive because of the huge amounts needed for producing jams, syrups and liqueurs. In former times these time-intensive activities were done by children, because older members of the family had to dedicate their labour to more important farm activities.

Harvesting mushrooms is a time-intensive and very popular local activity too. But *Boletus* sp., *Cantharellus* sp. and others are not only appreciated by local people, but also by tourists. Local people often mentioned the overharvesting of mushrooms by 'tourists' in the region. These 'tourists' are actually well organized collectors. They travel in big groups in buses, communicate with walky-talkies and often do not only harvest for their own needs but also for selling fresh or dried mushrooms in local markets in neighbouring regions of Eastern Tyrol.

Regardless of whether people prefer to gather alone or in groups, they usually like to share the products derived from gathered wild plants. For example, women take pride in providing delicacies for their families, serving them to their guests or sharing homemade products with their relatives or neighbours. Interviews were often sweetened by the women themselves, offering us homemade jams or liqueurs.

Shared Traditions and Beliefs Concerning Wild Plant Use

Barter and gift giving, for instance of gathered plant species and products, are shared traditions in the daily life of local people in Kartitsch. Other shared traditions related to gathering can be observed in local customs that are strongly based on the Catholic beliefs of the respondents and mainly practised on Catholic holidays. The blessing of the so-called *kräuterbuschn* ('bunches of herbs') on the Catholic holiday *Mariä Himmelfahrt* (Assumption Day, 15 August), and the blessing of bundled branches of *Salix* sp. on Palm Sunday were the only customs mentioned that use wild gathered plants (Assumption Day: 68 per cent of the respondents; Palm Sunday: 25 per cent).

Gathering herbs and flowers for Assumption Day is exclusively women's work (except for one male informant who keeps the tradition of his mother alive). Wild herbs are gathered one or two days before Assumption Day. Women make bunches combining these gathered wild herbs with cultivated plants that are collected from the gardens.

Our participation in a family's preparation of bunches helped to elucidate criteria for the composition of bunches. Women focused on tradition, on the *conditio sine qua non* of medicinal herbs, on the blaze of colour, and on the availability of certain herbs and flowers. For example, one woman replaced *Hypericum* sp., which she knew was part of every *kräuterbuschn* in former times, with *Solidago virgaurea*, because the latter is also a yellow flowering species but is easier to find near the farmhouse.

The respondents of Kartitsch did not mention symbolic meanings of specific plant species during interviews. However, an older woman from a neighbouring village explained in an informal conversation that in former times each herb represented a location, such as an alpine hut or arable plot, which should be protected from negative influences, such as lightning or hail.

Through our participation in the celebration of Assumption Day, it was confirmed that the diversity of wild plants in bunches carried by women varied significantly. Herbs and flowers are made into small bunches, the *kräuterbuschn*, which are sprinkled with holy water during the ceremony to give them the Virgin Mary's blessing.[8] After the blessing, bunches are brought home for decorative and ritual purposes.

Older informants remembered from their childhood times when every girl carried her own bunch of herbs to the blessing. Ten to fifteen years before our survey, this tradition nearly disappeared because of the decreasing number of people attending church services. Having recognized the potential effect of losing local religious practices, a few engaged women in Kartitsch revitalized the ritual of gathering and bundling herbs and flowers for Assumption Day. In the course of these efforts, the local custom was modified by these women, by making smaller bunches than was usual in the past. Nowadays, bundling of one bunch for each family household is not very common, and so those that are made are distributed after the blessing among the congregation, and are also desired souvenirs for tourists. Most respondents emphasised that, most importantly, the bundling and blessing of herbs and flowers is a beautiful and old custom for them, which is worth keeping alive.

The use of these herbs is well known among local people. The burning of a few herbs from these blessed bunches in the stove during heavy thunderstorms should bring protection against lightning and other dangers for humans and farm animals. Nearly all interviewees repeatedly confessed that they did not really believe that the burning of blessed herbs during heavy thunderstorms could keep away danger. Nevertheless, 25 per cent of respondents (mainly the older generation among them) still practise this ritual from time to time, probably – as some of these respondents supposed – because it lessens their fear in situations of emotional tension.

Another ritual use of these herbs is the 'smoking' of farmhouses and stables in the nights before Christmas, New Year and Epiphany (mentioned by nearly one-quarter of the respondents). As in the past, fireproof vessels are filled with glowing coals and blessed incense is strewn on them. The rising smoke is said to be purifying, blessing rooms and stables for the following year. The incense is mixed not only with blessed herbs, but also with branches of *Salix* sp. gathered before the holidays of Assumption Day and Palm Sunday. Blessed branches of *Salix* sp. are said to have a similar function of protecting humans and farm animals from danger. The blessing of bundled branches of *Salix* sp. on Palm Sunday was listed by more than 25 per cent of informants, asked to name local customs associated with the gathering of wild plants. One informant adds to this mixture a few roots of *Peucedanum ostruthium* that have a balmy scent – following his parents' and grandparents' tradition.

Conclusions

In other studies in Europe that take into account the social aspects of gathering, this activity is regarded as a woman's domain, where gift-giving and exchange of wild plants contribute to extending women's social space and strengthening social relations (Forbes 1976; Pieroni 1999; Agelet, Bonet and Vallès 2000; Ertuğ 2000, 2003; Pieroni 2003; Pieroni et al. 2003). Our network of plant gatherers in Kartitsch shows that women are also the local experts on wild plant gathering. The densely connected 'knots' shown in Figure 3.2 represent 'knowledge pools' of women who are considered by local people to be very knowledgeable of wild plants. Their knowledge has also been confirmed by the results of our interviews. These women share their knowledge and the majority of their exchange partners were persons from within their own municipality. This suggests that the gathering of wild plants is still a traditional element of the local knowledge system of Kartitsch. The women's knowledge, as measured by the free listing, and their actual behaviour in terms of gathered species, are highly correlated. Those plants that are best known are also those that are most frequently used. However, nonmaterial uses are still important, and several plants no longer collected are still known.

Nevertheless, this living part of local culture has changed in the last few decades particularly with regard to the motivation and uses of gathered wild plants and the practice of gathering. Motivation for gathering wild plants is now dominated by emotional rather than economic reasons. Women gather because they enjoy doing it, rather than because of a need to survive. A similar trend has been observed in Anatolia (Ertuğ 2000), while in Italy and Spain health- and nutrition-related aspects predominate (Pieroni 1999, 2000; Tardío, Pascual and Morales 2002; Pieroni et al. 2005a; Rivera et al. 2005), and in Bulgaria income generation is still very important (Ploetz and Orr 2004).

Plant species that are not gathered any more, or only gathered to a lesser extent, are mostly species that were used for medicinal purposes in former times. Although there are still some medicinal plant species that are very popular and in common use (e.g., *Arnica montana*), the majority are in decline (e.g., *Cetraria islandica, Thymus* sp., *Equisetum* sp., *Tussilago farfara*).

Studies in southern Italy (Pieroni 2003) and Greece (Forbes 1976) point out that people rarely go to the fields solely for the purpose of gathering of wild plants. They would rather combine gathering with other work or do it on their way home from the fields. This is also true for Kartitsch, but with less frequency and less importance now than in former times. The decrease in gathering of wild plants that are collected en route is due to fewer hikes done on foot (trips are more often taken by car), replacement of manual work by machinery and changes in agricultural land use. However, for some species of berry and mushroom, gathering has become a joyful social event explicitly planned for the whole family.

The custom of blessing herb bunches on Assumption Day is an example of change. The overall importance of the custom remains the same, but it has shifted from an activity performed by every family to one carried out by a few women, though their intention is to benefit anyone who attends the ceremony. For these women, the perpetuation of the tradition is what is most important and they hold only a general interest in wild plants themselves. They make a large quantity of small bunches and share them with the visitors at the church. While the blessing of herbs on Assumption Day remains embedded in the fixed structure of this Catholic festive day and follows a strict protocol, the participants are social actors who shape customs according to their liking. Although people no longer expect consecrated plants to protect them from danger, they practise the burning ritual from time to time: the plants continue to soothe anxiety, even if it be for reasons other than the threat of nature. Similar qualities of blessed herbs are known from the French Alps or the Spanish Pyrenees (Villar et al. 1987; Brüschweiler 1999). Also, in northern Italy several wild plant species protect against the 'evil eye', bad influences or spirits, and illnesses (Pieroni 2000).

The observation that herbs that are more readily available in the proximity of the houses or in homegardens (e.g., *Hypericum* sp.) are substituted for others in the Assumption Day bunches cannot be generalized, but it may indicate the impact of changes in agricultural production.

In free listing wild plants, respondents also listed plants they cultivated in homegardens. This is not a misunderstanding of the question asked, but a reflection of the history of plant management in the area, whereby wild plants that are no longer easily available, either due to habitat loss or lack of time for gathering, have been slowly moved into nearby habitats and therefore come under greater management by farmers. This process was confirmed by participant observation during fieldwork and is supported by other ethnobotanical studies on homegardens in Eastern Tyrol (Vogl-Lukasser 1999). This has also been reported as a very common practice in other European countries (Greece, Turkey, several mountain regions in Catalonia, and Italy: Forbes 1976; Agelet, Bonet and Vallès 2000; Ertuğ 2003; Pieroni 2003).

The change process from a predominantly subsistence- to a predominantly market-oriented livelihood system, paired with improvements in the national healthcare system and the ongoing secularization of rural societies in the Austrian Alps in the twentieth century, has resulted in a noticeable shift in meanings ascribed to wild plants by local people in Eastern Tyrol. Wild food plants have lost their importance as vital components of the diet, and are nowadays perceived as well appreciated delicacies. Wild medicinal plants for self-medication in farmers' families are no longer indispensible, but rather an alternative for people that prefer natural remedies to synthetic ones. It is still very popular to use wild plants in local customs because of their decorative function. However, belief in their sacred character and in the effects of ritual use of wild plants has almost disappeared.

Acknowledgements

We are grateful to all the people of Kartitsch, who shared their knowledge and experiences about wild plants with us, for their patience and hospitality. Dr Rajindra K. Puri supported us with advice during fieldwork.

Notes

1. Some of these countries are: Turkey (Ertuğ 2003, 2004); Greece (Forbes 1976; Iatrou and Kokkalou 1997; Lambraki 2000; Savvides 2000); Albania (Pieroni, chapter 2 in this volume); Italy (Pieroni 1999; Nebel, Pieroni and Heinrich 2006); Germany (Pieroni et al. 2005c); Portugal (Carvalho and Morales, chapter 7 in this volume); and Spain (Bonet and Vallès 2002; Tardío, Pascual and Morales 2005; Rigat, Garnatje and Vallès 2006).
2. Salience is a tool to recognize the culturally most important items in a domain (Quinlan, Quinlan and Nolan 2002). The Smith's Salience Index, developed by Jerry Smith (1993), takes into account the frequency of an item and the average rank of the item across multiple free lists, where each list is weighted by the number of items in the list.
3. This chapter uses Bye's (1993: 717) definition of *wild* and *cultivation*: 'Wild plants survive and reproduce naturally without the necessity of human intervention' (in our study: species that grow outside farmers' gardens and outside arable plots). *Cultivation* refers to 'where peoples' special actions modify environmental conditions to promote optimal production and reproduction' (in our study: gardens and arable plots, because only cultivated species from these agroecosystems were listed by the respondents. Whenever wild species were cultivated in gardens they were not included in the analysis presented here).
4. *'Aanfach als Hobby und wegen der Liebe zur Natur und zu die Pflonzen und weil des olls Leben isch. Und weil mir ollm firkimp, man konn reidn mit de, mit die Pflonzen, und de reidn foscht hinter.'*

5. *'I hon a Freide. In der Natur zu sein, a bissl kimp fa dr Groassmuater, a bissl fa dr Mamm und mir mochts selber Freide, wenn i gea und olls bliaht .... Und i eppas honn, wenn aans Bauchwea oder eppas hot, nor hot man wos.'*
6. *'Wenn man oft in die Täler zum Vieche geat, nor isch automatisch, dass man eppas mitnimp... I gea holt olm boll i ins Tol gea, bring i irgend eppas ham, wos i unterwegs sieg, nimm i holt mit.'*
7. *'Und eigentlich hot man lei gsommelt, wenn man zufällig do gongen isch und vorbeigongen isch, nor hot man holt eppas genomm, man isch nitt also extra gongen.'*
8. Out of consideration for the religious celebration, details of the bunches could not be studied during the service nor afterwards.

References

Agelet, A., M.A. Bonet and J. Vallès. 2000. 'Homegardens and their Role as a Main Source of Medicinal Plants in Mountain Regions of Catalonia (Iberian Peninsula)', *Economic Botany* 54(3): 295–309.

Bernard, H.R. 2002. *Research Methods in Anthropology: Qualitative and Quantitative Approaches*. Walnut Creek, Calif.: AltaMira Press.

Bonet, M.A. and J. Vallès. 2002. 'Use of Non-crop Food Vascular Plants in Montseny Biosphere Reserve (Catalonia, Iberian Peninsula)', *International Journal of Food Sciences and Nutrition* 53(3): 225–248.

Borgatti, S.P. 1996a. *ANTHROPAC 4.0.* Natick, Mass.: Analytic Technologies.

——— 1996b. *ANTHROPAC 4.0 Methods Guide.* Natick, Mass.: Analytic Technologies.

Bouby, L. and Y. Billaud. 2005. 'Identifying Prehistoric Collected Wild Plants: A Case Study from Late Bronze Age Settlements in the French Alps (Grésine, Bourget Lake, Savoie)', *Economic Botany* 59(3): 255–267.

Brugger, R. 2001. 'Landwirtschaft', in Katholischer Tiroler Lehrerverein (ed.), *Bezirkskunde Osttirol*. Innsbruck: Löwenzahn, pp.132–136.

Brüschweiler, S. 1999. *Plantes et Savoirs des Alpes: L'exemple du val d'Anniviers.* Sierre: Editions Monographic SA.

Bye, R. 1981. 'Quelites – Ethnoecology of Edible Greens – Past, Present, and Future', *Journal of Ethnobiology* 1(1): 109–123.

——— 1993. 'The Role of Humans in the Diversification of Plants in Mexico', in T.P. Ramamoorthy, R. Bye, A. Lot and J. Fa (eds), *Biological Diversity of Mexico: Origins and Distribution*. New York and Oxford: Oxford University Press, pp. 707–731.

Christanell, A. 2003. *Wildsammlung in Kartitsch, Osttirol: Eine ethnobotanische Untersuchung des Sammelns, der SammlerInnen und der von ihnen genutzten Pflanzenarten,* M.A. dissertation. Vienna: University of Vienna.

Cunningham, A.B. 2001. *Applied Ethnobotany: People, Wild Plant Use & Conservation*. London: Earthscan Publications Ltd.

Ertuğ, F. 2000. 'An Ethnobotanical Study in Central Anatolia (Turkey)', *Economic Botany* 54(2): 155–182.

——— 2003. 'Gendering the Tradition of Plant Gathering in Central Anatolia (Turkey)', in P.L. Howard (ed.), *Women & Plants: Gender Relations in Biodiversity Management & Conservation*. London & New York: Zed Books, pp. 183–196.

——— 2004. 'Wild Edible Plants of the Bodrum Area (Muğla, Turkey)', *Turkish Journal of Botany* 28: 161–174.

Etkin, N.L. (ed.). 1994. *Eating on the Wild Side: The Pharmacologic, Ecologic, and Social Implications of Using Noncultigens*. Tucson, Ariz., & London: The University of Arizona Press.

Forbes, M.H.C. 1976. 'Gathering in the Argolid: a Subsistence Subsystem in a Greek Agricultural Community', in M. Dimen and E. Friedl (eds), *Regional Variation in Modern Greece and Cyprus: Toward a Perspective on the Ethnography of Greece*. Annals of the New York Academy of Sciences 268, pp. 251–264.

Gärtner, G. 2001. 'Pflanzenwelt', in Katholischer Tiroler Lehrerverein (ed.), *Bezirkskunde Osttirol*. Innsbruck: Löwenzahn, pp. 110–114.

Goller, A. and A. Kofler. 2001. 'Kartitsch', in Katholischer Tiroler Lehrerverein (ed.), *Bezirkskunde Osttirol*. Innsbruck: Löwenzahn, pp. 243–248.

Gruber, E.S. 2005. *Sammelarten und Wegwissen: Von Frauen, die Wildgemüse und Heilkräuter Sammeln*, M.A. dissertation. Vienna: University of Natural Resources and Applied Life Sciences Vienna.

Hanf, M. 1998. *Farbatlas der Wildkräuter und Unkräuter*. Stuttgart: Ulmer.

Hatzer, A. 2001. 'Die Mundarten', in Katholischer Tiroler Lehrerverein (ed.), *Bezirkskunde Osttirol*. Innsbruck: Löwenzahn, pp. 75–77.

Iatrou, G. and E. Kokkalou. 1997. 'Rarity, Conservation, Importance and Ethnopharmacological Knowledge of the Greek Flora', in V.H. Heywood and M. Skoula (eds), *Identification of Wild Food and Non-food Plants of the Mediterranean Region*. Chania: CIHEAM-IAMC, pp. 65–75.

Ingruber, R. 2001. 'Landeskunde kurz und bündig', in Katholischer Tiroler Lehrerverein (ed.), *Bezirkskunde Osttirol*. Innsbruck: Löwenzahn, p. 8.

Köck, K. 2001. 'Musikkapellen und Chöre', in Katholischer Tiroler Lehrerverein (ed.), *Bezirkskunde Osttirol*. Innsbruck: Löwenzahn, pp. 70–71.

Lambraki, M. 2000. *Ta Xopta*. Athens: Ellinika Grammata [in Greek].

Nebel, S., A. Pieroni and M. Heinrich. 2006. '*Ta chorta*: Wild Edible Greens Used in the Graecanic Area in Calabria, Southern Italy', *Appetite* 47(3): 333–342.

Netting, R.M. 1981. *Balancing on an Alp – Ecological Change & Continuity in a Swiss Mountain Community*. Cambridge: Cambridge University Press.

Pardo-de-Santayana, M., J. Tardío and R. Morales. 2005. 'The Gathering and Consumption of Wild Edible Plants in the Campoo (Cantabria, Spain)', *International Journal of Food Sciences and Nutrition* 56(7): 529–542.

Pieroni, A. 1999. 'Gathered Wild Food Plants in the Upper Valley of the Serchio River (Garfagnana), Central Italy', *Economic Botany* 53(3): 327–341.

——— 2000. 'Medicinal Plants and Food Medicines in the Folk Traditions of the Upper Lucca Province, Italy', *Journal of Ethnopharmacology* 70: 235–273.

——— 2003. 'Wild Food Plants and Arbëresh Women in Lucania, Southern Italy', in P.L. Howard (ed.), *Women & Plants: Gender Relations in Biodiversity Management & Conservation*. London & New York: Zed Books, pp. 66–82.

———, B. Dibra, G. Grishaj, I. Grishaj and S.G. Maçai. 2005b. 'Traditional Phytotherapy of the Albanians of Lëpushë, Northern Albanian Alps', *Fitoterapia* 76: 379–399.

———, M. E. Giusti, H. Münz, C. Lenzarini, G. Turković and A. Turković. 2003. 'Ethnobotanical Knowledge of the Istro-Romanians of Žejane in Croatia', *Fitoterapia* 74: 710–719.

———, H. Münz, M. Akbulut, K.H.C. Başer and C. Durmuşkahya. 2005c. 'Traditional Phytotherapy and Trans-cultural Pharmacy among Turkish Migrants Living in Cologne, Germany', *Journal of Ethnopharmacology* 102(1): 69–88.

———, S. Nebel, C. Quave, H. Münz and M. Heinrich. 2002. 'Ethnopharmacology of Liakra: Traditional Weedy Vegetables of the Arbëreshë of the Vulture Area in Southern Italy', *Journal of Ethnopharmacology* 81: 165–185.

———, S. Nebel, R.F. Santoro and M. Heinrich. 2005a. 'Food for Two Seasons: Culinary Uses of Non-Cultivated Local Vegetables and Mushrooms in a South Italian Village', *International Journal of Food Science and Nutrition* 56(4): 245–272.

Plenderleith, K. 1999. 'Traditional Agriculture and Soil Management', in UNEP, *Cultural and Spiritual Values of Biodiversity*. London: Intermediate Technology Publications, pp. 285–323.

Ploetz, K and B. Orr. 2004. 'Wild Herb Use in Bulgaria', *Economic Botany* 58(2): 231–242.

Quinlan, M.B., R.J. Quinlan and J.M. Nolan. 2002. 'Ethnophysiology and Herbal Treatments of Intestinal Worms in Dominica, West Indies', *Journal of Ethnopharmacology* 89: 75–83.

Rigat, M., T. Garnatje and J. Vallès. 2006. *Plantes i gent: Estudi etnobotànic de L'Alta Vall del Ter*. Ripoll: Centre d'Estudis Comarcals del Ripollès.

Rivera, D., C. Obon, C. Inocencio, M. Heinrich, A. Verde, J. Fajardo and R. Llorach. 2005. 'The Ethnobotanical Study of Local Mediterranean Food Plants as Medicinal Resources in Southern Spain', *Journal of Physiology and Pharmacology* 56, Suppl. 1: 97–114.

Savvides, L. 2000. *Edible Wild Plants of the Cyprus Flora*. Nicosia: Private publication.

Smith, J.J. 1993. 'Using ANTHROPAC 3.5 and a Spreadsheet to Compute a Free-list Salience Index', *Cultural Anthropology Methods Newsletter* 5: 1–3.

Staller, M. 2001. 'Das Klima', in Katholischer Tiroler Lehrerverein (ed.), *Bezirkskunde Osttirol*. Innsbruck: Löwenzahn, pp. 107–109.

Statistik Austria. 2001. *Volkszählungsergebnisse 2001*. Retrieved 10 June 2006 from http://www.statistik.at

Strasser, L. 2001. 'Die Schützen', in Katholischer Tiroler Lehrerverein (ed.), *Bezirkskunde Osttirol*. Innsbruck: Löwenzahn, p. 72.

Tardío, J., H. Pascual and R. Morales. 2002. *Alimentos silvestres de Madrid: Guía de plantas y setas de uso alimentaria tradicional en la Comunidad de Madrid*. Madrid: Ediciones La Librería.

———, H. Pascual and R. Morales. 2005. 'Wild Food Plants Traditionally Used in the Province of Madrid, Central Spain', *Economic Botany* 59(2): 122–136.

Till, S. 2001. *Wildkräuter-Delikatessen: Wildpflanzen und Pilze aus Wald und Wiese*. St. Pölten, Austria: NP-Buchverlag.

Trum, B. and P. Lotter. 1998. *Wildkräuter-Kochbuch: Sammeln – zubereiten – genießen*. Kempten, Germany: Dannheimer.

Villar, L., J.M. Palacín, C. Calvo, D. Gómez and G. Montserrat. 1987. *Plantas medicinales del Pirineo Aragonés y demás tierras oscense*. Huesca: Diputación de Huesca & CSIC.

Vogl, C.R. and B. Vogl-Lukasser. 2003. 'Lokales Wissen von Biobauern über ausgewählte Elemente der Agrarbiodiversität im Bezirk Lienz (Österreich): Zur Bedeutung, Anwendung und Weiterentwicklung ethnobiologischer Forschungsfragen und Methoden in der Forschung im Ökologischen Landbau', in B. Freyer (ed.), *Beiträge zur 7. Wissenschaftstagung zum Ökologischen Landbau: Ökologischer Landbau der Zukunft, 24–26 Feb., Wien*. Vienna: Institute for Organic Farming, University for Natural Resources and Applied Life Sciences Vienna, pp. 403–406.

Vogl-Lukasser, B.N. 1999. *Hausgärten: Studien zur funktionalen Bedeutung bäuerlicher Hausgärten in Osttirol basierend auf Artenzusammensetzung und ethnobotanischen Analysen*, Ph.D. dissertation. Vienna: University of Vienna.

Weller, S.C. and A.K. Romney. 1988. *Systematic Data Collection*. London and New Delhi: Sage Publications.

CHAPTER 4

Local Innovations to Folk Medical Conditions

Two Major Phytotherapeutic Treatments from the Maltese Islands

TIMOTHY J. TABONE

Introduction

Trade and travel have linked various parts of Europe for millennia, especially between the southern European countries, the Mediterranean islands and North Africa. Over the centuries biological and cultural exchanges have been an important part of this dynamic movement. Malta is one such location, and shows very startling cultural influences to folk medicine categories and treatments, marvellous adaptations to local flora, local innovations etc. In this chapter, the results of long-term ethnobotanical work in Malta are described, focusing on two phytotherapies that quite probably have their origins elsewhere, but demonstrate local adaptation through processes of plant substitution and unique mixtures. The findings support the general proposition that current plant use in any particular location in Europe today needs to be examined in light of the historical ecology of much wider geographical spaces and longer temporal scales.

Malta

The islands of the Maltese archipelago – Malta, Gozo and Comino together with some minor islands – lie just south of Sicily in the central Mediterranean Sea (Figure 4.1). At 245.7 km^2, Malta is the largest of the

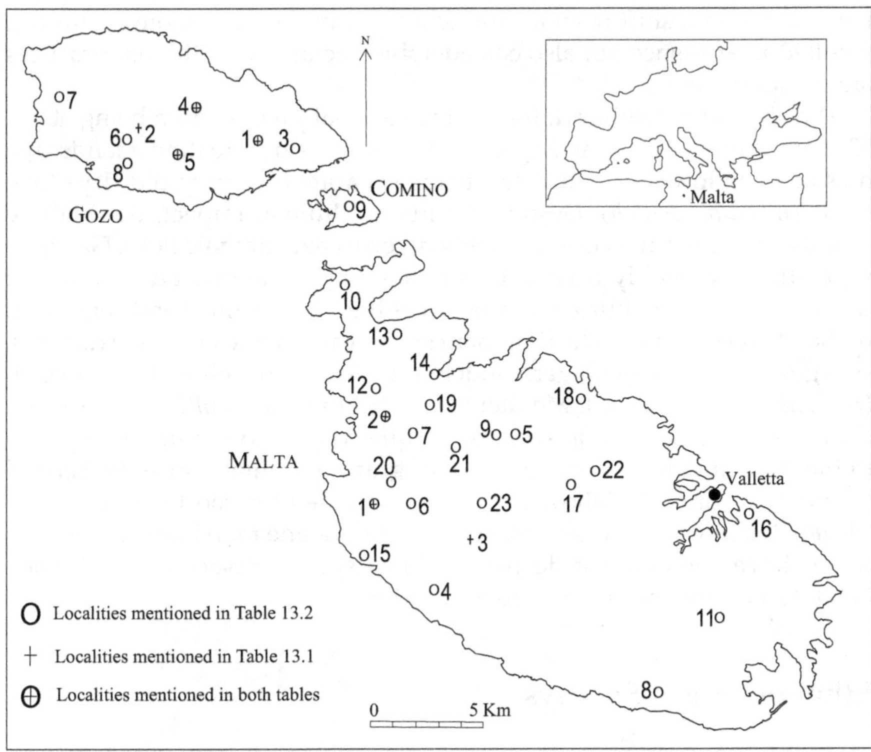

Figure 4.1. Map of the study site. *Localities in Malta*: 1. Baħrija; 2. Għajn Tuffieħa; 3. Rabat; 4. Dingli; 5. Għajn Riħana / Misraħ Għonoq; 6. Għemmieri, Rabat; 7. Mġarr; 8. Żurrieq; 9. Bidnijja; 10. Ċirkewwa; 11. Għaxaq; 12. Manikata; 13. Mellieħa; 14. St. Paul's Bay; 15. Mtaħleb; 16. Kalkara; 17. Lija; 18. Mghatab; 19. San Martin; 20. Bingemma; 21. Imselliet; 22. L-Iklin; 23. Mtarfa. *Localities in Gozo and Comino*: 1. Nadur; 2. Rabat; 3. Qala; 4. Xagħra; 5. Xewkija; 6. Fontana; 7. San Lawrenz; 8. Munxar; 9. Comino (adapted from Pedley, Clarke and Galea 2002: 20–21)

islands, while the total land area of the archipelago is about 316 km^2, and the coastline is roughly 190 km long. The resident population is approximately 400,000. Malta thus is one of the smallest countries of the world and yet has one of the highest population densities (> 1,100 pers/km^2), not including more than one million tourists that visit annually. The official languages are Maltese (*Malti*) and English. Maltese is closely related to Arabic and is the only Semitic language in the world to be written in the Latin script. Its vocabulary is heavily laced with Sicilian, Italian and English loan-words. The Maltese are predominantly Roman Catholic and the influence of religion is strong, with over 60 per cent of the Maltese being regular church attendees. Malta has witnessed massive social change since World War II: Political independence, compulsory education and a shift from a rural to a tourism-based economy. This has resulted in affluence but also considerable decline in traditional practices and belief systems.

The climate is semi-arid, the average annual precipitation being about 530 mm. Agroecosystems dominate the countryside, creating a landscape mosaic of fallow and cultivated fields separated by very old dry-stone walls (*ħitan tas-sejjieħ*). Despite the intense human impact, the Maltese Islands still harbour a comparatively diverse characteristic flora. Garrigue vegetation commonly occurs on coralline limestone plateaus, which are unsuitable for agriculture, and constitutes the chief semi-natural vegetation of the Maltese Islands. The dominant species are thyme (*Thymbra capitata*), spurges (*Euphorbia* spp.), germander (*Teucrium* spp), yellow kidney vetch (*Anthyllis hermanniae* L.), Mediterranean heath (*Erica multiflora*), rockrose (*Cistus* spp.) and *Fumana* species. Maquis vegetation is increasing due to the dramatic drop in the number of grazing animals since the Second World War, often in fallow fields in the shade of carob trees (*Ceratonia siliqua*). Freshwater habitats are scarce, while saline marshlands and sand dunes have been almost destroyed. Most species associated with such habitats are rare and threatened with extinction.

Ethnobotanical Surveys

The author has been conducting a systematic ethnobotanical survey of the Maltese Islands since 1995, though some data were sporadically collected before. A total of 213 men and 155 women were consulted using a standardized interview method. Most interviewees were farmers and/or herders, who have spent all their lives in the countryside. The chief age group targeted were people over seventy years of age, the pre–Second World War generation, who are undoubtedly the most knowledgeable about traditional Maltese pharmacopoeia. Interviews were carried out in Maltese, in which the author is fluent, and which is often the only language used by people in the countryside . Most interviews took place in the informants' private homes or while sitting in village squares, which are situated around the parish church parvis *(iz-zuntier*). These are the centre of village life, where many of the older people spend their mornings (after having heard Mass) and afternoons sitting on benches and chatting.

A list of all Maltese plant names recorded in the botanical literature was compiled and informants were asked what each was used for. Also, farmers and herders were accompanied by the author while out in the countryside tending herds of goats or while working in their fields, and asked the names of any plants they saw and to point out medicinal plants. Many previously unrecorded vernacular names were discovered and new informants asked about them.

'The Jaundice Medicine' (Id-Duwa tas-Suffejra)

Much traditional Maltese phytotherapy is focused on the treatment of shock/fright (*qatgħa*). One important treatment method is drinking bottles of herbal mixtures called 'the shock/fright medicine' (*id-duwa tal-qatgħa*). The following are examples of shock/fright incidents related by people who actually took this medicine and believed that they were cured by it, or had a family member who did.

- Bombs falling during the war
- Being surprised at night by a strange woman dressed in white
- Being pounced on by a dog
- Being charged at by a heifer
- Being surprised by a rat while sitting on a toilet
- Witnessing a woman faint
- One man is sitting on a mule-driven cart; a dog attacks the mule; the mule panics, causing the cart to overturn and its rider to fall down into a field ten feet below
- One fourteen-year-old girl catches her uncle stealing animal fodder from their field; he rebukes her and she is upset very badly
- Witnessing somebody's death or injury by car accident or narrowly escaping being hit by a car

Symptoms of Shock/Fright

The symptoms resulting from the shock/fright incidents listed above include the following.

- General feeling of physical weakness
- Legs feeling limp
- Loss of appetite
- Feeling broody
- Feeling shaken, disturbed (*ħadd daħdiħa ġo fija*)
- Fever, running a temperature which doctors seem unable to cure
- Skin rashes, boils
- Severe itching to the point where one cannot stay in the sun and where one cuts oneself with a knife in despair for relief
- Amenorrhoea
- Jaundice (*suffejra*), yellowing of eyes and skin

While all these are treated with 'the shock / fright medicine', jaundice is the most frequently mentioned symptom and so the medicine is often referred to as 'the jaundice medicine' (*id-duwa tas-suffejra*). However, the association between shock/fright and jaundice needs further examination.

While some informants said that jaundice was a symptom of shock/fright (*bill-qatgħa taqbdek is-suffejra*), others asserted that the two words were in fact synonymous (*qatgħa u suffejra l-istess*). One woman from Xagħra, Gozo, made a rather puzzling statement, implying a conceptual belief in two types of jaundice: 'The jaundice we are talking about is when one gets a fright, not the yellowing of eyes and skin. That is another kind of jaundice.'

Diagnosis

When wondering whether illnesses that one is experiencing are consequences of some past shock/fright, the following two measurements are made (see Figure 4.2):

1. The length of the fathom, from the tip of the middle finger of one hand across to the tip of the middle finger of the other hand.
2. The height of the body.

These measurements must be equal. If the fathom is less than the height, one is being affected by past frights and treatment with herbal medicine is required. 'The more your arm has shrunk inwards [i.e., the shorter the fathom], the more the shock has advanced and set in.'

Leaving the shock untreated results in increased susceptibility to fright, by even the most trivial accidents in daily life, such as someone calling your name. Two informants believed that relatives of theirs had actually died of a succession of frights.

The jaundice medicine used to be prepared by people with special knowledge. In Gozo, there used to be one or two such individuals in most villages. The most famous is the faith healer Francesco Mercieqa (1892–1967) (Bonnici 1985), popularly known as Frenċ tal-Għarb. He is still much revered to this day and the author has met people who actually

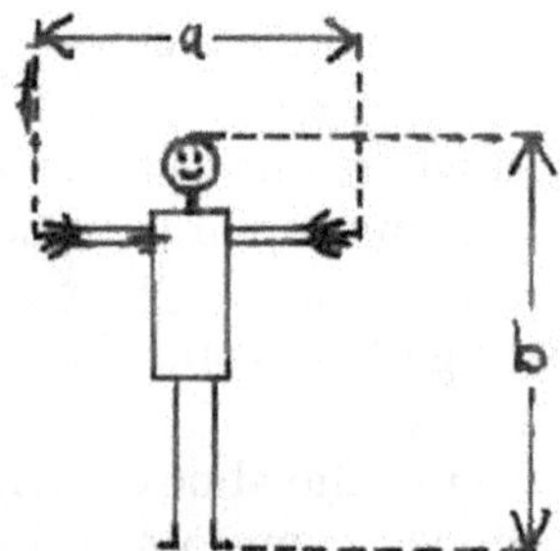

Figure 4.2. Measurements for the diagnosis of shock/fright syndrome. a, fathom; b, height.

pray to him, even though he has never been beatified. Another healer was the woman known as Karmni d-Dongra, long dead and only known to the author from one woman in Rabat, Gozo, who used her medicine; Tat-Trent, the nickname of a man from Nadur, Gozo, was only known to the author from one priest whose uncle had known him.

Table 4.1 lists the plant species that constitute the jaundice medicine. Beside these thirty-two species, one other species was recorded but remains unidentified: *sensaperilja,* used by a woman from Xaghra. Other ingredients of the recipe were wine, honey and whiskey. A number of the species listed are also used individually to treat the same folk conditions (fright/jaundice) throughout the country, especially *Cynodon dactylon, Equisetum ramosissimum* and *Verbena officinalis.*

Out of the twelve informants listed in Table 4.1, only two had actually prepared the medicine themselves. The rest remembered their mothers preparing it and had only partial knowledge of the recipe. There is substantial variation in the number and species of herbs used. For example, the man from Qala mentioned thirteen ingredients, while one woman from Xagħra suggested that it would be 'too thick and too strong with more than five herbs'. Attention is also paid to the taste of the mixture, so some of the plants used in the recipe are present for organoleptic as well as therapeutic purposes. Ingredients like black horehound (*Ballota nigra*) and little wormwood (*Artemisia arborescens*) are terribly bitter even if they are used in small quantities, so sweet herbs like rhizomes of Bermuda grass (*Cynodon dactylon*) are used to counteract the bitter taste. On the other hand, Sulla (*Hedysarum coronarium*) has a sickly sweet taste which can overwhelm, and thus, unlike other herbs, is used dry, not fresh.

Preparation Method and Dosage

The herbal mixture is usually boiled in water or wine and left to simmer on a low flame for hours, until its volume reduces by one half to three quarters. After cooling it is sieved. Sometimes the process of boiling and cooling is repeated three times in an attempt to extract as much of the plants' juices as possible.

The medicine must not be taken immediately after the shock incident. Two or three days must pass. According to Frenċ tal-Għarb, one must take a glass a day. The women from Għajn Tuffieħa suggested that one must not drink too much of it: just a tot (*grokk*), first thing in the morning.

Side-effects

Menstruating women cannot take this medicine as 'it stirs the blood', resulting in excessive issue of blood; nor can pregnant women, since some

Table 4.1. Plants used in the jaundice medicine. Status: W: wild; C: cultivated; SC: semicultivated. Localities and informants: 1: Baħrija, woman; 2: Għajn Tuffieħa, mother and daughter; 3: Rabat, woman; 4: Nadur, husband and wife; 5: Rabat, woman; 6: Qala, man; 7: Xagħra, woman; 8: Xagħra, woman; 9: Xagħra, woman; 10: Xagħra, woman; 11: Xagħra, husband and wife; 12: Xewkija, woman

Botanical taxon	Botanical family	Maltese name	Status
Artemisia arborescens L.	Asteraceae	Erba bjanka	C
Ballota nigra L.	Lamiaceae	Marrubja	W
Beta vulgaris L.	Chenopodiaceae	Selq	W, C
Borago officinalis L.	Boraginaceae	Fidloqqom	W
Centaurea melitensis L.	Asteraceae	Xewka tad-deni	EW
Ceratonia siliqua L.	Fabaceae	Ħarruba	SC
Cichorium spinosum L.	Asteraceae	Qanfuda	W
Citrus aurantium L.	Rutaceae	Larinġ tal-bakkaljaw	C
Citrus limon (L.) Burm. f.	Rutaceae	Lumi	C
Citrus sinensis (L.) Osbeck	Rutaceae	Larinġa helwa	C
Cuminum cyminum L.	Apiaceae	Kemmun	C x
Cynodon dactylon (L.) Pers.	Poaceae	Niġem	W
Equisetum ramosissimum Desf.	Equisetaceae	Demb iż-żiemel	W
Erica multiflora L.	Ericaceae	Xpakkapietra	W
Eugenia caryophyllata L.	Myrtaceae	Imsiemer tal-qronfol	*
Fumaria spp.	Papaveraceae	Daħnet l-art	W
Hedysarum coronarium L.	Fabaceae	Silla	C, SC
Hyoseris spp.	Asteraceae	Ċikwejra salvaġġa	W
Laurus nobilis L.	Lauraceae	Rand	W, C, SC
Lavatera spp. and *Malva* spp.	Malvaceae	Ħubbejża	W
Lippia triphylla L'Hér	Verbenaceae	Wiża	C
Matricaria recutita L.	Asteraceae	Kamumilla	W, C
Mentha spicata L.	Lamiaceae	Nagħniegħ	C
Origanum majorana L.	Lamiaceae	Merqtux	C
Pallenis spinosa (L.) Cass.	Asteraceae	Xewka tad-deni	W
Plantago weldenii Reich., *P. lagopus* L., *P. serraria* L.	Plantaginaceae	Salbet l-art	W
Rheum rhabarbrum L.	Polygonaceae	Abarbru	*
Rosmarinus officinalis L.	Lamiaceae	Klin	C, W
Ruta chalepensis L.	Rutaceae	Fejġel	W
Salvia fruticosa Mill.	Lamiaceae	Salvja	C
Thymbra capitata (L.) Cav.	Lamiaceae	Sagħtar	W
Verbena officinalis L.	Verbenaceae	Buqoxrom	W

* Not found growing in the Maltese Islands; bought from the grocer or chemist
x No longer cultivated

Part used	Localities											
	Malta			Gozo								
	1	2	3	4	5	6	7	8	9	10	11	12
Aerial parts		•		•		•	•	•		•	•	•
Aerial parts						•	•	•	•	•	•	•
Leaves					•							
Flowers		?				•						
Aerial parts				•								
Pods									•		•	
Aerial parts								•	•			
Leaves											•	
Leaves or fruit	•	•	•			•	•		•			
Leaves			•	•			•					
Fruits (*żerriegħa*)			•									
Underground rhizomes, roots			•		•				•			•
Aerial parts						•	•	•				
Aerial parts						•		•				
										•		
Whole plant			•									
Flowering tops									•			
?					•							
Leaves									•			
Leaves			•									
Leaves							•		•			•
Flowering tops		?				•		•				
Leaves									•			•
Leaves												•
Whole plant				•								
Whole plant					•							
									•			•
Leaves	•	•		•		•	•		•	•	•	
Leaves				•			•		•	•	•	•
Leaves	•											
Aerial parts		?										•
Whole plant	•	•	•			•			•		•	•

of the herbs used are reputed to be abortifacients, especially *Ballota nigra*, and also *Artemisia arborescens* and *Matricaria recutita.*

The Xpakkapietra Treatment for Kidney Stones

The *xpakkapietra/xkattapietra* is a very popular medicinal plant generic and one of the first that older villagers mention during interviews. An infusion made from it is used to remove stones from the kidney and the bladder (*il-ġebla tal-kliewi* and *il-ġebla fill-bużżieqa tal-awrina*). Even orthodox medicinal practitioners have been known to recommend it and many of those who use it do so with the hope of avoiding painful surgery in hospital. Maltese migrants to Australia take bunches of it to use in their adopted country.

Ethnotaxonomy

In past botanical literature, the *xkattapietra* has always been identified as *Satureja microphylla* [= *Micromeria microphylla* (d'Urv.) Benth.] (e.g., Borg 1927; Haslam, Sell and Wolseley 1977; Lanfranco 1989). However, throughout this research project it has become clear that the name *xpakkapietra* refers to at least six unrelated plant species (Table 4.2).

The taxa listed in Table 4.2 were identified by the author from actual specimens from bunches that informants had stored in their homes for future use, or to give to city dwellers who do not have easy access to wild habitats. In a few cases where specimens were not available, identification was made from a verbal description given by the informant. However, this has only been done where the taxon in question is so distinctive that errors in identification are easily avoided and where that taxon has been confirmed as a *xpakkapietra* by other informants. For example:

> *Xpakkapjietra* [*Erica multiflora*] .... It grows a lot on the rocky karstlands [*xagħri*] .... [I]ts leaves are thin and stiff. It is full of dense clusters of purplish flowers; it flowers at Christmas time .... Long ago ... they used to pick it and the narcissus [*narċis*] together, to sell as ornamental bunches. ... [I]t grows this tall [two feet]. Its leaves are very short and thin, like lavender.
>
> (A woman from St Paul's Bay)

> We call it *leħjet ix-xiħ*[1] [*Erica multiflora*]. It is abundant in the Wardija and San Martin area. ... purplish flowers. I have heard that it is a *xpakkapietra*, that it has the same medicinal use.
>
> (A man from Bidnijja)

In Malta 52 informants knew two species of *xpakkapietra* or had heard that there are two; 7 knew of three; 3 knew of seven; 2 just stated 'there are many!' Out of these, 13 state that there is both 'the male *xpakkapietra*' and 'the female *xpakkapietra*'. However, there is much variation as to which

Table 4.2. Plant species under the name *xpakkapietra*

Botanical taxon	Botanical family	Part used	Localities (number of mentions)			Total no of mentions
			Malta	Gozo	Comino	
Asperula aristata L.	Rubiaceae	Whole plant	Dingli Għajn Riħana / Misraħ Għonoq (2), Għemmieri (1), Rabat (2), Mġarr (1,2 +); Żurrieq (1)			9
Erica multiflora L.	Ericaceae	Aerial parts	Bidnijja (1 *), Ċirkewwa (2 *), Dingli (2), Għaxaq (1 *), Manikata (1), Mellieħa (1 +), St. Paul's Bay (1 +)	Fontana (1), Munxar (1), Nadur (3,3 +), Qala (5), San Lawrenz (1), Xagħra (3), Xewkija (2)	Comino (1)	29
Fumana arabica (L.) Spach and *F. thymifolia* (L.) Spach	Cistaceae	Aerial parts	Ċirkewwa (1), Mellieħa (3), Mġarr (1 +), Mtaħleb (1)			6
Satureja microphylla (d'Urv.) Guss.	Lamiaceae	Whole plant	Baħrija (2,1 *), Bidnijja (2), Ċirkewwa (2), Dingli (5), Għajn Tuffieħa (1), Għajn Riħana/ Misraħ Għonoq (1), Għaxaq (1), Kalkara (1), Lija (1), Manikata (2), Mellieħa (2), Mġarr (1), Mgħatab (1), Mtaħleb (1), San Martin (1), Żurrieq & Ħal Far (2)	Nadur (2)	Comino (1)	31
Valantia muralis L.	Rubiaceae	Whole plant	Baħrija (4), Dingli (1), Imselliet (1), L-Iklin (1), Mtaħleb (2)			9

+ Identification made from verbal description; no actual specimen seen by author
* Taxon not recognized under the name *xpakkapietra* by informant. However, informant has heard of it being recognized as such in other parts of the country

species is the male or the female. For some, the male is *Asperula aristata*; for others it is *Fumana* spp. *Satureja microphylla* is usually identified as the female, rarely as the male.

One man from Manikata stated that the male *xpakkapietra* had its own special name: *rijdnu*. The *rijdnu* consists of one main elongated prostrate stem whereas the female is caespitose, acquiring the domed habit of thyme bushes (*tissagħtar*). When this same informant showed specimens to the author, the male and the female turned out to be both *Satureja microphylla*. After some discussion, the author concluded that they are both the same species, and are only differentiated by their habit.

Erica multiflora is the main *xpakkapietra* in Gozo. Only 7 informants knew of two species under the name *xpakkapietra*. In each case, one of the two was *Erica multiflora* while the other was *Satureja microphylla* in two cases and remains unidentified in the rest. One man from Nadur knew of three, two of which are 'the male' and 'the female'.

The classification of *xpakkapietra* species into male and female herbs does not necessarily imply different uses depending on the sex of the patient. Apart from a man from Bidnija who stated that 'the larger one [25 cm] is for women; the shorter one [10 cm] is for men', there is nothing in the author's collection of citations to suggest that plant use is gendered. However, two informants did suggest that the two different species have different medicinal uses. One *xkattapietra* 'changes the blood [usually meaning cleanses the blood from shock and/or anxiety], while the other is for the urine' (a woman from Għamieri). 'The male is for the urine and the female is for stones' (a woman from Għajn Tuffieħa).

The classification into male and female is, it seems, a way of distinguishing two identically named species, and sometimes, specifying different uses of the medicines. Of course it is always possible that the symbolic meaning of the distinction has been lost in transmission over the generations.

The Xpakkapietra *in Phytotherapeutic Treatment*

Opinions vary widely among informants as to which of the several *xkattapietra* is the most curative. Only six informants described the two different species as equally effective. Several assert that 'the female' is better than 'the male'. The roots' ability to penetrate fissures and split rocks is seen as indicative of its power to break kidney stones. Hence a plant that grows in harder, upper coralline rocks (*blat samm* or *blat qawwi*) and is more difficult to uproot is believed to be more effective. 'It [*Satureja microphylla*] is very good, it grows among the rocks as it can split. … It [*Valantia muralis* and *Fumana* spp] grows where the soil is a little deep [hence not as good]' (a man from Mellieħa).

Many informants named a particularly familiar rocky patch of land where *xpakkapietra* is abundant from where they regularly collect. In

Gozo, Ta' Ċenċ cliffs and the upper coralline plateaus around Nadur were frequently cited as the best collecting areas. Areas of the softer globegerina limestone (*franka*) are avoided. Two informants asserted that the most effective *xpakkapietra* grows on coastal rocks, 'so that it absorbs sea spray and its grade improves, becoming very strong!' Hence farmers in the Mtaħleb area cited the sea-cliffs of Miġra l-Ferħa as the best collecting area.

Preparation Method and Dosage

An infusion is made 'like making tea' by pouring one-and-a-half pints of boiling water onto two generous handfuls of the herb. Some say that a decoction is made by boiling the herb for a short time, 'for half an hour, or less'. One drinks a tot once to three times a day, most importantly first thing in the morning on an empty stomach. 'The stone is crushed' and expelled in the urine within a week, sometimes even within three days.

Possible Side-effects

The *xpakkapietra* is widely reputed to be an abortifacient. One informant from Mġarr, Malta, said that two pregnant women once visited him and he was reluctant to give them *xpakkapietra,* as he was afraid that it might induce abortion. However, when they insisted, he relented and gave it to them, and they did not suffer a miscarriage. 'One of them could hardly go up the stairs, she was in such pain.'

Several informants cautioned against taking more than a shot of *xpakkapietra* infusion at a time. A woman from Baħrija was advised by the doctor not too take too much of it, as 'it even tears the flesh'. A man from Nadur, Gozo, stated that the smaller of the two *xpakkapietr*a species 'is hard on who drinks it. It attacks you. As soon as you drink it, you feel it has made a dent inside you. You can feel it working inside you.' Another man from the same locality said that *Satureja microphylla* is so strong, that it even dissolves the limescale inside the kettle in which it is boiled. Two informants, one from Manikata and one from Mġarr, Malta, stated that *xpakkapietra* diminishes one's vision if taken in excess (*id-dawl tal-għajnejn … tnaqqsu; taqta' id-dawl mill-għajnejn*).

A man from Nadur, Gozo, stated that one of the three *xpakkapietra* species is too strong for humans and is given to horses. He recounted the story of a man that was accidentally given 'the horse's *xpakkapietra* … and after two days, he had to return to hospital, suffering from a terrible headache'.

The Origins of Maltese Phytotherapies

These two phytotherapies illustrate the dynamic origins of ethnobotanical knowledge in Malta, and by extension elsewhere in the Mediterranean.

The causes and symptoms described for the Maltese *qatgħa* (shock/fright) folk condition closely resemble those documented for the Spanish *susto* folk condition: a state of fear resulting from a sudden jolt/crash/impact or sudden impression/surprise, first causing psychological problems such as nervousness, later resulting in the weakening of physical equilibrium, making one more susceptible to several maladies or their carrying agents (Kuschick 1995). A succession of such sudden incidents results in loss of appetite, vomiting, dizziness, trembling and sadness (García Barbuzano 1981). This *susto* concept is widespread in Latin America, also occurring in the Canary Islands and evidently in the central Mediterranean, indicating that the Iberian Peninsula is the epicentre of the fright syndrome, whence it has spread to several regions under Spanish dominion, including the Maltese Islands.

Here it is appropriate to outline historical periods of Spanish influence through which the *susto* concept may have permeated Maltese culture. Malta first came within the Spanish sphere when the Aragonese took Sicily and Malta from the Angevins in 1282. According to Luttrell (1975: 44), there then emerged

> a single economic and strategic unit, a Western Mediterranean common market in which merchants of Valencia, Barcelona and Perpignan could buy and sell in the Balearics, Sardinia and Sicily, while at the same time controlling those islands and the safe harbours they needed along their routes to the lucrative markets in North Africa and the Levant.

Malta was a minor trading post within this Siculo-Aragonese context where 'the Catalans were importing … Maltese cotton, either directly from Malta or from Sicily … they carried cloth, oil, sardines and dried fruits to Malta in exchange' (Luttrell 1975: 53). Some Catalans may have settled in Malta during this period. The Maltese Islands were ceded to the Order of the Knights of St John in 1530, but contacts with Spain did not cease. According to Vassallo (1997), trade with Spain flourished during the eighteenth century, largely because of Brigantine trade expeditions: Maltese merchants would purchase goods from several Mediterranean ports including Naples, Leghorn, Genoa and Marseilles, beach their ships in Iberian ports and spend months in Spain and Portugal selling goods door to door, in streets and squares. In the early 1760s one in ten of all foreign merchants in Spain was Maltese, and one German traveller wrote: 'the Maltese do a lot of trade in Cadiz and you won't find any important city all over Spain where you cannot find them. They have more privileges and rights than Spanish shopkeepers who only sell small amounts …. They … arrive with entire loads of all kinds of goods … and take large sums of cash back home from Spain' (Vassallo 1997: 28). Barcelona was an important buyer of cotton – then Malta's only cash crop. As the Brigantine expeditions began to die out, Maltese merchants began setting up retail firms in Spain, opening shops and residing there, usually for periods of seven to ten years, enough time for them to interact with Spaniards and absorb ideas from them.

It is also possible that the *susto* concept permeated Malta from Spain via Sicily, which remained under Spanish rule until the eighteenth century. The Sicilian influence has been the most prominent and most enduring foreign influence on Malta. From the ancient Roman period up to 1530, the Maltese Islands were ruled together with Sicily. Even after the arrival of the Knights of St John, Malta remained dependent on Sicily for most of its grain. Most commerce and foreign contacts were with Sicily; for example, more than 80 percent of vessels leaving Malta's harbours between 1564 and 1600 were directed to a Sicilian destination (Cassar 2000). There have also been population exchanges between the two countries.

The following two nonphytotherapeutic cures for shock/fright documented by Pitre for Sicily have also been documented by myself and past folklorists for Malta, indicating Sicilian influence in the development of the Maltese shock/fright belief system.

1. After a sudden fall, one must immediately urinate on a new broom in order to offset the development of symptoms of the fall and the resultant state of fear.
2. After one has had a fright, one immediately extinguishes a smouldering ember in wine and drinks the wine.[2]

However, the Spanish and Latin American *susto* concept lacks the association with jaundice which is so prominent in the Maltese conception of it. The Maltese shock/fright – jaundice syndrome seems to have resulted from syncretism of the Spanish *susto* with the *mal d'arco* folk condition found in the Vitalba valley and Lucania in Southern Italy. While *mal d'arco* 'illness of the rainbow', does not occur in the Maltese Islands as a separate folk condition and is not caused by shock/fright, but by urinating outdoors in the direction of a rainbow, its symptoms and diagnosis, as described by Pieroni and Quave (2005: 86–88) in southern Italy, are identical to those described from Malta for shock/fright: mainly hepatitic symptoms, especially jaundice, and also weakness in the arms and knees. Diagnosis of *mal d'arco* is by comparing the lengths of the affected person's fathom and height, as described above.

Such blending of folk conditions is not unique to Malta, as throughout Latin America the Spanish *susto* concept has syncretized with the Native American 'loss of soul' folk condition, where the soul is shocked out of the body and may only be retrieved with the aid of shamans (Berlin and Berlin 1996; Kuschick 1995).

Although a similar phytotherapeutic treatment for *susto* – consisting of an infusion of three aromatic herbs (*Mentha* sp.; *Citrus sinensis, Chenopodium ambrosioides* L.) – is used in the Canary Islands (García Barbuzano 1981), the *duwa tas-suffejra* medicine described above seems to have been developed only in the Maltese Islands. Two informants from Gozo stated that the basic principle for combining a number of different herbs in a single recipe was that 'if one herb does not cure, another will', indicating that the described medicine possibly evolved through a trial-and-error approach, where the most commonly used medicinal plants were eventually mixed into one medicine in an attempt to ensure effective healing. Even the variation in the number and species of plants used indicates local innovation, depending on the personal experience, tastes, beliefs and idiosyncrasies of the healer.

The Italian etymology of the Maltese name *xpakkapietra* (Italian *spaccare* – to split; and *pietra* – stone) indicates a Siculo-Italian lineage for Maltese treatment of kidney stones. However, the *xpakkapietra* species used in the Maltese Islands are not known to be used in Sicily and Italy and the species from Sicily recorded by Pitre under the name *spezza-petri* ('stones-splitter', *Saxifraga rotundifolia* L.) do not occur in the Maltese Islands. Pitre (2004) lists other species used in Sicily for the same purpose: *Betonica officinalis* L., *Eryngium campestre* L. and *Ranunculus asiaticus* L., which also do not occur in Malta; *Satureja nepeta* (L.) Scheele (syn. *Calamintha nepeta* (L.) Savi) commonly occurs but is not used medicinally; *Laurus nobilis, Cynodon dactylon, Allium cepa* L., *Lycopersicon esculentum* Mill., *Malva* spp. and *Petroselinum crispum* (Mill.) Fuss, all of which are used medicinally in Malta but not for kidney stones. Thus, it is likely that the Maltese peasants applied the *spezza-petri* concept learnt from the Sicilians to the Maltese flora, selecting the species to use through the doctrine-of-signatures approach: the plant's ability to split hard rocks hinting at its effectiveness in splitting kidney stones (see above). *Asperula aristata, Satureja microphylla* and *Valantia muralis* are of similar habit, bearing a superficial resemblance to each other, and often occur together in rock crevices; the attempt to distinguish the most effective among them through male/female classification (see above) indicates much local experimentation. Yet until further data on kidney-stone phytotherapeutic treatments in Sicily are published, caution is necessary when asserting that the male/female *xpakkapietra* classification concept is a Maltese particularity.

Conclusion

These two Maltese phytotherapies exemplify the evolution of ethnobotanical knowledge, which reflects and embodies the history of people, their ideas and the environment in which their society has developed. In the context of globalization and climate change today, we can expect that ethnobotanical knowledge in Malta will continue to evolve, incorporating influences from around the world, inventing new responses to these challenges that might involve both local and introduced flora, and thus continuing the dynamism characteristic of its past.

Notes

1. *Leħjet ix-xiħ* (old man's stubble) is a documented vernacular name for *Erica multiflora* (eg., Borg 1927; Haslam, Sell and Wolseley 1977; Lanfranco and Lanfranco 2003), and the author has confirmed it from several localities.
2. Siculo-Spanish elements in other aspects of Maltese culture have been recorded, providing parallels for the Siculo-Spanish lineage of the Maltese shock/fright folk condition. The few surviving Medieval church wall-paintings and frescoes show strong Siculo-Catalan style influences (Luttrell 1975); the Good Friday processions with ornately dressed life-sized statues representing different stages in the passion and death of Jesus Christ show great resemblance to such processions in Sicily, southern Italy and Spain (Cassar-Pullicino 1992); and there are also linguistic links, for example the word for 'gun': *xkubetta* (Maltese); *escopeta* (Spanish); *scupetta* (Sicilian).

References

Berlin, E.A. and B. Berlin. 1996. *Medical Ethnobiology of the Highland Mayas of Chiapas, Mexico.* Princeton, New Jersey: Princeton University Press.

Bonnici, A. 1985. *Frenċ Ta' L-Għarb, il-bidwi li sthajluh tabib.* Malta: Imprint Ltd., Qormi.

Borg, J. 1927. *Descriptive Flora of the Maltese Islands.* Malta: Government Printing Office.

Cassar, C. 2000. *Society, Culture and Identity in Early Modern Malta.* Malta: Mireva Publications.

Cassar-Pullicino, J. 1992. *Studies in Maltese Folklore.* Malta: Malta University Press.

García Barbuzano, D. 1981. *Prácticas y creencias de una santiguadora canaria.* Tenerife: Centro de la Cultura Popular Canaria.

Haslam, S.M., P.D. Sell and P.A.W. Wolseley. 1977. *A Flora of the Maltese Islands.* Malta: Malta University Press.

Kuschick, I. 1995. *Medicina popular en España.* Madrid: Siglo XXI.

Lanfranco, E. 1989. 'The Flora', in P.J. Schembri and J. Sultana (eds), *Red Data Book for the Maltese Islands*. Malta: Interprint, pp. 8–70.
——— and G. Lanfranco. 2003. *Kullana Kulturali Vol.47: Il Flora Maltija*. Malta: Publikazzjonijiet Indipendenza.
Luttrell, A.T. 1975. *Approaches to Medieval Malta*. London: The British School at Rome.
Pedley, M., M.H. Clarke and P. Galea. 2002. *Limestone Isles in a Crystal Sea, the Geology of the Maltese Islands*. San Gwann, Malta: Publishers Enterprises Group.
Pieroni, A. and C.L. Quave. 2005. 'Folk Illness and Healing in Arbëreshë Albanian and Italian Communities of Lucania, Southern Italy', *Journal of Folklore Research* 42: 57–97.
Pitre, G. 2004 [1894]. *Medicina popolare siciliana*. San Giovanni La Punta, Italy: Edizioni Clio.
Vassallo, C. 1997. *Corsairing to Commerce: Maltese Merchants in XVIII Spain*. Malta: Malta University Publishers.

Appendix 4.1. Maltese pronunciation

Ċ ċ like 'ch' in *ch*air
Għ considered a single consonant and usually silent
Ġ ġ like '*j*' in *j*ar
G g like '*g*' in *g*ate
H h silent
Ħ ħ like '*h*' in *h*at
J j like '*y*' in *y*ellow
X x like '*sh*' in *sh*ip
Z z like '*ts*' in *ts*unami
Ż ż like 'z' in zoology

Chapter 5

Local Awareness of Scarcity and Endangerment of Medicinal Plants in Roussenski Lom Natural Park in Northern Bulgaria

Hugo J. de Boer

Introduction and Background

The tradition of herbal remedies is deeply rooted in Bulgarian society – one of the oldest Bulgarian translations of Dioscorides' *De Materia Medica* dates to the fourteenth century – and still has a central role in Bulgarian daily life (Antonova 2007). A majority of people, young as well as old, know what herbs to use to cure or prevent common illnesses such as colds, fevers, coughs, stomach disorders, wounds etc. In fact, the use of alternative medicine or traditional medicine has been increasing in Bulgaria, as in many European countries today, and some suggest this is due to a growing scepticism of modern biomedicine (Council of Europe 1999). The increasing demand for plant-based alternative medicine as well as low prices for raw materials from Eastern Europe has led to strong growth in the herbal trade in Europe. As a result, Bulgaria has grown, in the last two decades, to become Europe's second largest net exporter of medicinal plants, with about 75 per cent sourced from the wild. This trade has also become an important source of income for many people in the countryside. While the collection of medicinal plants was previously regulated through quota systems, during the harvest boom of the 1990s these laws were only enforced in protected areas, and then with little vigour (Kathe, Honnef and Heym 2003). As one might expect, this intense harvesting of wild medicinal plants has put a considerable number of species at risk of becoming endangered (Lange and Mladenova 1997). According to Hardalova (1997), only 20 per cent of medicinal plants in Bulgaria are not threatened by wild collection.

This chapter presents the results of a study conducted in two villages within the area of Roussenski Lom Natural Park, to determine the current levels of collection and use of wild medicinal plants, with a particular focus on the awareness of endangerment of medicinal plants by inhabitants. Homegardens were inventoried to determine whether medicinal plants are cultivated and people were interviewed about the collection and use of plants from the Park. Concurrently, people were asked whether they had noticed a decrease in the availability of medicinal plant species from the Park during their lifetimes.

General Background

Bulgaria has a population of over 7 million of which 84 per cent are Bulgarians. The largest minority groups are Turks (9.4 per cent) and Roma (4.6 per cent), with the latter having one of its largest populations in the country. Throughout history several empires have occupied Bulgaria: Roman, Byzantine and Turk occupations have had a strong influence on Bulgarian culture and traditions. Before the Second World War, Bulgaria was predominantly rural and agricultural, but after the war industrialization took place at a very fast pace, with a focus on heavy industry, and today agriculture is practised by only 17 per cent of the population. Between 1946 and 1989, Bulgaria was a communist-dominated 'people's republic'. During the 1980s increasing inflation, unemployment and a decline in economic development caused the welfare system to collapse. The standard of living deteriorated, many people became homeless and crime flourished. Following the 1989 revolutions in Eastern Europe, free elections and a market economy were introduced. Currently, the economy is showing signs of recovery, and Bulgaria is now a functioning market economy and an EU member since 1 January 2007. The population has decreased from 9 million in 1990 to 7.3 million in 2007 (World Facts US 2007), mainly caused by a slowing birth rate and high emigration rates to Western Europe, and today Bulgaria has the most rapidly ageing population in Europe (Hossmann et al. 2008). The consequences of this demographic transition, coupled with an expanding economy, for the transmission of traditional knowledge of medicinal plants are potentially devastating, with knock-on effects for the conservation of the plants themselves. The villages studied for this research exemplified well these processes and their consequences for old traditions and wild plants.

Flora

The topography and geography of Bulgaria show great variation, contributing to a rich flora and fauna. The flora contains about 3600 species, of which 5–10 per cent are endemic (Peev et al. 1998; Nature Protection Directives 2001). Along the northern border runs the river Danube, alongside which the Danube plain spreads and covers about one third of the Bulgarian land area. Here the climate is continental and the vegetation is of central European type. The study villages, Koshov and Červen, and the Roussenski Lom Natural Park are found on a limestone plateau overlooking the eastern part of the Danube plain (see Figure 5.1). The south is dominated by the Rila-Rodopi massive which includes the highest peak on the Balkan Peninsula, 2925 metres above sea level. The climate is Mediterranean and the vegetation is dominated by deciduous forest, with beech and sometimes coniferous forest at higher altitudes. Across central Bulgaria run the Central Balkan Mountains with a climate as well as vegetation that take an intermediate form between north and south. Along the Black Sea coastline is a narrow plain of originally steppe-land with grass and annual plants.

Traditional Herbal Medicine

In the early days of Bulgarian history, when Thracian and Slavic tribes lived in the Balkan area, diseases were strongly connected to animistic belief systems and supernatural forces. In Thracian mythology, animals such as wild boar, deer and dogs are the bearers of bad health and disease. The bearers of good health are snakes, eggs, river waters, springs, dew on flowers and green twigs. Bad omens, harmful objects, beings or phenomena cause disease whereas herbs, minerals and some animals bring about

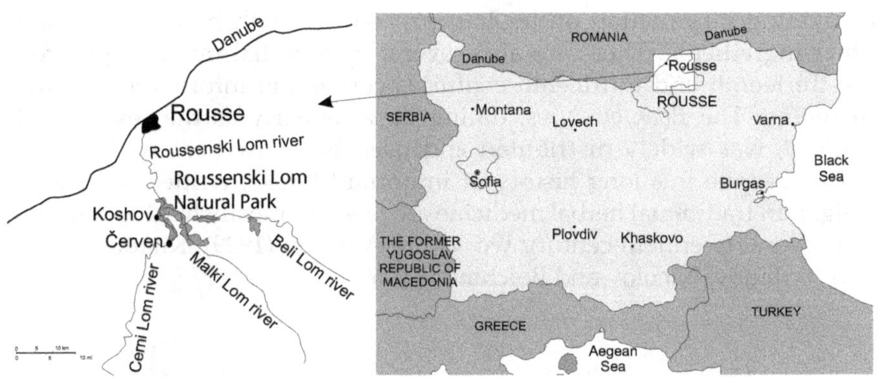

Figure 5.1. Map of the study area: Roussenski Lom Natural Park, Bulgaria

health. In Slav mythology, disease is seen as a bad demon entering the body. To be cured, Slavic tribes turned to natural phenomena like storms and lightning, the sun and moon, stars, rivers, springs and trees. They believed in occult powers of objects and divided plants and animals into either 'disease-breeding' or 'health-giving'. These beliefs are preserved in Bulgarian folklore and form the basis for many of the herbal remedy practices in Bulgaria today. The *vechtitsas*, or Bulgarian witch doctors, are specialists in the use of magic to ward off or cure illnesses caused by supernatural forces, and can be said to maintain this old tradition of magical, sometimes demonic, medicine (Georgiev 1979).

The five centuries of Ottoman rule left Bulgaria isolated from new developments in scientific medicine occurring in the rest of Europe. This led to the continued use and empirical development of herbal medicine in Bulgaria (Georgiev 1979; Petkov 1986). This isolation also protected the long tradition of herbal medicine from the purges of so-called witches, often women healers, carried out by the Catholic Church during this period. In fact, the *vechtitsas* were honoured due to their knowledge of medicine. According to Petkov (1986), the Greek Hellenic naturalist and philosopher Theophrastus mentions in his collected work, *Enquiry Into Plants*, that 'Trache [one of two provinces of the Roman Empire that later became Bulgaria] was the region richest in medicinal plants'. Dioscorides, author of *De Materia Medica*, served as a doctor in the Roman Army from 41 to 68 AD, and thus travelled extensively, recording herbal remedies from many areas, perhaps including Trache.

Throughout Bulgarian history, there have been many famous healers. In a glossary included in John the Exarch's *Hexaemeron* (857 AD, see Christov 1996), he describes the healing effect of several medicinal plants. Ivan Rilsky, a famous healer of the turn of the tenth century, was later made a saint in Bulgaria. Vassilii Vratch and his followers, a tenth- and eleventh-century anti-feudal movement known as the Bogomils, were also famous as healers. They produced a collected work called *Zeleinik* containing information on healing procedures with herbs, honey and other ingredients. There are also several apocryphal manuscripts from the thirteenth and fourteenth centuries containing information on folk medicine. The *Book of Cures*, dating back as early as the seventeenth century, was widely distributed and used by many healers (Georgiev 1979). Despite this long history of important books, written records on Bulgarian traditional herbal medicine are few and mostly hard to come by, but several twentieth-century works are Ahtarov (1939), Kitanov (1953), and Iordanov, Nicolov and Boichinov (1969).

Trade and Commercialization of Medicinal Plants

Some 20 per cent of the 3600 or so Bulgarian plant species are rich in bioactive secondary metabolites, making them suitable for use as medicinal and aromatic plants (MAPs) (Ivancheva and Stantcheva 2000; Leporatti and Ivancheva 2003). Today about 750 MAPs are collected and used in Bulgaria (Kathe, Honnef and Heym 2003).

The diversity of MAPs as well as a cheap, and otherwise unemployed, workforce has led to the growth of an industry utilizing medicinal plants. Southeastern Europe in general has become an important source of MAPs, with more than 300,000 tonnes, worth more than US$ 800 million, traded annually in 1995 and 1996 (Lange and Mladenova 1997). Bulgaria is the second largest net-exporter of pharmaceutical plants in Europe, with an average annual export of 10,050 tonnes in the period 1991–2000 (Kathe, Honnef and Heym 2003).

The basis of the trade is an industry that is present at several levels in the community: from large-scale cultivation to collection of plants in the wild by the local population for household use. Although relatively large in comparison with neighbouring countries, the fraction that is cultivated only comprises 20–25 per cent of the total harvest of MAPs (Hardalova 1997). Thus the local gathering of wild plants is of great importance for both the industry and the local collector as a means of income (Kathe, Honnef and Heym 2003). Many of those who collect medicinal plants, for instance students, do so to generate some extra income in the summer and have no or very little education in how to intensify production in a sustainable manner. With regard to this and other threats to medicinal plants it is thus important to concentrate on the local collector. Today work is being done to start cultivation of more species and on a larger scale; initiatives are coming from both the Bulgarian government and the EU (Peev et al. 1998; Nature Protection Directives 2001; Kathe, Honnef and Heym 2003).

The generally escalating environmental pressure of today is of great concern for medicinal plant abundance, because of its economic importance. Several species are acutely endangered, and numerous more are threatened. Therefore there is a need for action towards conservation of plants in general and medicinal plants in particular. Such actions were taken during the last century, starting with the establishment of the first nature reserves in 1934. Thereafter more protected areas were established and laws for the preservation of natural resources passed so that today about 5 per cent of the Bulgarian territory is protected.

A report from the German Federal Agency for Nature Conservation (Kathe, Honnef and Heym 2003) comparing Bulgaria with Albania, Bosnia-Herzegovina, Croatia and Romania found that Bulgaria has probably the best system for regulation of MAPs. Since 1989, Bulgaria's Ministry of Environment and Water has used a quota system to manage the gathering of MAPs. Regional quotas are issued annually for endangered or threatened MAP species, which can rotate between regions from year to year (Lange and Mladenova 1997). The 2000 'Law on Medicinal Plants' dictates that the Regional Forestry Service and Regional Inspectorates of Environmental Protection monitor, control and regulate wild MAP-collecting activities. Regarding implementation, it generally seems to work out well, but there have been reports of collection of endangered or rare plant species (Kathe, Honnef and Heym 2003). On the other hand, Ploetz's countrywide survey of wild herb use (Ploetz 2000; Ploetz and Orr 2004) suggests that the laws are rather ineffective.

A Study of Endangered Medicinal Plants in Roussenski Lom Natural Park

Bulgaria currently ranks eighth on the list of the leading countries involved in global trade of botanical drugs (Lange and Mladenova 1997). Local harvesting of wild-growing plants is of great importance for both this industry and the local collector as a means of income, as only 20–25 per cent of the total amount of exported medicinal plant material is cultivated.

At the national policy level much is being done to protect medicinal herbs, through new legislation and establishing protected areas, but little is known about the awareness of the problem at the level of communities. Based on a questionnaire survey, Ploetz (2000: 34) states that local people in Bulgaria are aware of overexploitation of medicinal plants, and according to her study more than 50 per cent of the surveyed people 'indicated that they believe the abundance of Bulgaria's herbs is threatened'. In a similar manner, but using a direct research method with semistructured interviews, our research focused on the question of whether inhabitants in villages within the area of Roussenski Lom Natural Park are aware of any threat to medicinal plants in the area.

Roussenski Lom Natural Park

Roussenski Lom Natural Park is a limestone plateau eroded over millions of years by tributaries to the Danube (see Figure 5.1). In the transition from the Pliocene to the Pleistocene the 2500 m thick limestone layers in the area were slowly horizontally uplifted. That forced the river to cut its bed up to 100 m into the rock. The outcome is a remarkably beautiful meandering gully. The limestone caves in the area have been used as monasteries and underground churches, but also as dwellings and storage. The rock churches at Ivanovo are UNESCO World Cultural and Natural Heritage sites. The limestone escarpments and hillsides are rich in flora and fauna, and many uncommon plants are found here. An area of 3259 ha was declared a Natural Park in 1970.

The small villages of Koshov and Červen, with populations of around five hundred, lie within the park. Many of the younger generation have moved to the nearby city of Rousse, so the villages are currently inhabited by older people and commuters. In recent years some of the houses in and outside the villages have been bought and restored as holiday houses for people from Rousse and elsewhere. People live off small-scale farming, frequently using donkey-drawn carts as their main mode of transport and horses for ploughing and working the fields.

Studying in Koshov and Červen

Through interviews with the inhabitants of Koshov and Červen (Table 5.1, Figure 5.2), inventories of seventeen gardens (Table 5.2), and reports of the collection and use of medicinal plants from the park (Table 5.3), we tried to determine whether people still gather and use medicinal plants and whether any of these plants have become scarcer during their lifetime. The interviews and inventories were done together with either Sveza Filipova, a Bulgarian ethnographer, or Maya Vasileva, a botanist, who also assisted with the identification of plant species as well as the

Table 5.1. Informants for awareness of endangerment of medicinal plants study

Name	Sex	Age	Village
Danka Milanova	Female	85	Koshov
Krastina Atanasova	Female	~65	Koshov
Pena Bozhinova	Female	~70	Koshov
[§]Anna Avramova	Female	~50	Koshov
[§]Demitr Avramov	Male	~50	Koshov
[§]Margarita Garvaloa	Female	~45	Koshov
Ginka Kostova Manova + *	Female	~60	Koshov
Bogomil Bogdanov +	Male	61	Červen
Ivana Ilieva +	Female	7	Červen
Deltzo Mintskev –	Male	81	Červen
Desha Ivanova –	Female	75	Červen
Dana Dimova + *	Female	83	Červen
Ivanka Jordanova + *	Female	83	Červen
Elena Stojonova +	Female	73	Koshov
Nadejda Eneva +	Female	71	Koshov
Snejana Marinova + *	Female	57	Koshov
Stojan Petrov Stojanov + *	Male	64	Koshov

Grey shades indicate that informants were interviewed together. Key: – does not gather medicinal plants; + gathers plants; * gathers medicinal plants; [§]No photograph available in Figure 5.2

Figure 5.2. Informants for the study on awareness of diminishing plant species. See Table 5.1 for additional information on the informants. A: Krastina Atanasova; B: Elena Stojanova; C: Danka Milanova; D: Dana Dimova; E: Pena Bozhinova; F: Ivanka Jordanova; G: Ginka Kostova Manova; H: Snejana Marinova; I: Stojan Petrov Stojanov; J: Deltzo Mintshev and Desha Ivanova; K: Bogomil Bogdanov and Ivana Ilieva; L: Nadejda Eneva

Table 5.2. Homegarden species with reported folklore or medicinal use. A total of seventeen home gardens were inventoried, and only species found in more than one garden are listed here

Species	Number of gardens	Use
Aloe vera (L.) Burm.f.	2	Decoration, medicinal
Anethum graveolens L.	15	Spice, medicinal
Buxus sempervirens L.	13	Decoration, folklore, magic
Calendula officinalis L.	3	Decoration, medicinal
Cydonia oblonga Mill.	6	Food, medicinal
Geranium macrorrhizum L.	12	Decoration, folklore, magic
Juglans regia L.	5	Food, medicinal
Mentha spicata L.	7	Spice, medicinal
Satureja hortensis L.	5	Medicinal, spice
Sempervivum tectorum L.	2	Decoration, medicinal
Vitis vinifera L.	14	Food, beverages, medicinal

Table 5.3. Use and preparation of medicinal plants by informants in Koshov and Červen, both cultivated in homegardens and wild-harvested from Roussenski Lom Natural Park

Scientific name	Main source	Disease treated[a]	Part used	Preparation[b]	Informant(s)[b]
Allium cepa L.	Cultivated	Cold, cough (7), rhinitis (10)	Bulb	Syrup with *A. cepa*, *J. regia*, *O. basilicum*, *M. chamomilla*, *S. ebulus* and *M. spicata* (7), fresh juice in nose (10), tea with *P. radula*, walnut, apple and *lokum* (turkish sweets)	7, 10, 16
Allium sativum L.	Cultivated	Blocked nose, rhinitis	Corm	Fresh juice dropped in nose	7
Artemisia absinthium L.	Wild harvested	Cold	Leaves	Infusion	9
Calendula officinalis L.	Cultivated	Cold	Flower	Infusion with *T. vulgaris* and *Tilia* spp.	12
Fragaria x *ananassa* Duchesne ex Rozier	Cultivated	Cold, fever	Leaves, berries	Infusion; infusion with *Salix alba* bark for fever	9
Hedera helix L.	Wild harvested	Cough	Leaves	Infusion	2, 13
Hypericum perforatum L.	Wild harvested	Cold	Entire plant	Infusion with *T. vulgaris*, *S. virgaurea*, *O. vulgare* and *M. chamomilla*	10
Juglans regia L.	Cultivated	Throat pain, cold, cough, breathing difficulties	Young nut	Crushed in tea (2), syrup with *A. cepa*, *J. regia*, *O. basilicum*, *M. chamomilla*, *S. ebulus* and *M. spicata* (7), tea with *A. cepa*, *P. radula*, apple and *lokum* (turkish sweets) (16)	2, 7, 16
Malus domestica Borkh.	Cultivated	Cough	Fruits	Infusion with *P. radula*, *J. regia*, *A. cepa* and *lokum* (turkish sweets)	16
Malva sylvestris L.	Wild harvested	Throat problems, cold	Flowers	Infusion	1, 9

Table 5.3. *continued*

Scientific name	Main source	Disease treated[a]	Part used	Preparation[b]	Informant(s)[b]
Matricaria chamomilla L. (syn *M. recutita* L.)	Wild harvested	Cold, throat pain, cough, to prevent cold	Flowers	Dried, infusion to drink (2) or to gargle with (5), with sodium bicarbonate as an inhalation (3) or as syrup with *A. cepa*, *J. regia*, *O. basilicum*, *Tilia* spp., *S. ebulus* and *M. spicata* (7), tea with *O. vulgare*, *T. vulgaris*, *H. perforatum* and *S. virgaurea* (10)	2, 3, 5, 7, 9, 10, 14
Melissa officinalis L.	Cultivated	Cold	Leaves	Infusion	3, 4
Mentha spicata L.	Cultivated	Swollen throat glands	Entire plant	Infusion	1
Menyanthes trifoliata L.	Wild harvested	Cold	Leaves	Infusion	9
Ocimum basilicum L.	Cultivated	Cough	Entire plant	Infusion (2), syrup with *A. cepa*, *J. regia*, *O. basilicum*, *M. chamomilla*, *S. ebulus* and *M. spicata* (7)	2, 7, 14
Ononis spinosa L.	Wild harvested	Cold	Leaves	Infusion	9
Origanum vulgare L.	Wild harvested	Cold	Leaves	Infusion with *T. vulgaris*, *S. virgaurea*, *H. perforatum*, *M. chamomilla*	10
Pelargonium radula L'Hér	Cultivated	Cold (3), cough (16)	Leaves	Mixture with *A. cepa*, *J. regia*, *M. spicata* (3); infusion with apple, *A. cepa*, *J. regia* and *lokum* (turkish sweets) (16)	3, 16
Pinus sylvestris L.	Wild harvested	Cough (1, 14), cold (14)	Spring shoots	Syrup with sugar (1), infusion (14)	1, 14
Robinia pseudoacacia L.	Wild harvested	Cold	Flowers	Infusion of dried or fresh flowers	3, 12
Salix alba L.	Wild harvested	Fever	Bark	Infusion with strawberries	9
Sambucus ebulus L.	Wild harvested	Breathing difficulties, throat pain, cold, cough	Flowers	Infusion of dried flowers (1) or syrup with *M. chamomilla*, *O. basilicum*, *A. cepa*, *J. regia* and *M. spicata* (7)	1, 5, 7, 13

Table 5.3. *continued*

Scientific name	Main source	Disease treated[a]	Part used	Preparation[b]	Informant(s)[b]
Sinapis arvensis L.	Wild harvested	Cough	Seeds	Mixture with honey and butter	10
Solidago virgaurea L.	Wild harvested	Cold	Leaves, flowers	Infusion with *O. vulgare, T. vulgaris, H. perforatum* and *M. chamomilla*	10
Symphytum officinale L.	Wild harvested	Lung problems/ diseases	Roots, leaves	Infusion	14
Thymus vulgaris L.	Wild harvested	Cold	Entire plant	Infusion (3, 9), infusion with *O. vulgare, S. virgaurea, H. perforatum* and *M. chamomilla* (10), infusion with *Tilia* spp. and *C. officinalis* (12)	3, 9, 10, 12, 14
Tilia spp.	Wild harvested	Cold, cough, breathing difficulties, fever (8)	Flowers	Syrup with *A. cepa, J. regia, O. basilicum, M. chamomilla, S. ebulus* and *M. spicata* (7, 8), infusion with *T. vulgaris* and *C. officinalis* (12)	7, 8, 12, 14
Tribulus terrestris L.	Wild harvested	Cold	Entire plant	Infusion	9
Trifolium pratense L.	Wild harvested	Cold	Flower heads	Infusion	9
Tussilago farfara L.	Wild harvested	Cough, cold	Leaves	Infusion of dried leaves	1, 8, 10, 15
Viola tricolor L.	Wild harvested	Cold	Entire plant	Infusion	9

[a]The informants were asked specifically about whether they knew plants to treat colds, coughs and fever.

[b]The numbers in the columns refer to the informants: 1) Anna Avramovi; 2) Margarita Garvalova; 3) Ginka Kostova Manova*; 4) Anna Avramova, Demitr Avramov, Margarita Garvalova; 5) Velika Marinova; 6) Vera Jordanova: 7) Snejana Marinova*; 8) Danka Milanova*, Krastina Atanasova*, Pena Bozhinova*; 9) Stojan Petrov Stojanov*; 10) Maya Vasileva*; 11) Nikolai Kubratov; 12) Malinka Tsoneva; 13) Blagoi Bogdanov; 14) Group interview in Cerven; 15) Minka Mincheva; 16) Penka Goranova. For informants marked with (*) additional information can be found in Figure 5.2 and Table 5.1.

translation of Bulgarian plant names. Research was conducted by graduate students from Uppsala University during April and May 2005. Given the short period for research, the findings presented here are necessarily preliminary, but still offer insights into the way local Bulgarians think about and use medicinal plants.

We were very fortunate to come across a local expert, Mr Stojan Petrov Stojanov, who had a better understanding of medicinal plant populations than the other informants we interviewed, concerning both their sizes and localities, gained from nearly forty years of excursions in the area. He claimed to be the only person in Koshov who presently had this extensive knowledge and that over the years he had taught some of it to the rest of the village's inhabitants. He also claimed that previously people had not been very interested in the wild medicinal plants of the area and did not think much of his ambles, but that lately townspeople and tourists had begun to ask him to guide walks in the area. Mr Stojanov had much to say about local exploitation of the Natural Park, as is reported below, and his contribution to the study was crucial. This raises an important methodological point: the domain of medicinal plant knowledge is usually restricted to specialists, with the general public typically having some but often just a very shallow understanding of these plants and their uses. Sometimes there may be only a few people, or even just one person, in a community with the in-depth, accurate knowledge that researchers might find useful. It is therefore essential that these local experts are sought out, especially nowadays as younger generations are less and less likely to take an interest in these traditional practices.

Wild-harvesting of Medicinal Plants

Interviews with residents (see Table 5.1) suggest that collection of MAPs for personal use is fairly common among the adults and elder residents living in Roussenski Lom Natural Park. The majority of informants collect a few species of medicinal plants, in the right season when these are available, but none reported to do it on a commercial basis (see Table 5.3). Local people use medicinal plants cultivated in their homegardens (see Table 5.2), as well as collected from the surrounding Park. In this preliminary study more than 25 per cent of informants reported *Matricaria chamomilla* L., *Sambucus ebulus* L., *Thymus vulgaris* L., *Tilia* spp. and *Tussilago farfara* L. as important medicinal plants collected from the Park, all of which are among the seventeen most important MAP species harvested from the wild in Bulgaria (Kathe, Honnef and Heym 2003). These species are all used to treat common ailments, such as colds, coughs, fever and sore throats. These species, as well as the other species reported (see Table 5.3), are generally fairly common in the Park according to a local botanist and none are prohibited (Maya Vasileva, pers. comm.; see also Table 5.4).

Table 5.4. Awareness of diminishing plant species

	Mentioned by informants (No. #)	Species listed as endangered[1] found in home gardens	Restricted for picking in RL NP (Vasileva, pers. comm.)	Most important exported MAP species[2]	Endangered MAP species in regions or entire country[1]	Monitored in the Annual Reports of the Regional Forestry Directorates[3]
Achillea clypeolata Sm.	2					
Asplenium trichomanes L.	1		•			•
Atropa belladonna L.		•			•	•
Centaurea cyanus L.	1					
Centaurium erythaea Rafn.	1					•
Convallaria majalis L.		•			•	•
Glaucium flavum Crantz		•			•	
Hypericum perforatum L.	2			•		•
Linaria vulgaris Mill.	1					
Matricaria chamomilla L.	1			•		
Papaver rhoeas L.	1					
Rhamnus cathartica L.	1					
Sambucus ebulus L.	1			•		•
Stipa spp.	1					
Valeriana officinalis L.	1		•	•		

[1]Kathe, Honnef and Heym 2003, Appendix D.1. Data from the Red Data Book of Bulgaria and official orders from the Ministry of Environment and Water.
[2]Kathe, Honnef and Heym 2003, p. 47.
[3]Kathe, Honnef and Heym 2003, Appendix C.1.

In addition, people collect those herbs that are plentiful in the area or buy them from local pharmacies that stock dried medicinal plants, which are always readily available. One resident suggested that the present generation rely on these dried plants and plant extracts, and have very little knowledge of wild medicinal plants, neither how to find them nor how to use them.

Cultivating Medicinal Plants

Our inventory of homegardens also showed some interest in and use of medicinal plants. While most of the 130 different species found in the gardens are for food and decorative purposes, some do have medicinal uses such as *Aloe vera* (L.) Burm.f., *Geranium macrorrhizum* L., *Juglans regia* L. and *Sempervivum tectorum* L. (see Table 5.2). Some informants mentioned giving dried cultivated or collected medicinal plants to neighbours and friends, and we also accumulated quite a few samples during our research, but no commercialization of homegarden-cultivated or wild-harvested medicinal plants was reported. Growing medicinal plants in homegardens could be a sustainable alternative if demand grows or availability in the wild decreases, although both are unlikely as the rural population decreases through emigration and senescence.

Local Awareness of Scarcity

The general impression is that the area is, in fact, rich in plants in general, so that it is rarely a problem to find medicinal plants. Most respondents had little knowledge of whether medicinal plants in the area have become scarcer, but three species were mentioned: *Achillea clypeolata* Sm., *Asplenium trichomanes* L. and *Hypericum perforatum* L. However, our local expert, Mr Stojanov, noted diminishing populations of several species, those that are still relatively common and those that seem to have become locally extinct. He mentioned that eleven species have become scarcer during his lifetime: *Achillea clypeolata* Sm., *Centaurea cyanus* L., *Centaurium erythaea* Rafn., *Matricaria chamomilla* L., *Hypericum perforatum* L., *Linaria vulgaris* Mill., *Papaver rhoeas* L., *Rhamnus cathartica* L., *Sambucus ebulus* L., *Stipa* spp. and *Valeriana officinalis* L. He notes that *M. chamomilla* and *S. ebulus* populations are declining in the Park itself, but not at an alarming rate. Mr Stojanov's impression was that most plants gathered for personal use are generally easy to find, which is supported by our survey results (see Table 5.3). However, he claimed that Roma, living in a village near the Park, often illegally collect MAPs in the area for commercial purposes, and stressed that collection was not always done in the most sustainable manner. Kathe, Honnef and Heym (2003) identify commercial collection of MAPs as an activity predominantly performed by poor, ethnic minority and/or retired people in rural areas. Regrettably, in Bulgaria, as elsewhere in eastern Europe, marginalization and segregation of Roma is common (European Commission on Racism and Intolerance 2004), which may explain why the Roma would be involved in the wild-harvesting of MAPs in the Park. However, it is also the case that poorer and marginalized peoples and minority groups may be blamed for such illegal activities. Research with the Roma would be necessary to confirm this view, and also explore how and why this collection occurs in an unsustainable way, if it occurs at all.

Of the twelve plants mentioned by all informants, *A. trichomanes* and *V. officinalis* are listed as medicinal species restricted for picking in Roussenski Lom Park (Maya Vasileva, pers. comm.). It should also be noted that among cultivated medicinal plants in the gardens there were those that are either scarce in Bulgaria (Kathe, Honnef and Heym 2003) or under some kind of restriction in the Natural Park (Maya Vasileva, pers. comm.).

We conclude that the people of Koshov and Červen, in general, are not aware of a decrease in the availability of medicinal plants collected in the wild. These results clearly contradict the results found in the wider Bulgarian survey by Ploetz and Orr (2004). Why is it that, except for a few species, people do not think that medicinal plants are diminishing or threatened? The easiest explanation could simply be that they do not gather many wild plants, for medicinal or other purposes. People in general mostly use plants from their gardens or buy them dried, and only a few, quite common species are gathered from the wild. An alternative explanation could be that the abundance of collected species in the Natural Park is so high that no significant problems of overharvesting occur. This is probably only true for a few abundant species, such as *M. chamomilla*, which the specialist (Mr Stojanov) notes to be decreasing even though it is still widely available. Our local expert also listed several other plants that he believes to be in decline, so perhaps species are vulnerable, even in the Natural Park (see Table 5.4).

The Future of Medicinal Plants in Bulgaria

Small-scale collection of medicinal plants for personal use forms an important part of Bulgarian biocultural diversity, which has come under threat after the collapse of communism due to commercial overexploitation of wild medicinal plants. In addition, increasing urbanization and increased welfare have led to diminishing interest and knowledge transfer on medicinal plant use and collection, as modern medicine is both readily available and affordable. Overexploitation and related scarcity of medicinal plants is an acute problem. The main culprit is commercial picking for export to mainly western European countries. Bulgaria became the second largest net-exporter of raw medicinal plants in the 1990s, with the bulk of this export, 75–80 per cent, from wild-crafted medicinal and aromatic plants that are collected all over the country (Evstatieva and Hardalova 2004).

Future directions for the Bulgarian medicinal plant market would be to focus on increasing and intensifying cultivation of medicinal plants, as well as developing a domestic processing industry that exports to western Europe. Extensive cultivation would allow for a more predictable annual production, which could in turn lead to stabilization of prices and increased security for small-scale commercial collectors. Experimental cultivation, specifically targeted at threatened and endangered medicinal species, with EU or national funds could lead to an intentional undercutting of the market for these vulnerable species, which fetch high prices but are currently difficult to obtain because of their scarcity. Directed research into cultivation of endangered MAPs is being conducted already (Kupke, Schwierz and Niefind 2000). Increased commercial cultivation could promote sustainability on a countrywide scale, but it could also undermine small-scale collecting by local people, including minority populations such as the Roma. The earnings generated by small-scale collecting of medicinal plants are probably significant for the subsistence of many rural people that have limited income. In addition, small-scale collection of medicinal plants for personal use is not necessarily a threat to floral biodiversity, and sustainable collection forms a traditional cultural practice in Bulgaria.

A transition from an uncontrolled overexploitation to a sustainable system of regulated, commercial wild-harvesting and cultivation is already under way, and Bulgaria has probably the best system of regulation regarding MAPs in Eastern Europe (Kathe, Honnef and Heym 2003). There is a quota system for commercial harvesting, where quotas are issued annually for specific species in defined regions. Wild MAP-collecting activities are then monitored, and the implementation seems to work out well in general, though there have been reports of collection of endangered or rare plant species (Kathe, Honnef and Heym 2003). The quota system makes it difficult to trade regulated species over the quota maxima, and thus limits overexploitation to a certain degree, but the species and volumes subject to the quota system are still quite limited (Mladenova 1998).

The average age of the informants in this study reflects in part a bias toward selecting elderly informants because of their presence and willingness to participate, but in part also the extensive migration of the youth to urban centres. The latter is a trend of concern for the preservation and continuation of this part of the overall biocultural diversity of Bulgaria. Elderly informants represent knowledge repositories of biocultural diversity and documenting their knowledge is a crude way of preserving their skills beyond the extent of their lives. Ideally their knowledge would be taught directly to a younger generation of Bulgarians.

Conclusion

In conclusion, then, in Roussenski Lom Natural Park the loss of biocultural diversity through general aging of the population of holders of medicinal plant knowledge appears more immediate than the loss of medicinal plant biodiversity. Identifying community members with expert knowledge of medicinal plants and involving them in participatory monitoring of wild harvesting could benefit both Park rangers and the younger generation, and could be a sustainable way of maintaining today's and reviving yesterday's traditions.

Acknowledgements

Our gratitude goes to all the Bulgarian informants who were all so willing and kind to spend time sharing their knowledge with our research team. Thanks also to our colleagues Dr Lars Björk, Anneleen Kool and Marie Melander, without whose help and organization this project would never have been realized; and also to our Bulgarian colleagues: Maya Vasileva, Dr Rousi Rousev (Rousse University), Dr Sveza Filipova, and Dr Ljuba Evstatieva (Bulgarian Academy of Sciences). I would like to acknowledge the efforts of the students who collected the raw data for this chapter: Annika Bengtsson, Erik Carlsson, Karolina Eriksson, Päivi Hennola, Rebecka Strandberg and Marie Wiberg Andersson. The Swedish Institute (SI) is acknowledged for granting travel stipends to all students.

References

Ahtarov, B. 1939. *Materiali za bulgarski botanitcheski retchnik.* Sofia: BAN.

Antonova, J. 2007. 'Pharmacy in Medieval Bulgaria', *Pharmazie* 62: 467–469.

Christov, I. 1996. 'The Creative Logos, the Nature of Things, and Their Uniqueness in the ''Hexaemeron' of Joannes Exarchus', in J. Aertsen and A. Speer (eds), *Individuum und individualität im Mittelalter. Miscellanea Mediaevalia* 24: 99–110.

Council of Europe. 1999. 'A European Approach to Non-Conventional Medicines: Resolution 1206', in *Official Gazette of the Council of Europe.* Retrieved 6 September 2008 from http://assembly.coe.int/Mainf.asp?link=/Documents/AdoptedText/ta99/ERES1206.htm

European Commission on Racism and Intolerance. 2004. 'Third Report on Bulgaria'. Council of Europe, CRI 2.

Evstatieva, L. and R. Hardalova. 2004. 'Conservation and Sustainable Use of Medicinal Plants in Bulgaria', *Medicinal Plant Conservation* 9/10: 24–27.

Georgiev, M. 1979. 'Bulgarian Folk Medicine', *Medico-biologic Information* 4: 25–30.

Hardalova, R. 1997. 'The Use of Medicinal Plants in Bulgaria and Their Protection', in J. Newton (ed.), *Planta Europaea: Proceedings of the First European Conference on the Conservation of Wild Plants*, Hyères, France, 2–8 September 1995. London: Plantlife, pp. 184–187.

Hossmann, I., M. Karsch, R. Klingholz, Y. Koehnke, S. Kroehnert, C. Pietschmann and S. Suetterlin 2008. *Europe's Demographic Future: Growing Imbalances (Summary).* Berlin: The Berlin Institute for Population and Development.

Iordanov, D., P. Nicolov and A. Boichinov. 1969. *Phytotherapy in Bulgaria.* Sofia: Medicina i Phyzcultura.

Ivancheva S. and B. Stantcheva. 2000. 'Ethnobotanical Inventory of Medicinal Plants in Bulgaria', *Journal of Ethnopharmacology* 69: 165–172.

Kathe, W., S. Honnef and A. Heym. 2003. *Medicinal and Aromatic Plants in Albania, Bosnia-Herzegovina, Bulgaria, Croatia and Romania.* Bundesamt für Naturschutz-Skripten 91. Bonn: Federal Agency for Nature Conservation.

Kitanov, B. 1953. *Bulgarian Folk Medicine.* Vol. 3.

Kupke, J., A. Schwierz and B. Niefind. 2000. *Arznei- und Gewurzpflanzen in Osteuropa. Anbau, Verarbeitung und Handel in 18 ausgewahlten MOE-Landern. Materialien zur Marktberichterstattung* 34: 1–95. Zentrale Markt- und Preisberichtstelle GmbH. Bonn, Germany.

Lange, D. and M. Mladenova. 1997. 'Bulgarian Model for Regulating the Trade in Plant Material for Medicinal and Other Purposes', in G. Bodeker, K.K.S. Bhat, J. Burley and P. Vantomme (eds), *Medicinal Plants for Forest Conservation and Health Care; Non-wood Forest Products* 11: 135–146. Rome: FAO.

Leporatti, M.L. and S. Ivancheva. 2003. 'Preliminary Comparative Analysis of Medicinal Plants Used in the Traditional Medicine of Bulgaria and Italy', *Journal of Ethnopharmacology* 87: 123–142.

Mladenova, M. 1998. 'The Management System of Harvesting Medicinal Plants in Bulgaria', in *Medicinal Plant Trade in Europe: Conservation and Supply.* Proceedings of the First International Symposium on the Conservation of Medicinal Plants in Trade in Europe. Brussels: TRAFFIC-Europe, pp. 85–99.

Nature Protection Directives. 2001. *The Benefits of Compliance with the Environmental Acquits for the Candidate Countries, Part E: Nature Protection Directives.* ECOTEC, European Union.

Peev, D., S. Kozuharov, M. Anchev, A. Petrova, D. Ivanova and S. Tzoneva. 1998. 'Biodiversity of Vascular Plants in Bulgaria', in C. Meine (ed.),

Bulgaria's Biological Diversity: Conservation Status and Needs Assessment (2 volumes). Washington DC: Biodiversity Support Program, World Wildlife Foundation. Retrieved 20 May 2008 from http://www.worldwildlife.org/bsp/publications/europe/bulgaria/bulgaria.html

Petkov, V. 1986. 'Bulgarian Traditional Medicine: a Source of Ideas for Phytopharmacological Investigations', *Journal of Ethnopharmacology* 15: 121–132.

Ploetz, K.L. 2000. *An Ethnobotanical Study of Wild Herb Use in Bulgaria*, M.Sc. dissertation. Houghton, MI: Michigan Technological University.

——— and B. Orr. 2004. 'Wild Herb Use in Bulgaria', *Economic Botany* 58: 231–241.

World Facts U.S. 2007. 'Facts about Bulgaria'. Retrieved 8 September 2008 from http://worldfacts.us/Bulgaria.htm

CHAPTER 6

'My Doctor Doesn't Understand Why I Use Them'

Herbal and Food Medicines amongst the Bangladeshi Community in West Yorkshire, U.K.

ANDREA PIERONI, HADAR ZAMAN,
SHAMILA AYUB AND BREN TORRY

Urban Ethnobotany

There has been a slow but gradual shift in ethnobotanical research during the last twenty years, from exploring exotic places and rainforests to investigating back yards and urban environments. Critics of ethnobotany have seen this change as the reaction of ethnobotanists to the increasing difficulties in negotiating permissions with regional and national authorities of developing countries (especially Meso-American) for conducting classical bioprospecting-based research. However, there has been a move away from access to classical funding routes, which in turn has resulted in a change in research aims from a mere documentation of plant uses to more complex, hypothesis-driven research.

In reality, change has been forthcoming since the late 1990s when North American ethnobotanists began to explore the plant uses by people migrating and settling in urban environments (Balick, Kronenberg and Ososki 2000), possibly driven by the increasing interest of a number of stakeholders in minority ethnic health and issues related to the use and perception of Traditional Medicine (TM) and Complementary and Alternative Medicine (CAM) within Western societies. Although this shift may have been initially the result of many undesirable contingencies, it is nowadays widely accepted within the scientific community that 'urban' ethnobotanical studies may offer the unique possibility of a better understanding of how ethnobotanical knowledge changes over space and time.

In recent years a number of ethnobotanical studies have highlighted interesting cases amongst minority ethnic communities in urban environments (Corlett, Dean and Grivetti 2003; Nguyen 2003; Waldstein 2006; Pieroni and Vandebroek 2007, and chapters therein). The growing interest of medical anthropologists and other social scientists in ethnicity and health studies in multicultural societies parallels an increasing awareness that the meaning of health is often broader and more complex than what is understood in classical Western biomedical terms. In order to understand human well-being we need to take into consideration emic or insiders' health perceptions, beliefs and practices, instead of focusing merely on the etic or outsiders' approach. Emic approaches to minority-ethnic health studies represent important turning points in public-health discourses aimed at improving interventions devoted to minority ethnic groups in Western countries.

Strengthening or Adapting Cultural Identities?

As highlighted in the introduction of the recently edited book *Traveling Cultures and Plants: The Ethnobiology and Ethnopharmacy of Human Migrations* (Pieroni and Vandebroek 2007), there are scientific questions which remain open for investigation in the field of migration and ethnobiology. These include:

- Do people who migrate still depend on their own healthcare strategies within the domestic domain, including the continued use of food and medicinal plants brought over from their home countries or purchased in local shops in the new society, for common, chronic and/or culturally important health conditions? If so, why is this?
- In what ways do minority ethnic healthcare-seeking strategies change over time, in response to 'internal' dynamics of identity and representation within the minority ethnic community, and to external environmental, cultural, social and political changes in the country to which they migrate, including public healthcare policies?
- What are the existing articulations between minority ethnic groups' own healthcare systems and the institutionalised biomedical system? To what extent are institutional health actors in their new country of residence aware of these strategies?

Biocultural adaptation, cultural negotiation and identity are key issues for anthropological discourses on displacement and migrations.

Research in culturally homogeneous and/or nonurban environments has shown that to follow the pattern of change in traditional knowledge and use of plants among people who migrate to a new country implies the analysis of acculturation processes (Nesheim, Dhillion and Stolen 2006). Acculturation has been discussed in communication sciences as being the result of two simultaneous processes: one involving deculturation from the original culture, and the other involving enculturation towards the adopted culture (Kim 2001). The old model of cultural adaptation is, however, quite problematic, since it is highly unlikely that a culture, after moving, simply 'adapts' to the new, autochthonous culture. There are reasons for this; adaptation does not represent a sort of 'destiny'. On the contrary, it is only one of many diverse possibilities that minority ethnic groups have in their interface with a new culture. Adaptation is, in fact, the result of cultural negotiations with a new environment, which is not always culturally homogeneous. Indeed, Western metropolitan milieus are never quite culturally homogeneous. Often people who have migrated may choose other strategies; for example, those aimed at strengthening their cultural identities in a process that is similar to what other scholars in ethnoecology define as 'resilience' at ecological and cultural edges (Turner, Davidson-Hunt and O'Flaherty 2003). Strengthening their own identities in the midst of autochthonous/new populations means that people who migrate to a new country may want to deliberately retain their traditional knowledge and practices, in order to affirm their distinct cultural identity. What minority ethnic groups do in reality probably lies somewhere in between the two. The exact location depends very much on the dynamics in the migrated groups' changing interface with their new cultural context. Moreover, since ethnicity is also the complex result of social processes (Barth 1969), and cultural boundaries are very dynamic and may even be seen as constructs that are created by our own processes of representation (Clifford and Marcus 1986; Marcus 1998), plant uses and especially their representations may change rapidly in response to continuously shifting cultural negotiations.

Migrants' Ethnobiology and Bangladeshis in the U.K.

Health policies in many European countries and in the U.K. have been based and are still largely based upon 'assimilationist' approaches, where the assumption is that inequalities among ethnic groups will disappear once newly migrated families adopt belief systems and practices of the 'new society'. Assimilationist ideas are based upon simplistic explanations (Mason 2000) and they do not take into account the variability of well-being and health beliefs and perceptions, which exist among diverse ethnic groups. Biologists and health specialists have paid attention to the issue of the use and perception of TMs and other 'nonorthodox' practices of different medical systems among minority ethnic groups in Western European urban environments (Pieroni et al. 2005; Sandhu and Heinrich 2005; Ceuterick et al. 2007; Pieroni et al. 2007; Pieroni and Torry 2007; Pieroni et al. 2008) and even nonurban environments (Pieroni and Gray 2008), while other researchers have also analysed the dietary habits of people who have migrated (Jonsson, Hallberg and Gustafsson 2002; Burns 2004).

Too little is still known about the uses and perceptions of homemade herbal medicines and food in self-medication practices and the provision of healthcare within migrated populations' domestic arenas. Moreover, more research is needed to better understand how these practices among minority ethnic groups change over time. A deeper understanding of the patterns of use of TMs and of the non-Western healing strategies among people who migrate, and their dynamics, could be especially crucial for health practice and policies in Europe.

Studying Herbal and Food Medicines in Bangladeshi Communities in West Yorkshire

The primary aim of this study was to explore the range of TMs in use within the Bangladeshi community in West Yorkshire. We wanted to identify the purpose of their use and to capture how lay healthcare knowledge has changed over time.

This ethnobotanical study was undertaken between September and December 2007. Semi-structured interviews, focus groups and the free-list technique were used to collect data. The majority of data were qualitative; however, some data were converted into a quantitative form. Qualitative data were analysed using the Grounded Theory technique (Strauss and Corbin 1998), statistical data were analysed using SPSS and all graphs were produced in Microsoft Office Excel.

Research Participants

The U.K. 2001 National Census provides a useful insight into cultural diversity in England. In 2007 there were 324,000 people of Bangladeshi ethnic origin living in England, of which the majority (62 per cent) were aged between sixteen and sixty-four years; only 4 percent were aged sixty-five years and over (Self 2008). This minority ethnic group make up 0.6 per cent of the population of England (Phillpotts and Causer 2006: 102). The majority of families of Bangladeshi origin live in London. However, in Yorkshire and the Humber there were estimated to be 12,330 people living in the region (ONS 2001a); this represents just 0.2 per cent of people living in Yorkshire. Figure 6.1 provides a visual snapshot of the distribution of the Bangladeshi population within the U.K.

In comparison to other minority ethnic groups, and excluding the white majority ethnic group (who make up 92.1 per cent of the total population in the U.K.), the Bangladeshi ethnic group makes up just 6.1 per cent of all minority ethnic groups (ONS 2001b). The 2001 national census in the U.K. indicates that South Asians are the largest minority ethnic group, including Indian (22.7 per cent) and Pakistani (16.1 per cent) people. There has been a past tendency to report upon the South Asian ethnic groups as one single ethnic group. Consequently, early investigations into the health of the nation failed to recognize important cultural differences between different South Asian groups, and subsequently the diversity within different groups. The specific needs of people of Bangladeshi ethnic origin, therefore, have not been identified until more recently. In this study we wanted to investigate only the Bangladeshi group.

Figure 6.1. Distribution of the Bangladeshi population in the U.K. *Source*: ONS 2001b

A report into the health perceptions of minority ethnic groups in the U.K. found that Bangladeshi men and women were three to four times more likely than the general population to rate their health as bad or very bad (White 2002: 14). This is not surprising given that the study also found that Bangladeshi men and women living in England were nearly six times more likely than the general population to report having diabetes (White 2002: 14). These latter figures are considerably higher than those for Indian men and women in general. The validity of studies which rely upon self-perception scales or self-reporting techniques has long been questionable; however, these particular findings do raise serious consequences for public health policies in the U.K. The annual report by the national statistician in 2007 found that the proportion of people who reported a longstanding illness or disability, including mental health problems, remained stable across the whole of the U.K. However, the highest rates reported were for Bangladeshi women; Bangladeshi men were second highest, just behind Pakistani men (Dunnell 2008).

Bangladeshi households were also found to be the largest in the U.K., with an average of 4.5 people per household (Self and Zealey 2007: 15). Of course, this does not mean that such families are living in larger houses; on the contrary it is quite possible that many larger families living in deprived communities are living in overcrowded conditions. Not surprisingly, therefore, a report by the Joseph Rowntree Foundation in 1998 entitled 'Ethnic Minorities in the Inner City', which analysed data collected from 5,196 people from minority ethnic groups, found that Bangladeshis were most likely to live in the most deprived areas within the U.K. (Dorsett 1998).

In 2005 Bangladeshi men were amongst a small number of minority ethnic groups who were least likely to be in professional careers in comparison to the majority ethnic groups or most other minority ethnic groups; no Bangladeshi women werereported to be in such employment. In fact, in 2004/05, 86 per cent of children in Pakistani/Bangladeshi households in the U.K. were in the bottom 40 per cent of households ranked by disposable income, compared with 49 per cent of all children (Self and Zealey 2007: 47). These statistics reveal a number of social factors known to contribute towards poor health in families.

Sampling Method

To best meet the aims of this study it was necessary to recruit participants with a good history of traditional knowledge of lay healthcare; thus the inclusive criteria was to engage older people of Bangladeshi origin who fell into the category of first- or second-generation migrants. This participant profile was selected because it could provide the greatest opportunity to explore original lay-healthcare practices with TM as well as to obtain factual details about what information has been shared with younger generations within the Bangladeshi community.

Initially participant communities were randomly selected from four different towns in West Yorkshire, including Bradford, Leeds, Keighley and Halifax. Figure 6.2 is a map of the location of West Yorkshire within the U.K.

Members of these four Bangladeshi communities were approached on the basis of convenience and accessibility, and were invited to take part in the study. After initial participants were selected within each community a nonparametric snowball sampling method was used to recruit further members to the study. This snowball approach has the disadvantage that participants often recommend their friends or relatives to take part, who arguably share similar views and healthcare practices; this can introduce bias in the data and consequentially concerns about the validity of the

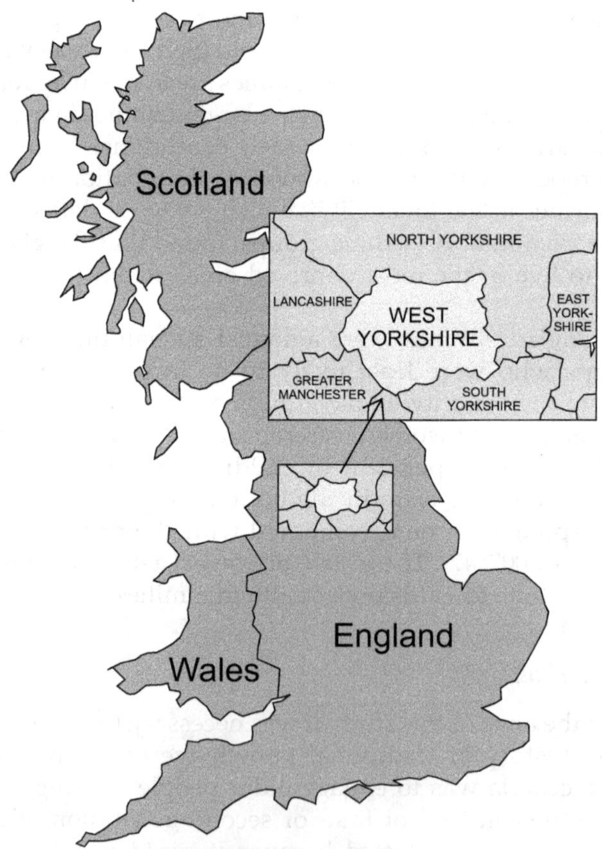

Figure 6.2. Location of West Yorkshire in Great Britain

results. However, in West Yorkshire the Bangladeshi communities are relatively small in comparison to other ethnic groups and so in reality it is possible that any method of participant recruitment could result in a degree of bias because of the strong likelihood that participants may know each other through friendship, relations or through shared community values. Children and young people were excluded from the study.

In total, seventy-nine participants were recruited; thirty-seven agreed to be interviewed and forty-two took part in the focus groups. To promote the most favourable setting and atmosphere during data collection, female participants were interviewed by female researchers and male participants interviewed by male researchers at a venue and time convenient to the participants; this also ensured strict Muslim beliefs regarding gender segregation could be respected in situations where participants engaged in religious practice. To reduce the risk of bias as a result of using different interviewers a fixed schedule of questions was used.

Ethical guidelines outlined by the American Anthropological Association (1998) and the International Society of Ethnobiology (2006) were observed throughout this study. We obtained ethics approval from Bradford University Ethics Committee. Informed consent was sought from participants in advance of signing them up to the study, and names and specific locations were made anonymous in order to preserve confidentiality. Permission was sought to use cameras and audio equipment.

Data Collection

In addition to a review of literature we used interviews, a free-listing method and focus groups to collect data.

Interviews and Free Listing

In total thirty-seven semistructured interviews were undertaken. Sessions were recorded. Participants were interviewed by two researchers; this ensured access to an interpreter because one researcher in each pair was fluent in South Asian languages. If a separate interpreter was required the researchers had access to family members. At the request of participants the interviews with men were always undertaken in a local mosque immediately after prayers, and interviews with women were undertaken either at home or in a Bangladeshi community centre around midday.

During interviews participants were invited to free-list plants, herbs, vegetables, spices, animal-derived products and ritual practices used for the maintenance of health. The free-list approach enabled the researchers to explore the participants' knowledge further as well as collect relevant botanical data; including names and parts of medicinal plants used or known to them, range of uses for the plants, the preparation and administration of the plants, their origin, frequency of use and any observed effects when used for the treatment of health. In addition, demographic data were also collected, including age, gender, place of birth and self-perceived identity.

Waldstein (2006) suggests that listing a remedy is not a proof of the actual use of it; in fact, the cultural importance of the plant is more likely to be evident from the number of participants who list it. The analysis, therefore, took into account this quantitative data.

Focus Groups

Three separate focus groups were undertaken with forty-two women participants in different parts of West Yorkshire (eight, fifteen and nineteen participants). We used focus groups because it allowed a forum where participants could share commonalities and diversities in practice. We believed this setting to be the best to expose such differences.

Botanical Identification

Quoted plant items were collected and identified using a standard work on Bangladeshi medicinal flora (Ghani 2003) and *Mansfeld's Encyclopedia of Agricultural and Horticultural Crops (Except Ornamentals)* (Hanelt and IPK 2001). Voucher specimens of the recorded wild taxa only were collected and deposited at the Herbarium of the Laboratory of Pharmacognosy at the University of Bradford (PSGB).

Biomedicine versus Traditional Medicine

The majority of the participants prefer TMs instead of Western biomedicines. Figure 6.3 shows that over twice as many participants rely upon TM for the treatment of illness and the maintenance of health. A number of concerns and remarks about the Western biomedical approach were raised. Male participants appeared to lack knowledge in the use of TMs. Female participants offered useful comments, which could be analysed into the following three categories:

1. There appeared to be a failure by some Bangladeshis to find biomedical approaches helpful. This was coherently expressed by one participant: 'If I use modern medication, it just leads to another problem. It never cures the root problem; eventually I end up using traditional medicine to cure the root cause' (Female, 31 years old).
2. The biomedical system (National Health Service, NHS in the U.K.) was used, but if so only as a last resort. For example: 'I only go to my

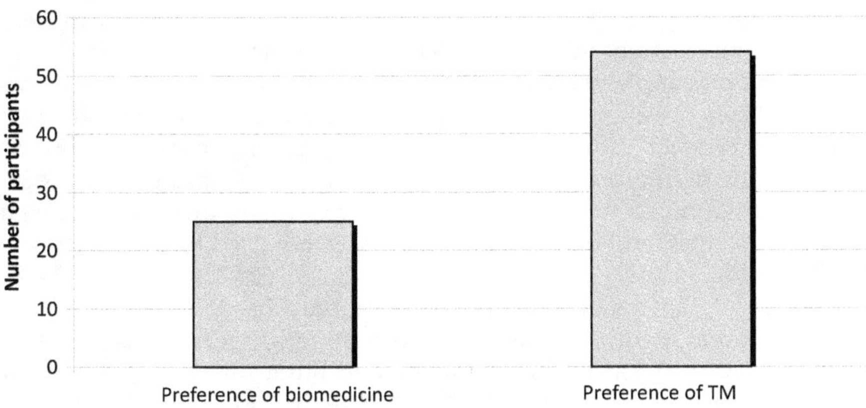

Figure 6.3. Participants' preference of TMs versus biomedical products

doctor when I'm very ill, and the TMs I have used have not helped me' (Female, 47 years old).

3. Communication problems between Bangladeshis and U.K. doctors appeared to raise two important issues:
   (i) It appeared to prevent healthcare staff from developing a greater understanding of healthcare behaviours and strategies used by minority ethnic families. One comment made suggested: 'My doctor doesn't understand why I use TMs and tells me just to take what he has given, without explaining how and why I need it. If I knew why I needed a certain medicine, I then could maybe find a traditional alternative' (Female, 53 years old).
   (ii) It appeared to act as a barrier to utilizing biomedical healthcare. For example: 'I have difficulty explaining myself to my doctor. I would like to know why I have [a] certain problem and what has caused it first and then be prescribed medication. I get frustrated because my English is not very good. (Female, 36 years old). 'Doctor doesn't care about patients. He doesn't spend enough time with me so I can't explain my problems' (Female, 61 years old).

Food-medicines

Participants in this study quoted approximately 150 preparations based on seventy-five folk taxa (sixty-one taxa were identified). Appendix 6.1 provides an outline of these findings. Most of these quoted remedies are still in use today within the U.K., whilst only a small number were said to have been prepared and used in the past in Bangladesh.

The data suggests that the majority of plants cited are used to treat only one or two minor ailments; for example, beet leaves are used to treat burns, and pineapple is used to cure threadworm. However, some plants have multiple uses; for example, mustard seeds and garlic are both used to treat up to seven different health problems, ranging from heart disease and arthritis (garlic) to eczema and constipation (mustard seeds). Neem leaves are found to treat up to six health disorders, including chicken pox and diabetes; and onion and olive oil are both used to treat up to five health problems.

We found that psyllium husk is commonly used in the treatment of constipation (n=45) and as a diuretic; garlic has a majority use as a general treatment for all illness (n=31); and bitter melon is popular for the treatment of diabetes (n=32). Other TMs included olive oil for the treatment of pain (n=23), rice to treat diarrhoea (n=22) and lemon to aid digestion.

Participants also listed thirteen other natural remedies, which are outlined in Appendix 6.2, and ritual practices for various illnesses based on chanting specific verses of the Qu'ran (see Appendix 6.3 for specific quotations).

Figure 6.4 shows the proportion of plant, animal and spiritual remedies used within TM by participants. Most of the quoted TMs were represented by food-medicines, whose natural ingredients are widely available in South Asian shops, but not so accessible in the U.K. The overlap between the food and medicine domains is well known in ethnobiology (Etkin 2006; Pieroni and Price 2006) and in urban ethnobotany as well (Pieroni et al. 2005; Sandhu and Heinrich 2005). However, it still appears that many health professionals are frequently unaware of minority ethnic medicinal perceptions of specific food items. A thorough and detailed knowledge

Figure 6.4. Proportion of plant, animal and spiritual remedies used within TM by participants

of traditional dietary patterns in South Asian households within the U.K. would certainly facilitate the design of therapeutic profiles for those patients affected by chronic and/or metabolic diseases (i.e. diabetes). The U.K. Government health policy white paper 'Saving Lives: Our Healthier Nation' resulted in the introduction of a range of initiatives to tackle health inequalities, including the '5 a Day' fruit and vegetables campaign, the introduction of Healthy Living Centres within deprived areas, and local Health Improvement Programmes (Department of Health 1999); central to the policy is the belief that diet is core to the maintenance of health throughout life. Research investigating the food and plant uses in health is therefore paramount if such initiatives are to be successful.

Availability of Ethnobotanicals

Although only a few participants raised this as an issue, the access to products and concern for the continuous availability of products was an interesting finding. Comments could be categorized into the following two themes:

1. There are limited numbers of products imported into the U.K. As participants point out: 'In the local Bangladeshi shops, herbs and products are only available to an extent. Some products are hard to import from Bangladesh' (Female, 47 years old). 'I can find most things that I need from either the local Bangladeshi shops or larger supermarkets, because I only use basic traditional food medicines. I don't think that there is any need to try to find foods that are not readily available here, it's just a big hassle' (Female, 24 years old).
2. Names of products are problematic when translated into English for sale in U.K. One interviewee summarizes the point: 'It is hard to try to explain to shopkeepers as to what exactly we are looking for, especially in the larger stores and especially since we do not know the English names of some products' (Female, 23 years old).

In Bangladesh healthcare services are considered scarce. Some Bangladeshis can access allopathic treatment, however. Alongside this exists the traditional system, which includes Ayurvedic (Kabiraji), Unani (Hakim), homeopathic and folk medicine. According to Akhter (2006) there are approximately six thousand registered and ten thousand unregistered practitioners (Kabirajis and Hakims). There are also approximately twenty-four registered herbal pharmaceuticals of which four of the big companies (Sadhana, Sakhti, Kundeshwari and Hamdard) are producing 80 per cent of traditional remedies (Akhter 2006). It is not surprising, therefore, that there continues to be a reliance upon TM use by migrant populations within the U.K. A deeper understanding of the TMs in use, together with an understanding of how they complement other healthcare systems in the U.K., could help improve access to those remedies currently unobtainable within the U.K. market.

Use of Ethnobotanicals among Different Age Groups

The findings showed that middle-aged to older participants were able to identify and quote the use of TMs more reliably than younger participants (Figure 6.5). However, the average number of examples appeared to remarkably decrease for participants the longer they had lived in the U.K. (Figure 6.6).

The majority of older participants suggested that the 'younger generations seemed more reluctant to use TMs or to know how to prepare them'. There appeared to be three reasons for this:

1. Changing attitudes of younger generations.
   (i) Younger generations were reported to prefer to use Western biomedical healthcare. Comments suggested that: 'The younger generation are not interested, they prefer using conventional medicines' (Female, 61 years old). 'Traditional knowledge is decreasing, because our children want quick results when they are ill and so use modern medicines' (Female, 57 years old).

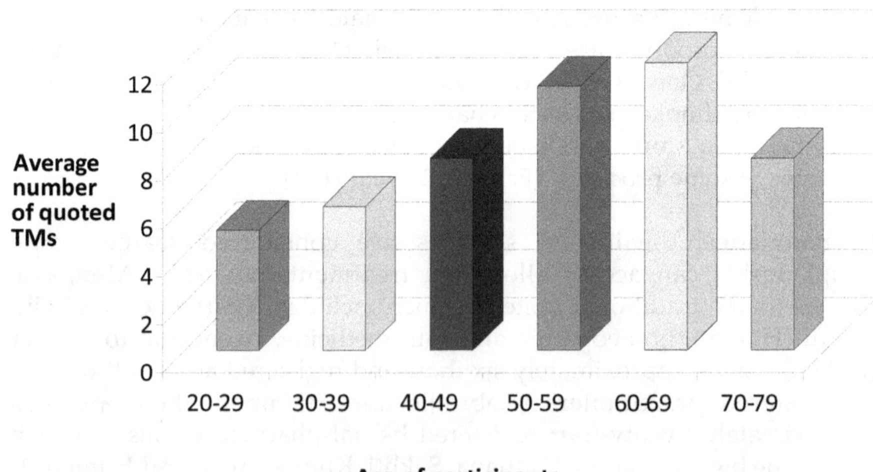

Figure 6.5. Average number of quoted TMs by different participant age groups

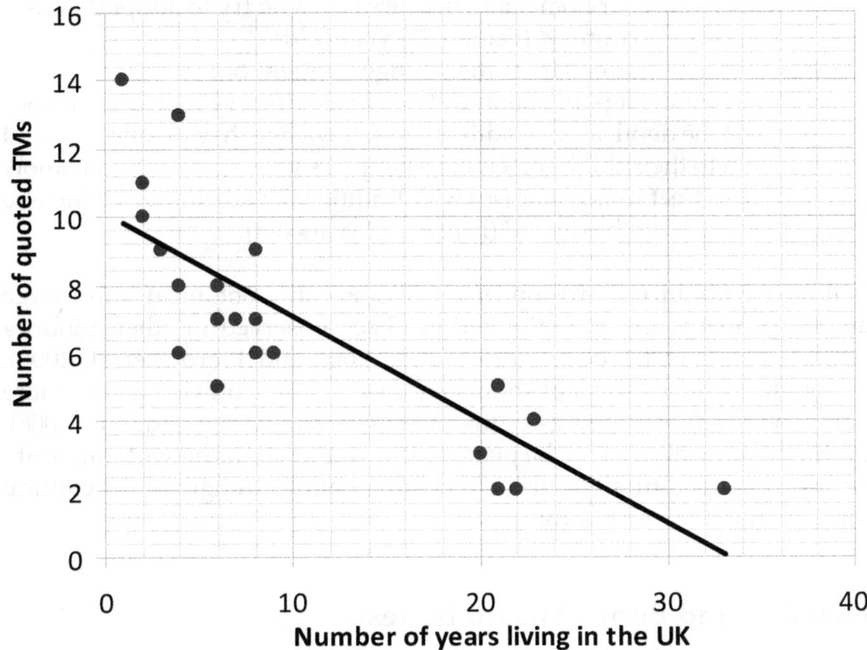

Figure 6.6. Number of quoted TMs versus the number of years participants have been living in the U.K.

(ii) It was reported that young people are not interested and do not understand TM. For example, respondents informed us: 'I have tried my best to tell my children about herbs and plants when cooking. However, they don't seem to very interested at times. They are too busy with watching the television or playing' (Female, 31 years old). 'Kids these days don't understand the importance of TMs; they don't understand the side-effects of modern medicines' (Female, 60 years old).

2. Knowledge is being lost within the home.
   (i) It seems that TM is no longer practised due to changing roles within the family: 'My knowledge has decreased because I do not cook as much as I did in Bangladesh. My daughters-in-law do most of the cooking' (Female, 75 years old).
   (ii) There appears to be a lack of time to educate the family: 'My children do seem to be interested in traditional knowledge. However, we never have the time to talk about it properly and educate them' (Female, 57 years old).
   (iii) It has been found to be difficult to practise the use of TM in a new country: 'Traditional knowledge is going to decrease if you move

away from your homeland, because now you try to adopt the ways of your new country' (Female, 32 years old).

3. Changing expectations from the younger generation.
   (i) A few younger participants did not appear to have high expectations about the depth and breadth of TM knowledge they planned to hand down to their children: 'I will be happy as long as my children know how to treat minor illnesses with traditional medicine, anything else they can visit the doctor' (Female, 22 years old).

Similar trends in the erosion of traditional ethnobotanical knowledge among people who have migrated has been observed in other minority ethnic groups in Europe (Pieroni et al. 2005 and Pieroni et al. 2008), although this process is not so much a case of being inevitable, but more likely the result of ongoing cultural negotiations (Pieroni and Gray 2008), in which a crucial role may be played by power relations between minority and dominant cultural groups and the fluctuating strength of the cultural identity of the minority group.

Diabetes and Public Health Issues

Participants were asked to identify common health problems being treated with TMs. Figure 6.7 shows the most important pathological conditions, for which participants mentioned using homemade TMs.

The management of diabetes with the use of TMs appears to be a popular, domestic health-seeking strategy within the Bangladeshi community in West Yorkshire. The data shows that seven different plants are in use for

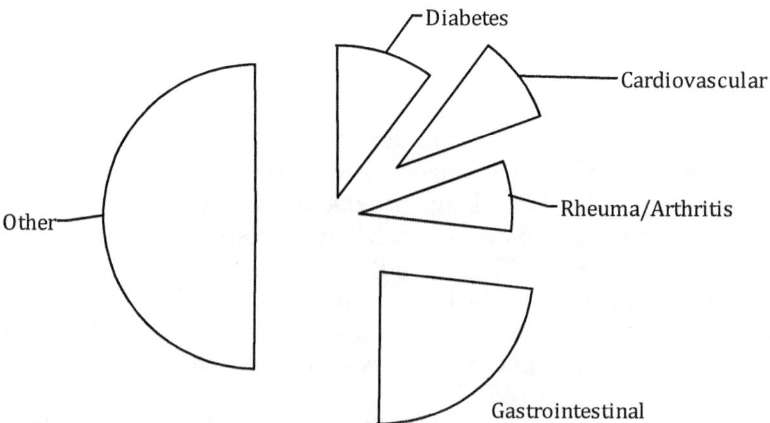

Figure 6.7. Most-quoted diseases treated by TMs

the treatment of diabetes; the most popular is bitter melon (n=32), and others include fenugreek leaves (n=7), white marudah bark (n=5), neem leaves (n=4), *sorrotha* leaves (not identified) (n=3), pointed gourd leaves (n=2) and Indian elm leaves (n=1). Specific details about their preparation are outlined in Appendix 6.1; all are prepared for ingestion as either liquid or food, and some are taken either in the morning or in the evening. Bitter melon has three different methods of preparation, including being taken as juice or in food.

The ethnobotany of Bangladeshis in Europe is completely unknown, although a few scholars in Bangladesh have paid attention in recent years to ethnobiological topics, especially concerning traditional knowledge on homegardens and medicinal plants in the home country (Alam 1992; Khan and Sen 2000; Millat-e-Mustafa, Teklehaimanot and Haruni 2002; Partha and Hossain 2007; Rahman, Uddin and Wilcock 007). The Bangladeshi community in the U.K. and its networks in Bangladesh, however, have been extensively studied in recent decades by diverse social scientists (Gardner 1993; Eade 1997; Phillipson, Ahmed and Latimer 2003; Gardner and Ahmed 2006). It has been widely reported that the Bangladeshi ethnic group has the poorest health status, especially considering the problems related to the high incidence of diabetes and its management (Kelleher and Islam 1996; Rhodes and Nocon 2003; Rhodes, Nocon and Wright 2003), and those connected to the practice of chewing betel nuts (Mannan, Boucher and Evans 2000; Prabhu et al. 2001).

The U.K. Government has published a National Service Framework for Diabetes in England and Wales (Department of Health 2001) which identifies specific national performance targets and actions that health professions are required to implement as a priority. Such targets include routine screening, access to information and education, and to work together with patients to an agreed care plan which involves regular health checks. The World Health Organisation has also identified diabetes as a significant cause for concern, estimating that 'diabetes deaths are likely to increase by more than 50 percent in the next 10 years without urgent action' (WHO 2008). Given the high relevance of this issue for public health policies in Europe, our data suggests that Bangladeshis in West Yorkshire have adopted their own strategies to attempt to manage this illness; this could have implications for their adherence as patients to biomedical care plans, if such plans fail to take into consideration lay health behaviours. It is quite possible that many health professionals involved in the management of diabetes among South Asians are not always aware of these practices.

Culturally sensitive approaches to South Asian patients affected by type 2 diabetes is more crucial than ever; studies using culturally sensitive methods have recently been documented for patients with type 2 diabetes from minority ethnic groups living in Glasgow (Baradaran et al. 2006), and other studies have stressed the importance of using a culturally sensitive approach for South Asians (Ahmed and Lemkau 2000).

Conclusion

There were four key findings in the study.

1. The majority of participants prefer to use TM rather than seek healthcare from the State (NHS), which operates a biomedical model of healthcare.
2. Healthcare practices appear to fall within two specific health-belief models, as outlined in Holland and Hogg (2001): the 'naturalistic' model, where the majority of participants are actively involved in the use of botanical specimens for prevention, maintenance and cure of health and illness; and the 'personalistic' model, where participants use rituals, such as chanting, to ward off illness.
3. Access to some of the natural products used by the participants was reported to be problematic.
4. The knowledge and use of the botanical specimens identified in this study appears to be greater within the older age group. This appears to result from the gradual change in roles, attitudes and expectations within families.

Our data provide valuable insights into concepts and views surrounding the health-giving properties of foods in the traditional diet within the domestic arenas of Bangladeshis living in West Yorkshire. This information could be crucial to both understanding dietary habits and improving the provision of healthcare through dietary consultation that takes into account emic views and concepts regarding 'healthy foods'. This would be especially beneficial for patients with type 2 diabetes.

Acknowledgements

Special thanks are due to all of the participants who agreed to take part in this study and to Ambreen Khalid, Raheela Majid, Tasneem Patel, Tayyabah Mirza, Ibrar Khokhar and Afeera Aleem, for having contributed to the study.

References

Ahmed, S.M. and J.P. Lemkau. 2000. 'Cultural Issues in the Primary Care of South Asians', *Journal of Immigrant Health* 2: 89–96.

Akhter, F. 2006. *Situation of Traditional Medicine in Bangladesh*. Retrieved 28 September 2007 from http://www.aifo.it/english/proj/traditional_medicine/Presentations/day1/Farida_Traditional%20Medicine_Presentation.pdf

Alam, M.K. 1992. 'Medical Ethnobotany of the Marma Tribe of Bangladesh', *Economic Botany* 46: 330–335.

American Anthropological Association. 1998. *Code of Ethics*. Retrieved 20 February 2008 from http://www.aaanet.org/committees/ethics/ethicscode.pdf

Balick, M., F. Kronenberg and A. Ososki. 2000. 'Medicinal Plants Used by Latino Healers for Women's Health Conditions in New York City', *Economic Botany* 54: 344–357.

Baradaran, H.R., R.P. Knill-Jones, S. Wallia and A. Rodgers. 2006. 'A Controlled Trial of the Effectiveness of a Diabetes Education Programme in a Multi-ethnic Community in Glasgow', *BMC Public Health* 6: 134.

Barth, F. 1969. *Ethnic Groups and Boundaries: The Social Organization of Culture Difference*. Long Grove, Ill.: Waveland Press.

Burns, C. 2004. 'Effect of Migration on Food Habits of Somali Women Living as Refugees in Australia', *Ecology of Food and Nutrition* 43: 213–229.

Ceuterick, M., I. Vandebroek, B. Torry and A. Pieroni. 2007. 'The Use of Home-remedies for Health Care and Well-being by Spanish Latino Immigrants in London: a Reflection on Acculturation', in A. Pieroni and I. Vandebroek (eds), *Traveling Plants and Cultures: The Ethnobiology and Ethnopharmacy of Human Migrations*. Oxford: Berghahn Books, pp. 145–165.

Clifford, J. and J. Marcus (eds). 1986. *Writing Culture: the Poetic and Politics of Ethnography*. Berkeley, Calif.: University of California Press.

Corlett, J.L., E.A. Dean and L.E. Grivetti. 2003. 'Hmong Gardens: Botanical Diversity in an Urban Setting', *Economic Botany* 47: 365–379.

Department of Health. 1999. *Saving Lives: Our Healthier Nation*. London: HMSO.

Department of Health. 2001. National Service Framework for Diabetes. Retrieved 20 June 2008 from http://www.dh.gov.uk/en/Healthcare/NationalServiceFrameworks/Diabetes/index.htm

Dorsett, R. 1998. *Ethnic Minorities in the Inner City*. London: JRF and The Polity Press.

Dunnell, K. 2008. *Diversity and Different Experiences in the UK*. Retrieved 15 June 2008 from http://www.mighealth.net/uk/images/4/4b/Divdiff.pdf

Eade, J. 1997. 'The Power of the Experts: the Plurality of Beliefs and Practices Concerning Health and Illness among Bangladeshis in Contemporary Tower Hamlets, London', in M. Warboys and L. Marks (eds), *Migrants,*

Minorities and Health: Historical and Contemporary Studies. London and New York: Routledge, pp. 250–271.

Etkin, N. 2006. *Edible Medicines: an Ethnopharmacology of Food*. Tucson, Ariz.: University of Arizona Press.

Gardner, K. 1993. 'Mullahs, Migrants and Miracles: Travel and Transformation in Rural Bangladesh', *Contributions to Indian Sociology* 27: 213–235.

——— and Z. Ahmed. 2006. *Place, Social Protection and Migration in Bangladesh: A Londoni Village in Biswanath. Working Paper T18*. Brighton: Development Research Centre on Migration, Poverty and Globalisation.

Ghani, A. 2003. *Medicinal Plants of Bangladesh with Chemical Constituents and Uses*. Dhaka: Asiatic Society of Bangladesh.

Hanelt, P. and Institute of Plant Genetics and Crop Plant Research (IPK) (eds). 2001. *Mansfeld's Encyclopedia of Agricultural and Horticultural Crops (except Ornamentals)* (6 volumes). Berlin: Springer Verlag.

Holland, K. and C. Hogg. 2001. *Cultural Awareness in Nursing and Health Care: an Introductory Text*. London: Arnold.

International Society of Ethnobiology. 2006. *Code of Ethics*. Retrieved 30 May 2008 from http://ise.arts.ubc.ca/_common/docs/ISECodeofEthics2006_000.pdf

Jonsson, I.M., H.R.M. Hallberg and I.B. Gustafsson. 2002. 'Choice of Food and Food Traditions in Pre-war Bosnia-Herzegovina: Focus Group Interviews with Immigrant Women in Sweden', *Ethnicity & Health* 7: 149–161.

Kelleher, D. and S. Islam. 1996. '"How Should I Live?" Bangladeshi People and Non-insulin-dependent Diabetes', in D. Kelleher and S. Hillier (eds), *Researching Cultural Differences in Health*. London: Routledge, pp. 220–237.

Khan, N.A. and S. Sen. 2000. *Of Popular Wisdom: Indigenous Knowledge and Practices in Bangladesh*. Dhaka: Integrated Action Research and Development.

Kim, Y. 2001. *Becoming Intercultural: An Integrative Theory of Communication and Cross Cultural Adaptation*. Thousand Oaks, Calif.: Sage.

Mannan, N., B.J. Boucher and S. J. Evans. 2000. 'Increased Waist Size and Weight in Relation to Consumption of *Areca catechu* (Betel-nut); a Risk Factor for Increased Glycaemia in Asians in East London', *British Journal of Nutrition* 83: 267–275.

Marcus, G. 1998. *Ethnography through Thick and Thin*. Princeton, NJ: Princeton University Press.

Mason, D. 2000. *Race and Ethnicity in Modern Britain*. New York: Oxford University Press.

Millat-e-Mustafa, M., Z. Teklehaimanot and A.K.O. Haruni. 2002. 'Traditional Uses of Perennial Homestead Garden Plants in Bangladesh', *Forest, Trees and Livelihoods* 12: 235–256.

Nesheim, I., S.S. Dhillion and K.A. Stolen. 2006. 'What Happens to Traditional Knowledge and Use of Natural Resources when People Migrate?', *Human Ecology* 34: 99–131.

Nguyen, M.L.T. 2003. 'Comparison of Food Plant Knowledge between Urban Vietnamese Living in Vietnam and in Hawai'i', *Economic Botany* 57: 472–480.

Office for National Statistics (ONS). 2001a. *Neighbourhood Statistics: Ethnic Group (UV09)*. London: Office for National Statistics. Retrieved 15 May 2008 from http://www.statistics.gov.uk

———. 2001b. *The National Census in England and Wales 2001*. London: Office for National Statistics. Retrieved 21 May 2008 from http://www.statistics.gov.uk

Partha, P. and A.B.M.E. Hossain. 2007. 'Ethnobotanical Investigation into the Mandi Ethnic Community in Bangladesh', *Bangladesh Journal of Plant Taxonomy* 14: 129–145.

Phillipson, C., N. Ahmed and J. Latimer. 2003. *Women in Transition: a Study of the Experiences of Bangladeshi Women Living in Tower Hamlets*. Bristol: The Policy Press.

Phillpotts, G. and P. Causer (eds). 2006. *ONS: Regional Trends 39 Report*. Hampshire: Palgrave Macmillan.

Pieroni, A. and C. Gray. 2008. 'Herbal and Food Folk Medicines of the Russlanddeutschen Living in Künzelsau/Taläcker, South-western Germany', *Phytotherapy Research* 22: 889–901.

———, L. Houlihan, N. Ansari, B. Hussain and S. Aslam. 2007. 'Medicinal Perceptions of Vegetables Traditionally Consumed by South-Asian Migrants Living in Bradford, Northern England', *Journal of Ethnopharmacology* 113: 100–110.

———, H. Muenz, M. Akbulut, K.H.C. Baser and C. Durmuskahya. 2005. 'Traditional Phytotherapy and Trans-cultural Pharmacy among Turkish Immigrants Living in Cologne, Germany', *Journal of Ethnopharmacology* 102: 69–88.

——— and L.L. Price (eds). 2006. *Eating and Healing: Traditional Food as Medicine*. Binghamton, NY: Haworth Press.

———, Q.Z. Sheikh, W. Ali and B. Torry. 2008. 'Traditional Medicines Used by Pakistani Migrants from Mirpur Living in Bradford, Northern England', *Complementary Therapies in Medicine* 16: 81–86.

——— and B. Torry. 2007. 'Does the Taste Matter? Taste and Medicinal Perceptions Associated with Five Selected Herbal Drugs among Three Ethnic Groups in West Yorkshire, Northern England', *Journal of Ethnobiology and Ethnomedicine* 3: 21.

——— and I. Vandebroek (eds). 2007. *Travelling Plants and Cultures: The Ethnobiology and Ethnopharmacy of Human Migrations*. Oxford: BerghahnBooks.

Prabhu, N.T., K. Warnakulasuriya, S. Gelbier and P.G. Robinson. 2001. 'Betel Quid Chewing among Bangladeshi Adolescents Living in East London', *International Journal of Paediatric Dentistry* 11: 18–24.

Rahman, M.A., S.B. Uddin and C.C. Wilcock. 2007. 'Medicinal Plants Used by Chakma Tribe in Hill Tracts Districts of Bangladesh', *Indian Journal of Traditional Knowledge* 6: 508–517.

Rhodes P. and A. Nocon. 2003. 'A Problem of Communication? Diabetes Care among Bangladeshi People in Bradford', *Health & Social Care in the Community* 11: 45–54.

———, A. Nocon and J. Wright. 2003. 'Access to Diabetes Services: the Experiences of Bangladeshi People in Bradford, UK', *Ethnicity & Health* 8: 171–188.

Sandhu, D.S. and M. Heinrich. 2005. 'The Use of Health Foods, Spices and Other Botanicals in the Sikh Community in London', *Phytotherapy Research* 19: 633–642.

Self, A. (ed.). 2008. *ONS: Social Trends 38*. Hampshire: Palgrave Macmillan.

——— and L. Zealey (eds). 2007. *ONS: Social Trends 37*. Hampshire: Palgrave Macmillan.

Strauss, A.L. and J.M. Corbin. 1998. *Basics of Qualitative Research: Techniques and Procedures for Developing Grounded Theory*. London: Sage.

Turner, N.J., I.J. Davidson-Hunt and M. O'Flaherty. 2003. 'Living on the Edge: Ecological and Cultural Edges as Sources of Diversity for Social-ecological Resilience', *Human Ecology* 31: 439–461.

Waldstein, A. 2006. 'Mexican Migrant Ethnopharmacology: Pharmacopoeia, Classification of Medicines and Explanations of Efficacy', *Journal of Ethnopharmacology* 108(2): 299–310.

White, A. 2002. *ONS Social Focus in Brief: Ethnicity 2002*. London: Office for National Statistics.

World Health Organisation. 2008. *Diabetes Programme*. Retrieved 20 June 2008 from http://www.who.int/diabetes/en

Appendix 6.1. Plant-based TMs quoted by the participants (in brackets the number of participants who have quoted each specific remedy). Quoted disappeared uses, which were common in Bangladesh, are indicated by an asterisk

Abelmoschus esculentus (L.) Moench. (Malvaceae). Okra or lady's finger/ *Bindi*. Fruit.

- Treatment for sore throat, coughs and hypoglycaemia. Cook as a vegetable and eat. (7)

Allium cepa L. (Alliaceae). Onion/*Payaaz*. Bulb.

- Used to relieve headache. Apply juice to forehead. (3)
- Helps to clear the eye in conjunctivitis. Bite onion and blow into eyes to produce tears. (7)
- Used for ear conditions in three ways: to treat earache (heat bulb and place in the ear), for its antiseptic properties, and after ear piercing. (3)
- Helps relieve phlegm in people suffering from chesty cough. Use bulbs and leaves in cooking. (14)
- Used to relieve/treat bruises. Cut onion in half and cover the surface in turmeric. Press it against the bruise for a couple of minutes. (4)

Allium sativum L. (Alliaceae). Garlic/*Roshun*. Bulb.

- Taken for heart diseases or hypertension. Eat raw garlic cloves in the morning. (9)
- Used in treatment of all illnesses. Use the bulb in curry or eat on its own. (31)
- Treatment for arthritis and rheumatism. Boil garlic in water and drink mixture. (3)
- To prevent nausea and sickness in pregnancy. Place cloves in a hot pan and blend with honey and sugar, and drink. (4)
- For relief of headache. Inhale the aroma from raw garlic cloves. (3)
- To treat sore throat. Crush the garlic and onion to make a paste, and swallow. (4)
- Used to treat stomach pain. Heat garlic and grate into hot water, and drink. (7)

Aloe barbadensis Mill. (Asphodelaceae). Aloe/*Kumari*. Gel and dried juice.

- Aloe is considered to be a powerful aphrodisiac. Mix the mucilaginous pulp of the leaf with sugar and make into a sweet preparation. (4)
- Used to treat eye irritations. Rub the dried juice that flows from the cut bases of the leaves onto the eyes. (8)

Amygdalus communis L. (Rosaceae). Almond/*Badham*. Kernel.

- Almond eaten raw is believed to help the brain. (22)

Ananas comosum (L.) Merrill (Bromeliaceae). Pineapple/*Anarosh*. Fruit.
- Pineapple is used to cure threadworm. Mix the juice of pineapple with '*chunah*' (calcium carbonate) and drink. (1)

Apium graveolens L. (Apiaceae). Celery/*Randhoni*. Root and fruit.
- Celery is good for tooth pain. Chew the root . (2)
- The fruit is believed to have antiseptic and sedative properties. (3)

Areca catechu L. (Arecaceae). Betel Nut/*Gua*. Nut.
- Used to aid digestion. Believed to promote relaxation. Can be taken for social reasons. Chew slices of the nut and lime inside betel (*Piper betle* L.) leaf. Spit out juice. Usually taken after meals. (2)
- Treatment for headache. Crush the nut, add honey and mix to form a paste. Place on forehead. (5)

Averrhoa carambola L. (Oxalidaceae). Starfruit/*Camranga*. Fruit.
- Used to heal sore throats. Cut fruit into small pieces and eat. (2)

Azadirachta indica Juss. (Meliaceae). Neem/*Neem patha*. Leaves.
- Treatment of indigestion. Cut leaves and boil in water to make herbal tea. (6)
- To treat diabetes. Grind and make into small balls and eat it on an empty stomach in the morning. (4)
- To relieve pain. Crush and make into a paste; make little balls out of it and apply to area of pain. (3)
- For skin infection. Blend leaves with turmeric and apply to infected areas. (5)
- To treat headaches and fever. Grind *neem patha* with garlic, onion and ginger and apply on head. (12)
- As a remedy for chicken pox. Boil neem in water, rub the solution on the body and bathe the child with the solution. (9)
- To cleanse skin and treat spots. Place the leaves in water, warm up the solution and apply to spots. (5)

Beta vulgaris L. (Chenopodiaceae). Beet/*Betsalang*. Leaves.
- Apply leaves to burns or bruises. (2)

Brassica sp. (Brassicaceae). Mustard/*Sureyo*. Seeds – Oil.
- Treatment of blocked nose. Mix fresh mustard paste and fresh garlic, add to boiled rice and eat. (2)
- To treat earache. Slightly heat the mustard oil and insert few drops into the ear. (8)
- For mouth ulcers. Gargle with mustard oil. (8)

- For dry lips and eczema. Apply mustard oil to affected area 3–4 times a day. (8)
- Used for the treatment of constipation. Roast *methi* (*Trigonella foenum-graecum* L.) seeds in mustard oil until black and rub in clockwise motion on stomach. (5)
- To treat a cough. Mix mustard oil with garlic and rub on chest. (1)
- A remedy for hypertension. Mix oil with water and pour onto head. (5)

Camellia sinensis (L.) Kuntze (Theaceae). Green tea/*Kava*. Leaves.
- Tea for treating coughs and colds. (3)

Carica papaya L. (Caricaceae). Papaya/*Khoyfol*. Fruit.
- To treat threadworm. Drink the juice of papaya. (1)

Citrus aurantifolia (Christm.) Swingle (Rutaceae). Green lime/*Adha lembu*. Fruit.
- Used to treat headaches and fever. Extract fruit juice and apply on forehead. Believed to bring down temperature. (8)

Citrus limon (L.) Burm. f. (Rutaceae). Lemon/*Lembu*. Fruit.
- Used for digestion. Drink lemon juice with meals, especially a diet of fish. (20)
- To treat stomach pains. Add lemon juice and salt to warm water and drink. (3)
- To prevent sickness. Breathe in the scent of a lemon or lemon leaves. (3)
- To cool down the body (reduce temperature). Massage lemon juice into the scalp. (8)

Citrus sinensis (L.) Osbeck (Rutaceae). Orange/*Zamir*. Fruit.
- For treating sore throat or indigestion or skin diseases. Eat fruit on its own. (7)

Cocos nucifera L. (Arecaceae). Coconut/*Naikol*. Nut – Oil.
- Used in the treatment of dry hair, the prevention of hair loss, and to treat chapped lips. Apply coconut oil to hair on a weekly basis. Apply externally to lips. (8)
- To treat a rash. Burn shellfish and mix ash with coconut milk and apply to rash. (2)*

Crocus sativus L. (Iridaceae). Saffron/*Zafraan*. Stamen.
- To aid the expulsion of trapped wind (carminative action). Use in cooking. (3)

- To ward off evil and help people afraid of the dark. Saffron can be added to a small amount of oil and rubbed on skin before going to bed. (3)

Cucumis sativus L. (Cucurbitaceae). Cucumber/*Khira*. Fruit.
- To treat eye inflammations. Cut into slices and place over eyes. (7)
- To promote weight gain for people who are anorexic. Eat together with dried dates (*Phoenix dactylifera* L. fruits). (4)

Cucurbita pepo L. (Cucurbitaceae). Courgette/*Toree*. Fruit.
- The vegetable is cooked and eaten to relieve stomach problems. (6)

Cuminum cyminum L. (Apiaceae). Cumin/*Jeera*. Fruits.
- Crush jeera, add to olive oil and massage on stomach to relieve stomach pain. (4)

Curcuma longa L. (Zingiberaceae). Turmeric/*Haldi*. Rhizome.
- To prevent stomach pains. Mix turmeric powder with milk and drink. (6)
- To prevent cough. Mix with honey and eat. (8)
- To heal bruises. Crush fresh turmeric rhizome and apply to skin. (7)
- To treat nausea in pregnancy. Add Rhizome to cooking. (19)

Daucus carota L. (Apiaceae). Carrot/*Gajor*. Root.
- Carrot is eaten raw to improve eyesight. (15)

Elettaria cardamomum (L.) Maton (Zingiberaceae). Cardamom/*Elaichi*. Seeds.
- Chew cardamom to treat toothache. (7)

Foeniculum vulgare Mill. (Apiaceae). Fennel/*Jamaain*. Fruits.
- To aid digestion. Chew fennel and drink with water. It can also be used to freshen the taste in the mouth. (13)
- To treat a sore throat. Chew fennel. (8)

Ficus carica L. (Moraceae). Fig/*Dumur*. Fruit.
- Used as a laxative. Eat raw or as dried fruit. (2)
- To remove 'gravels' (stones) in the kidney or bladder. Eat the fig. It is easily digested. (4)

Holoptelea integrifolia (Roxb.) Planch. (Ulmaceae). Indian Elm/*Naali*. Leaves.
- Used in the control of diabetes or for muscular pain. Cut the leaves and stems into small pieces, boil in water and drink/swallow. (1)

- As a treatment for chicken pox. Cut the leaves and stems into small pieces, boil in water, allow to cool down and then apply to the skin (spots/blisters). (1)

Hordeum vulgare L. (Poaceae). Barley/*Jau*. Fruits.
- Used to treat sickness and loss of appetite and to promote a healthy heart. Make into a soup and eat. (4)

Hymenaea courbaril L. (Fabaceae). Brazilian copal. *Jatoba*. Leaves.
- Crush and make into a paste and apply to cuts. (3)

Lawsonia inermis L. (Lythraceae). Henna/*Mehndi*. Powdered leaves.
- Used to reduce hypertension. Also used to maintain healthy hair and for hair colour. Mix powder with water to create a paste. Apply to head/hair. (16)
- Used in the treatment of mouth ulcers or skin burns. Also used to treat pain in general. Apply wet to mouth or skin. (3)

Lavandula sp. (Lamiaceae). Lavender/*Jusna ful*. Flowering tops – Essential oil.
- Apply oil around the nose and eyes, 2–3 times a day, to relieve a blocked nose. (1).

Lens culinaris Medik. (Fabaceae). Lentil/*Daal*. Seeds.
- To help clear skin, blend the lentils with lemon juice and turmeric and apply on the skin. (5)
- Consumed, it is used to help 'soften the heart', making one more sympathetic and caring. (2)

Lycopersicon esculentum Mill. (Solanaceae). Tomato/*Bangoin*. Fruit.
- Eat raw tomatoes to help circulation. (9)

Malus domestica Borkh. (Rosaceae). Apple/*Aifol*. Fruit.
- Eat on its own to treat indigestion and constipation. (8)

Mangifera indica L. (Anacardiaceae). Mango/*Aam*. Fruit.
- Use ready-made or homemade mango pickle and chew skin of mango until headache symptoms improve. (1)

Momordica charantia L. (Cucurbitaceae). Bitter Melon/*Karela*. Fruit.
- For diabetes. Drink a glass of raw *karela* juice every morning or afternoon. Take two small *karela*, blend and take out the seeds before drinking. (30)

- For diabetes. Eat the raw, green, outer part of the fruits every morning. Eat raw *karela* 1–2 times a week. (15)
- Used regularly for diabetics. Chop up and fry in oil to make curry. (32)

Musa x *paradisiaca* L. (Musaceae). Banana/*Khola*. Fruit.
- To treat diarrhoea. Eat raw green banana. (11)
- To treat diarrhoea. Boil banana with water and eat. (1)

Nicotiana tabacum L. (Solanaceae). Tobacco/*Gool*. Leaves.
- Used to cleanse teeth. Rub tobacco leaves on the teeth and spit out. (4)

Nigella sativa L. (Ranunculaceae). Kalhi/*Zeera*. Seeds – Oil.
- Good for health and heart; also good for stomach pains. Put a few seeds in the mouth and chew, and then drink water after; take every morning. Eat seeds raw. (6)
- To relieve a blocked nose. Roast the nigella seeds, place into a cloth and inhale. (3)
- To strengthen liver and stomach functions. Mix the oil with honey then drink. (1)
- Oil can be regularly used as a cream for the treatment of rheumatism/arthritis and hair loss. (1)

Ocimum basilicum L. (Lamiaceae). Sweet Basil/*Tuk malanga*. Leaves
- Mix with water and drink the solution 2–3 times a day for treating urinary-tract infections. (6)

Ocimum sanctum L. (Lamiaceae). Holy Basil/*Tulsi*. Leaves.
- To cure headache. Grind leaves and place on head. (1)
- Used as a laxative for children. Grind leaves and drink the juice. (3)

Olea europaea L. (Oleaceae). Olive/*Zaitun*. Fruits – Oil.
- When taken internally it helps people with gastric ulcer. (9)
- External application helps soften the skin and crusts in eczema and psoriasis. (12)
- To relieve pain. Massage oil onto joints. (23)
- For improved texture and healthy hair. Massage oil into hair and leave overnight, then wash hair. (21)
- Used for stomach ache. Add one spoon of olive oil to warm water and drink. (3)

Oryza sativa L (Poaceae). Rice/*Chawl*. Fruits.
- Rice is boiled in milk to make *Kheer* which can be ingested to help stomach problems; it is best taken on an empty stomach for digestive problems. (4)

- For post-operative food or diarrhoea. Boil rice and lentils to make *kitchari*. (22)
- For treatment of irritable eye. Grains of rice are placed in the lower eyelid. The rice will fall out during the next few days, removing the irritant. (1)
- For red/sore eyes. Place warm boiled rice in a cotton cloth, wrap to form a pouch and hold over the eye. (3)
- To overcome minor heart trouble. Rice is boiled in water. Without draining the rice, eat it for seven days. (3)

Phoenix dactylifera L. (Arecaceae). Date/*Khejur*. Fruit.
- Eat three dates daily to have aphrodisiac effects.
- For mouth ulcers. Make into a paste and apply to affected area. (1)

Piper betle L. (Piperaceae). Betel leaf/*Panchi*. Leaf.
- Used to treat headache. Grind the leaf to extract juice. Mix with equal amounts of honey and apply paste on forehead 2–3 times a day. (3)

Plantago psyllium L. (Plantaginaceae). Psyllium/*Isphagullah* husk. Seeds.
- Used as a diuretic and for constipation. Leave to soak for 1–2 hours in water and then drink for a few days, depending on symptoms and until better. (45)
- Used for constipation and heartburn. Place in water and drink. (5)

Psidium guajava L. (Myrtaceae). Gujava/*Piara*. Fruit.
- The fruit is eaten raw to prevent infections and make the skin healthier. (1)

Punica granatum L. (Punicaceae). Pomegranate/*Daleem*. Fruit and bark.
- To allay thirst in diarrhoea and dysentery. Drink the juice of the fruit (4)
- For expulsion of worms, including tapeworms. Drink the decoction of the root bark of pomegranate. (3)
- For jaundice, high blood pressure, piles and arthritis. Drink the juice on its own or with honey. (6)

Raphanus sativus L. (Brassicaceae). Radish/*Mooley*. Leaves and roots.
- To aid digestion. Eat leaves and roots raw. Has no beneficial use if cooked. Should be avoided before sleep as it may cause a 'prickling' sensation. (2)

Salvadora persica L. (Salvadoraceae). Toothbrush tree/*Miswak*. Stems.
- For treating toothache. Use the stems to clean teeth or place on teeth. (6).

Spinacia oleracea L. (Chenopodiaceae). Spinach/*Saag* or *Paalak*. Leaves.
- Cooked as a vegetable curry for good health and to help improve eyesight over a long period of time. (9).

Swertia chirata Buch.-Ham. ex Wall. (Gentianaceae). Chiretta/*Cherata*. Stems.
- To relieve muscle pain. Place *cherata* in water and allow to boil. Drink when required. (5)

Tamarindus indica L. (Fabaceae). Tamarind/*Imlee*. Fruit.
- Used to refresh mouth. Also used as a stimulant. The fruit pulp is eaten fresh or in a pickle sauce. (2)
- In pregnancy tamarind is recommended to help avoid cravings. (1)
- To relieve constipation. Soak tamarind in water and drink solution. (5)
- For hypertension. Place tamarind in boiled water until dissolved and drink. (4)

Terminalia arjuna (Roxb.) Wight & Arn. (Combretaceae). White Marudah/*Arjun*. Bark.
- For treating diabetes. Place *arjun* in water and allow to boil. Drink daily with breakfast. (5)

Terminalia chebula Retz. (Combretaceae). Chebulic Myrobalan/*Hortoki*. Fruits.
- Fruit is chewed for stomach pains and 'clears' blood. (1)*

Trigonella foenum-graecum L. (Fabaceae). Fenugreek/*Methi*. Leaves.
- Used to treat diabetes. Soak in water for a few hours and drink before going to bed. (7)

Trichosanthes dioica Roxb. (Cucurbitaceae). Pointed gourd/*Potol*. Leaves.
- Used to treat diabetes. Take leaf of plant and turn into juice. Drink the juice. (2)

Triticum aestivum L. (Poaceae). Wheat/*Gom*. Fruits.
- Used to bring down a person's temperature. Wheat flour is mixed with water to make dough; this is then heated and ingested. (1)

Vaccinium sp. (Ericaceae). Cranberry/*Kanchee*. Fruits.
- Drink juice for constipation. (1)

Vitis vinifera L. (Vitaceae). Grape/*Angoor*. Fruit.
- Eat the fruit for a laxative effect. (2)
- Unripe fruit juice is used in throat infections and for thirst. (3)

Zingiber officinale Roscoe (Zingiberaceae). Ginger/*Adha*. Rhizome.

- To treat indigestion. Skin is peeled off and cut into little pieces before swallowing. (18)
- To treat a cough. Boil 2–3 pieces of ginger for a few minutes and then drink the juice. Drink 2–3 times a day. (9)
- For headache or fever. Grind ginger on its own, or with onions, to make a paste and apply on the forehead. (10)
- For sore throat. Boil cinnamon, cardamom and bay leaf with the ginger in water. Drain and drink the water 2–3 times a day until symptoms are relieved. (4)
- For headache and fever. Grind ginger with garlic, onion and neem patha and place on head. (12)
- To treat diarrhoea. Leave to soak in approximately 250 ml water overnight and then drink the next day. (2)
- For treatment of headache. Crush garlic and add to tea without milk to make *adhda cha*. (6)
- To relieve the pain in sore throats. Place in mouth and chew. (10)

Not identified mushroom. *Kumbhi*.

- The juice of mushrooms is used for improving eyesight. (5)

Not identified. *Shuaga*. Seeds.

- For mouth infection/ulcer. Roast and grind *shuaga*. Add honey and put in mouth. (1)

Not identified. *Sorrotha*. Leaves.

- For diabetes and rheumatism/arthritis. Soak leaves and collect juice. (3)*
- For rheumatisms/arthritis and diabetes. Boil in water and drink. (3)

Not identified. *Loja pati tree*. Fruit juice?

- Used for female watery discharge. Collect juice and drink every day until cured. (2)*

Not identified. *Kathe patha*. Leaves.

- Grind leaves and put on cuts. (1)

Not identified. *Tuka tree*. Seeds.

- For constipation. Soak seeds until swollen, then add to sherbet and drink. (2)*

Not identified. *Rifusi*. Leaves.

- Make a paste with a little water and apply on wounds, cuts, burns and eczema. Once dried, apply another layer until heals. (1)*

Not identified. *Tunni-man*. Leaves.

- Collect juice from leaf and drink or cook with fish, as a healthy beverage/food. (1)*

Not identified. *Dephol*. Fruit.

- Used to treat sore throat. Cut the *dephol* fruit into small pieces, boil in water, drain and drink the juice. (2)

Not identified. *Senfisal*. Leaves.

- Used as a diuretic. Place leaves into water and leave to soak 2–3 hours. Drink solution. (6)

Not identified. *Bubraaz*. Root.

- To relieve period pain. Root of tree pulled out on either Saturday or Sunday and placed in a small metal locket. This is then tied to a piece of black string around the stomach. (3)

Not identified. *Khosu*. Leaf.

- For treating fever. Crush the leaf, add salt and place on the forehead. Tightly wrap a piece of cloth over this. Leave for 6–8 hours. (4)

Not identified. *Kha*. Roots.

- For treatment of toothache. Blend the roots into a paste and apply to tooth. (1)

Not identified. *Shutika*. Mixture of plants.

- Used to treat hot flushes after pregnancy. Paste is placed over the head to cool down the body. (1)

Appendix 6.2. List of other natural TMs quoted by the participants

Calcium carbonate. *Chunah*. Add water to produce a white paste. Give to the person with threadworms (3). Mix *chunah* with water and leave overnight. Two layers will form. The water on the top layer mixed with coconut oil is placed on burns (5). For headache, crush *chunah* and garlic together to form a paste. Place on temple and add a layer of tobacco to form a shield (7).

Cotton material (*duppatta*). Roll the duppatta very tightly and thinly around the waist, lower abdomen area, and keep it in this way for a week; ideal to do this straight after childbirth (2).

Chicken soup. *Yakni*. Leave chicken to boil with herbs and spices for approximately two hours and once cooked drink for treating colds and fever (15).

Egg albumen. *Andu*. Beat the egg white and apply to wounds. Used for treating burns (2).

Fish. *Machee*. Cooked in spices or a sauce or grilled. Oil is thought to be good for the heart and health (10).

Hair. *Sool*. To treat warts; tie a piece of hair around the wart tightly to remove it (3).

Honey. *Modhu*. Take it on its own or mixed with turmeric to help cure dry or chesty coughs; take every day until cured (12). Boil milk or water and add one teaspoon of honey to help sleep and aid relaxation; also good for coughs, stomach ulcers and indigestion. Use occasionally when required (16). For coughs or to increase appetite add a tablespoon of honey and some lemon juice to hot milk or water and drink (5). For sore throat mix lemon juice and honey in hot water and drink (3).

Milk. *Duud*. Drink boiled milk everyday to prevent illness (3).

Petrol. Apply the petrol to hair to get rid of head lice (2).

Salt. *Loon*. Mix salt in hot water and drink 2–3 times a day to treat a sore throat (6).

Soot. *Surma*. Apply to the eye as eyeliner once daily when required; used by both sexes; it is said to be useful for treating red and watery eyes, for improving eyesight, and enhancing the shape of the eye (19).

Tibet snow cream. Apply cream around the nose and eyes to relieve a blocked nose. Use 2–3 times a day (2).

Water. *Pani*. Drink lots of water to heal a headache (3). Wet cloth with water and apply to forehead to treat a headache (5). Steam inhalation to treat a headache (2).

Appendix 6.3. Recorded spiritual treatments based on the Qu'ran's prayers

Labour pains.
Chapter 17, section 13, verse no: 30.
When a woman is in the throes of labour above verse should be recited and blown on the stomach or it can be written on paper and worn as *taweez*. This will cause the delivery to become swift and less painful.

Fever.
Chapter 9, section 14, verse no: 301.
The above should be recited and blown on a person who has fever. Otherwise it could be written with saffron on a plate, the plate washed with a cup of water and the patient be made to drink this water.

Depression.
Chapter 9, section 15, verse no: 38.
Should be written on a piece of paper and worn as a taweez and placed on the heart.

Palpitation of the heart.
Chapter 3, section 17, verse no: 82-86.
The verse should be written on the inside of a new clay utensil with saffron ink. The plate then should be washed with one or two cups of water and the patient be made to drink this water.

Disease of the spleen.
Chapter 22, section 17, verse no: 41.
Write the above verse on paper and tie it on the portion where the spleen is situated; then the disease shall be cured.

Piles.
Chapter 1, section 15, verse no: 127-129.

Nose bleeds.
Chapter 4, section 5.
The above verse should be written on paper and fixed between the two eyes.

For a specific pain anywhere on the body.
Chapter 15, section 12, verse no: 1.
Place the hand on the portion of the body where the pain is felt. Recite the above verse once and blow three times on the affected area.

Headaches.
Chapter 27, section 14, verse no: 19.
Recite the above verse three times and blow on the patient and the headache shall disappear.

Earache.
Chapter 11, section 9, verse no: 31.
Write the above verse with leek extract on the inside of the copper dish. Rub off the writing with a teaspoon of pure honey. Collect the honey in a teaspoon and heat it mildly and administer three drops in the affected ear.

Bone fracture.
Chapter 11, section 5, verse no: 139.
Recite the above verse on the affected region.

When bitten by a poisonous insect or snake.
Chapter 19, section 11, verse no: 130.
Circulate the finger around the bitten area and recite the above verse seven times in one breath. The patient shall recover shortly.

CHAPTER 7

Persistence of Wild Food and Wild Medicinal Plant Knowledge in a Northeastern Region of Portugal

ANA MARIA CARVALHO AND RAMÓN MORALES

Introduction

People tend to be strongly dependent on the landscape and natural environment in which they live and work, especially in extreme environments (e.g., high altitude, desert areas) and regions with geographical and social isolation. Beginning as children, these people have learned how to discover and understand the signs of nature and to observe changes in the landscape. However, they have also shaped that landscape according to their own beliefs and material needs. This adaptive knowledge is often a practical one, based on empirical observation and long experience, and transmitted through oral traditions. Such knowledge is not merely of academic or historical interest but is fundamental to maintaining cultural continuity and identity and, possibly, could play a role in achieving sustainable use of plant resources in the future (Svanberg and Tunón 2000).

Today, the geographical and social conditions that led to the isolation of some European regions are no longer prevalent, and so changes in the cultural, economic and political contexts of plant use in these areas are coming faster and faster. The current reduction in the human population, due to out-migration and a general drop in the birth rate, and the abandonment of agriculture have been critical for rural areas and ways of life in these regions and have promoted the loss of cultural traditions. How these changes are affecting the system of local knowledge of plant resources and the maintenance of traditional plant use practices are some of the questions addressed in this chapter, based on research in such a region of northeastern Portugal.

We documented local knowledge and uses of food and medicinal plants among rural people living inside and in the vicinity of the Natural Park of Montesinho, in the region of Trás-os-Montes, in northeastern Portugal. This area has suffered from geographic and economic isolation and a slow, steady decline of the population, as young people move out of the area, so we were particularly interested in seeing whether and how traditional practices would persist under these conditions. Do the people recognize these trends, and are they concerned about them? Are there mechanisms that people are employing to maintain their knowledge?

Our research specifically analyses the effects of geographic or cultural isolation on traditional knowledge and practices (for another example of this type of analysis, see Pieroni and Quave 2005). While there have been several recent ethnobotanical studies in Portugal (e.g., Camejo-Rodrigues et al. 2003; Novais et al. 2004), none have as yet addressed our research questions and, unfortunately, they are difficult to use for comparative analyses due to the different purposes and methodological approaches used by the authors (Carvalho 2005).

An Overview of Northeastern Portugal

> *Terra quente e terra fria. Léguas e léguas de chão raivoso, contorcido, queimado por um sol de fogo ou por um frio de neve. Serras sobrepostas a serras. ... É esta a terra de homens inteiros, saibrosos, que olham de frente e têm no rosto as mesmas rugas da terra* (Miguel Torga in *Um reino maravilhoso* (1941[2002]).
>
> Hot land and frost land. Miles and miles of feral and contorted ground, sometime sun-baked, sometime frozen. Mountains overlapped with mountains. ... This is the land of real men, carved in stone, men that can look you in the eye and have on their faces the same wrinkles the land has (Miguel Torga in *A Marvellous Rein* (1941[2002]).

This quotation characterizes well the remote ruggedness of northeastern Portugal and its inhabitants. This is a mountainous region with a highly diverse landscape, still with vegetation and flora (Aguiar and Carvalho 1994; Aguiar 2001) similar to the deciduous forest cover that dominated the north of the country for thousands of years (Costa et al. 1998; Aguiar 2001). Native fauna and flora have been sharing the same space, hard climate and soil conditions with humans and their activities for a long time, apparently without substantial changes. These factors supported the creation of a protected area in part of this territory during the 1980s, the Montesinho Natural Park (Figure 7.1), covering 734 km^2, comprising eighty small villages with about seven thousand inhabitants (INE 2001), and involving two municipalities, Bragança and Vinhais.

Figure 7.1. Map of the Iberian Peninsula. Arrow points to the Montesinho Natural Park, in northeastern Portugal

The rural character of this Portuguese province (called *Trás-os-Montes*, meaning 'behind the mountains') has been preserved mainly due to social, economic and land-use constraints (Rodrigues 2000). Moreover, the national road system is very inefficient or nonexistent in many cases (Carvalho 2002). Today, despite the emergence of a few new roads since the 1990s, physical communication between villages and nearest towns is still difficult.

For decades, the population of the Montesinho Natural Park has been isolated from other populations and dependent on natural resources. Agriculture and pastoralism were the main economic activities, sometimes complemented by income from mining and smuggling, as the Park borders the Spanish province of Castilla-Léon, Zamora. While most households were essentially self-sufficient and subsistence oriented, some products, such as grains, chestnuts, potatoes, livestock, textiles, handicrafts, charcoal and wood, could be sold or traded outside the communities to generate extra income. In former times, medical and veterinary services were far from these villages. People's food and medical needs relied on materials found in the natural surroundings, especially wild and cultivated plants, on small-scale animal breeding, and on fishing and hunting. Wild-gathered species were an important supplement and alternative to the regular diet – sometimes the sole source of micronutrients, minerals and vitamins (e.g., vitamin C) – and the source of raw materials for homemade remedies for primary healthcare and treatment of human and animal diseases.

Over time, this close relationship between people and their natural and agricultural environment has led to the development of a rich knowledge base on plants, plant uses and related practices. Some of this local ecological knowledge has been documented in old botanical or agricultural monographs (e.g., Coutinho 1877; Palhinha 1946), and in several historical and anthropological research projects conducted in northeastern Portugal (e.g., Alves 1934[1985]; Dias 1953[1984]), but there was no detailed or prolonged ethnobotanical study of such knowledge and practices until 2001 (Carvalho 2005).

In 1986, when Portugal joined the European Community (EC), the northeastern most interior region and the Trás-os-Montes province were some of the most depressed areas in all of Portugal. More than half of the workforce was committed to agricultural activities, but these were not profitable enough to support the population. Thus the area was quickly depopulating, as young people migrated to other European countries, or to the towns near the Atlantic coast, where it was easier to find employment.

New rural development policies and the Common Agricultural Policy (CAP) guaranteed farming prices, ensuring stable incomes for farmers, and the region benefited from European funds for sustainable development. According to national statistical and administrative data (Lima 2000), Trás-os-Montes province has received a quarter of the total European funds attributed to Portugal, for improving quality and safety and developing a farming sector in tune with contemporary ideas of environmental protection and animal welfare. The new policies were attractive for farmers, and middle-aged emigrants returning home, who were offered investments in agriculture (e.g., CAP agrienvironment measures) based on the development of regional farm products such as cheese, olive oil and meat products. These were guaranteed a somewhat protected and more lucrative market based on their certification as 'Protected Designation of Origin' (PDO) and 'Protected Geographical Indication' (PGI). Funds were also available if farmers took measures to promote environmental protection and preserve rural heritage, or decided to develop alternative activities such as agrotourism or the sale of wild, natural products like honey, herbal teas or mushrooms.

Participation in these EC-funded schemes favoured and reinforced the traditional knowledge system, and contributed to a balance between traditional and modern farming practices. Subsidies allowed farmers to maintain their cultural heritage and to adapt older and modern agricultural technologies to new and different purposes. The increasing demand for new and authentic rural goods and services encouraged the revival of evocative aromas, flavours and feelings from old, rural areas. The most desired products – edible plants, mushrooms, medicinal plants and PDO and PGI products, such as olive oil, cheese, veal and sausages – all required traditional knowledge and wisdom for their harvest and production. In fact, some communities were revived as a result of the infusion of cash, with people returning and traditional farming activities and festivals being conducted and celebrated with new enthusiasm. Besides Rio de Onor, already a well known community due to its particular ethnographic history (Dias 1953[1984]), Montesinho, Moimenta, França, Babe and Baçal also became popular through their local events celebrating the main rural daily or seasonal tasks, including *a segada* (the grain harvest), *o ciclo do linho* (the flax production and linen-making process), *o mata-porco* (the pig slaughter), and the rites of St John's Day (24 June) and St Stephen's Day (26 December).

Unfortunately this revival was rather short lived, as a succession of CAP reforms, beginning in 1992 and continuing for twelve years, introduced 'flanking' measures that cancelled or reduced some crop subsidies and imposed strict production conditions that affected traditional agricultural farming systems. These measures disappointed local farmers and caused the abandonment of some crops (such as cereals, potatoes and fodder). Moreover, these new policies constrained the ability of small farmers to diversify their activities, and agriculture was suddenly viewed as an impossible task without competitive advantages because of rising production costs versus low profits and uncertain wages. As some authors have pointed out (e.g., Batista 1993), there was little chance that development based on small-scale agricultural projects could be profitable in a European and global context because people enjoy and take pride in being productive, on a practical level, and disregard having a passive role as mere guardians of nature. Besides, landscape diversity depends on the ways in which land is used, especially agricultural and pastoral practices; the abandonment of these activities can lead to significant structural changes in landscape and vegetation communities. Nowadays, this northeastern region, and Portuguese rural society in general, are no longer primarily engaged in agricultural activities, instead people have abandoned the small villages or taken precarious jobs in the nearest towns. Community residents are mostly older people living on minor retirement pensions for farmers. There is a high level of out-migration among the youth and young adults, resulting in an ageing population and threatening the ability of the government to implement regional policies. Many leaders in the area are doubtful of the viability of economic alternatives due to the population decline.

Given this dynamic historical and political-economic context, we expected that traditional botanical knowledge would also have declined, though we hoped that amongst the population, primarily the older people, there would be sufficient memory of such knowledge even if it was no longer being practised. Thus we set out to document present knowledge and use of wild food and medicinal plants among several communities of the Montesinho National Park.

Studying Ethnobotany in Montesinho National Park

Out of eighty communities, we choose for our survey all the communities (thirty) that matched the following criteria: they had to be located inside the Park territory (e.g., communities on the southern Park border, under the influence of a nearby town, were not selected), have a population of over fifty residents, represent as many of the thirty-six parishes of the region as was possible, and have a history of agropastoral activities and homegardens until very recently. From 2002 to 2004, during all four seasons of the year, unstructured interviews with individual farmers, participant observation and group discussions were conducted in order to gather ethnobotanical information. Voucher specimens of all plants discussed and encountered were collected, identified and stored in the Escola Superior Agrária de Bragança Herbarium (BRESA). The flora and vegetation of Montesinho Natural Park studied by Aguiar (2001) was the basic botanical reference for this work. He records 1271 vascular plant taxa from the area, including 446 with potential medicinal uses. Plant use and management and other related technologies were also documented, and a photographic collection and an ethnobotanical database were created.

Our eighty interviewees were mostly women (85 per cent) and the average age was 62 years old. Seventy were chosen randomly and ten were key informants selected from those residents considered knowledgeable by their neighbours. After getting to know our informants and engaging in casual conversation, each participant's gardening and gathering activities were followed and observed. We reported information about all the species she/he would consider medicinal or edible, and recorded their local names, uses and relevant stories. Afterwards, in our informant's house or garden, the collected samples were discussed in greater detail. In other cases, participant observation of plant-collecting trips and the use of plants gave us insights into the knowledge required for plant collection, storage and preparation. We also recorded information on the means by which this knowledge is learned and transmitted, and their opinions on the extent to which the younger generation is maintaining these traditions.

Wild Food and Medicinal Plants in the Natural Park of Montesinho

Up until the 1970s, in particular during spring and autumn, the women, shepherds, hunters and fishermen of the area all used to collect a variety of plants for both food and medicine as they engaged in their daily activities. A few men, including some well known smugglers, were skilled in gathering mushrooms and some of the rarer medicinal plants, such as the bladderseed[1] (*Physospermum cornubiense*) or the mountain arnica (*Arnica montana* L.), from sites deep in the woods and from across the border in Spain.

Those were days of great activity in the villages; women and children were active plant gatherers and foragers, while most of the men cultivated field crops, worked in the woods or had jobs outside the community in farming, mining, roadworks and reforestation. However, those were also times of scarcity and hard work without breaks or holidays. As one respondent put it, '*ninguém sabia o que eram fins de semana ou férias ... mas também não havia pressas!*' ('no one knew about weekends or holidays ... but nobody lived under stress!'). Gathering and the consumption of wild food or medicinal plants are strongly correlated with hunger, insufficient (or nonexistent) medical care and formal education, and geographic isolation, as already shown in other ethnobotanical studies (Pardo-de-Santayana, Tardío and Morales 2005; Tardío, Pascual and Morales 2005; Grivetti 2006). All of these criteria appear to have been typical of communities in the Park up until the 1970s.

Gathering from the Wild

More than half of the 364 plant species used by people are considered wild (199 species), and 140 of these (70 per cent of the wild species) were reported to have been traditionally gathered from the wild and consumed either as food (10 per cent), as medicinal plants (58 per cent), or both (32 per cent). Most of these plants, as well as the remaining 59 species of the wild flora, have also been used in the area for other purposes, such as veterinary medicine, fuel, fodder, and handicraft or building materials (Carvalho 2005).

In the study area, wild plants are known as '*plantas bravas que crescem por aí*' (literally meaning 'wild plants growing around'). This popular characterization refers to species growing on their own in different environments. These include plants from woods, scrubland and riversides; from natural prairies or meadows; and from disturbed environments, such as roadsides, crop-field borders, fallow fields, or hedgerows and boundary walls. Others fit into the farmers' concept of 'weeds', which are plants out of context, as those weeds, locally named *ervas*, are closely associated with

Table 7.1. Some plants traditionally gathered from the wild, grouped according to the principal source or growing place

Source/Growing place	Number of species	Some examples: scientific name (English name)
Naturalized species in the village surroundings	16	*Chenopodium ambrosioides* L. (Mexican tea), *Cydonia oblonga* Mill. (quince), *Ficus carica* L. (fig tree), *Laurus nobilis* L. (bay tree), *Melissa officinalis* L. (lemon balm), *Salvia officinalis* L. (common sage)
Disturbed environments	63	*Borago officinalis* L. (borage), *Chelidonium majus* L. (celandine), *Malva neglecta* Wallr. (dwarf mallow), *Umbilicus rupestris* (Salisb.) Dandy (pennywort), *Rumex acetosella* L. (sorrel), *Urtica urens* L. (stinging nettle)
Weeds	43	*Capsella bursa-pastoris* (L.) Medik. (shepherd's purse), *Chondrilla juncea* L. (naked weed), *Plantago* spp. (plantains), *Portulaca oleracea* L. (green purslane), *Sonchus asper* (L.) Hill (sow thistle)
Prairies and meadows	11	*Filipendula ulmaria* (L.) Maxim. (meadowsweet), *Lolium perenne* L. (perennial ryegrass), *Ornithopus pinnatus* (Mill.) Druce (yellow bird's-foot), *Prunella vulgaris* L. (heal all), *Thymus pulegioides* L. (thyme)
Riversides	13	*Equisetum arvense* L. (common horsetail), *Fraxinus angustifolia* Vahl. (ash), *Hedera helix* L. (ivy), *Humulus lupulus* L. (wild hops), *Sambucus nigra* L. (elder), *Tamus communis* L. (black bryony)
Bushes/scrubland	18	*Arbutus unedo* L. (strawberry tree), *Cistus ladanifer* L. (gum rockrose), *Cytisus multiflorus* (L'Hér.) Sweet (white broom), *Halimium lasianthum* (Lam.) Spach (rockrose), *Lavandula stoechas* L. (topped lavender), *Tuberaria lignosa* (Sweet) Samp. (perennial herb similar to spotted rockrose)
Woods	15	*Arenaria montana* L. (mountain sandwort), *Crataegus monogyna* Jacq. (hawthorn), *Fragaria vesca* L. (wild strawberry), *Physospermum cornubiense* (L.) DC. (bladderseed), *Quercus pyrenaica* Willd. (oak), *Prunus spinosa* L. (sloe)

agricultural practices, homegardens and specific cultivated plants, like wheat, rye, potatoes and cabbages. Finally, some are naturalized species, not depending on human activities for survival, but established for at least five decades of local consumption and traditional use (see Table 7.1).

Many of these wild species are in fact found in more than one of the habitats described above, but when it comes to harvesting for particular uses, species tend to be gathered in only a few or just one of these several locations. According to several informants, the quality of the vegetal material depends on its particular location. An illustrative example is the very common mountain sandwort (*Arenaria montana*), whose leaves and flowers are still used to prepare a powerful infusion recommended as a diuretic and an anti-inflammatory. Mountain sandwort should be collected during spring along paths through the oak trees: '*vai-se pela melhor seixinha nos barrancos dos carvalhos*' ('one can find the best mountain sandwort under oak trees'). As stated by older informants, exposure to sunlight decreases the potency of mountain sandwort, so the best growing conditions (moisture, light and humus) are to be found on the shady forest floor, particularly under the oak trees. After having been dried in the shade, the medicinal potency of these plants will be preserved at least through the year and quite possibly for even longer.

The quality of gathered plant material may also vary according to season and time of day. Many of the green edible plants and some fungi should be picked in early spring, when the weather is still cold enough to avoid insects laying their eggs on these plants: '*até ao cantar do cuco por causa dos bichos*' ('before the cuckoos sing to avoid worms'). That event coincides with the first calls of the cuckoo, which is regarded as the harbinger of spring and rising temperatures. This is the case, for instance, for watercress (*Rorippa nasturtium-aquaticum*), red bryony sprouts (*Bryonia dioica*) and oyster mushrooms (*Pleurotus ostreatus*). Another interesting example is the collection of flower caps, and sometimes leaves or shoots, used for medicinal infusions and macerations (i.e., a process of soaking or steeping the plant material). It is said that the best time to harvest is in the early morning, before sunrise. From a phytochemical perspective, harvesting at dawn is probably a very good idea, as aromatics are very rich in volatile substances (essentials oils) and other compounds that easily vaporize with heat or are degraded with sunlight. This method of harvesting is applied to the flowering parts from chamomile, St John's wort, thyme and elder, as well as to fennel and mugwort (*Artemisia vulgaris* L.) shoots. Some of our informants find it preferable to combine time with a special day, such as St John's Day. People believe that mid-summer plants have miraculous healing powers and they therefore pick them on this precise date, '*melhor colher na madrugada de S. João, que é um dia bento*' ('It is better to gather during St John's dawn, which is a holy day').

Edible Plants and Fungi

At the beginning of our survey, one middle-aged man declared: *'ervas são medicinas, não se comem!'* ('herbs are remedies, they are not to be eaten!'), while someone else mentioned *'chupar ervas é brincadeira de crianças e tonteira de pastor!'* ('eating/sucking herbs are children's games and shepherds' foolishness'). These statements suggest a prejudice against wild edible plants. In fact, interviewees were initially not particularly enthusiastic about discussing wild plants traditionally used as food, even though they had consented to be interviewed on the subject! We speculated that there might be numerous reasons for such prejudice, including the widespread negative association between wild food and poverty and hunger, as some informants pointed out. Other reasons for the decline in interest in wild plants may derive from the increased use of homegardens in the 1970s and 1980s, where various staple products for daily meals were cultivated. Also, in the last twenty years, mobile shops have started to visit the villages once a week to sell groceries, vegetables, fruit and other goods, offering a bigger array of products and discouraging the maintenance of diverse gardens and wild gathering.

Against all expectations, then, we were surprised to find that fifty-seven vascular plants and sixteen fungi species from the wild flora were recognized and referred to as traditional edible plants by our informants. These species used to be consumed (some still are) in different and creative ways, such as raw in salads, boiled or fried, eaten with a potato-based, bean-based or chickpea-based soup, as spices, sweets and fruits, or as beverages (see Table 7.2 for some examples).

Compared with the number of cultivated food plants commonly available and also inventoried, the contribution of wild food plants to the total dietary intake is not very important, although they do provide important minor nutrients (such as vitamins and mineral salts) that are not present in daily meals based on bread and potatoes.

It is also interesting to highlight the importance of fungi diversity. About sixteen species belonging to fourteen genera have been gathered and were/are usually eaten, according to the informants. Even though we have not been able to compare this result with other areas, we assume that it is an unexpectedly high number of species because collecting edible fungi is rather difficult. Mushrooms are not easily found (i.e., they are usually hidden in the herb and ground layer or inside old tree trunks) and their collection is only possible for a short time, and extremely dependent on adequate climatic conditions.

Wild food was, sometime still is, commonly preserved and stored, for consumption during the long and hard winters. This was especially true for wild fruits, as well as cultivated products (e.g., cherries, pumpkins, tomatoes), and aromatic species used in herbal teas or spices. Marmalades

Table 7.2. Wild edible plants and fungi reported by more than 25 per cent of the informants, in descending order of most frequently mentioned

Scientific name	English name	Part(s) used and gastronomic use
Origanum vulgare L.	Oregano	Dried leaves and florets used as spices in salads, for seasoning olives, chopped meat and bread used in various types of sausages
Laurus nobilis L.	Bay	Dried leaves used as spices, for seasoning olives, chopped meat and bread used in various types of sausages
Bryonia dioica Jacq.	Red bryony	Sprouts, scalded in water, then cooked with garlic and olive oil, or with scrambled eggs or omelettes
Foeniculum vulgare Mill.	Fennel	Leaves, stems and seeds used as spice, liquor and herbal tea
Fragaria vesca L.	Wild strawberry	Fruit, raw, jam and liquor
Rorippa nasturtium-aquaticum (L.) Hayek	Watercress	Leaves, raw in salads and greens for soup
Glechoma hederacea L.	Ground ivy	Fresh leaves, for seasoning soups, herbal tea
Rubus ulmifolius Schott	Blackberries	Fruit, raw, jam and liquor
Chamaemelum nobile (L.) All.	Chamomile	Florets, herbal tea
Mentha pulegium L.	Pennyroyal	Fresh leaves, spice and herbal tea
Montia fontana L.	Water chickweed	Leaves, raw in salads
Borago officinalis L.	Borage	Leaves, greens for soup
Lavandula stoechas L.	Topped lavender	Leaves, for seasoning wild rabbit
Urtica dioica L.	Stinging nettle	Leaves, boiled then smashed and fried with garlic and olive oil or cooked in soup
Pterospartum tridentatum (L.) Willk.		Flowers and young shoots for seasoning wild rabbit, hare or chicken, or to prepare a juicy rice dish
Portulaca oleracea L.	Green purslane	Leaves, raw in salads and cooked in soups
Rumex acetosa L.	Common sorrel	Leaves, raw in salads; raw stems are sucked
Boletus edulis Bull.: Fr.	Cep	In both cases, caps are stewed with meat, or sautéed with onion and boiled potatoes, or cooked with rice
Boletus pinophilus Pilát & Dermeck	Pine tree porcine	
Macrolepiota procera (Scop.:Fr.) Singer	Parasol mushroom	Grilled intact or chopped with scrambled eggs
Pleurotus ostreatus (Jacq.: Fr.) Kummer	Oyster mushroom	Sliced and fried with garlic and olive oil or with scrambled eggs

were prepared from wild strawberries, blackberries and sloes (*Prunus spinosa* fruits), and a traditional jam was/is made from quince pulp. Quite exceptional and much appreciated *licores* (liqueurs) are made from wild hop florets, fennel and bladderseed (*Physospermum cornubiense*) fruits, sloes and wild strawberries. A wide range of herbs were/are gathered, dried and stored in different ways depending on their two main purposes: seasoning food or preparing beverages.

As reported in other Iberian Peninsula regions (Bonet and Vallès 2002; Pardo-de-Santayana, Tardío and Morales 2005; Tardío, Pascual and Morales 2005), some species (e.g., fennel, mints, lemon balm, chamomile, lesser calamint) are widely used in herbal teas that are drunk hot in winter and cold in summer, as a refreshment, for the pleasure of aroma and taste, daily or in social events. Most of these species and their resulting products such as marmalades, spirits and infusions are also deliberately consumed for their preventative or curative properties and are considered 'medicinal foods' (Pieroni 2000; Pieroni and Quave 2006).

From Famine Plants to Tasty and Fragrant Recipes

Hunger and starvation were common events in this corner of Portugal, mostly during the Spanish Civil War and the Second World War. For about ten years (1936–1945) the most important local crops, such as grain and potatoes, and also meat, were sent to armed forces allied to the Portuguese government (General Franco in Spain and to Germany). Thereafter in the 1960s, when the independence wars of the former African colonies started, many of the young soldiers were from Trás-os-Montes and thus the region suffered a lack of men to work the fields, a situation that entailed scarcity all over this area.

Some interviewees were able to remember and others had been told by their parents that in those days there was not enough flour for making bread. Therefore chestnut flour was used as a substitute, and topinambur (*Helianthus tuberosus* L.) tubers were consumed instead of potatoes and distilled for alcohol production. For some, buds, flowers or fruits from various plants (see Table 7.3) were consumed with rye bread during the working day to avoid hunger pains – '*para enganar a fome*'.

The sole hot meal of the day might be composed of rye bread and a soup made of boiled water with stinging nettle, or borage leaves, enriched with a tablespoon of rye, and seasoned with herbs, taking advantage of a variety of aromatic wild species (e.g., water mint, apple mint, peppermint, pennyroyal, catmint or ground ivy).

One fascinating episode was recounted in Rio de Onor (a communal agropastoral village until the 1970s) concerning the consumption of sorrel (*Rumex acetosa*). There was such a lack of food at the time that the

Table 7.3. Species quoted as famine food, in descending order of most frequently mentioned

Scientific name (part used)	English name
Mentha x piperita L. (leaves)	Peppermint
Rumex acetosa L. *and R. acetosella* L. (stems)	Common sorrel and sheep's sorrel
Rubus ulmifolius Schott (fruits)	Blackberries
Prunus avium L. (fruits)	Wild cherries
Fragaria vesca L. (fruits)	Wild strawberry
Lamium purpureum L. (flowers)	Red dead-nettle
Prunus spinosa L. (fruits)	Sloes
Halimium lasianthum (Lam.) Spach (flower buds and fruits)	Rockrose
Malva sylvestris L. and *M. neglecta* Wallr. (fruits)	Common and dwarf mallow
Conopodium majus (Gouan) Loret (tubers)	Pignut
Cytinus hypocistis L. (flowers)	
Crataegus monogyna Jacq. (fruits)	Hawthorn
Rosa canina L. (pseudofruits)	Dog rose

Conselho (council of elders) took over the management of sorrel harvesting by selecting the gatherers and the sites where they would collect, and then organizing the daily distribution to each family. '*Tocavam os sinos a avisar da distribuição das azedas*' (the church bells rang out to announce the distribution of sorrel).

Seasoning and preserving food are still common procedures that have an influence on the traditional cuisine and are fundamental to many regional recipes. Today, there is a particular interest in species with natural flavours suitable for enhancing the taste and smell of food. In former times, their use may have been related to nutritional needs, especially during those famine periods when wild edible plants were the main source of nourishment for the rural families, but nowadays they are specially used for flavour and aroma. There are a great number of use-reports (151) and plants (30 per cent of those cited) connected with the use of these condiments and the processes of food preservation. In one case, the use of the Fabaceae *Pterospartum tridentatum* as a wild green and as a spice appears to be unique in Europe.

Recently, people have brought some of the most popular plants used as food additives and beverages from the wild to grow in their homegardens, in order to make them easily available. However, according to most informants, the gathering and consumption of wild edible plants is in steady decline throughout the area.

Wild Medicinal Plants and Local Folk Phytotherapy

Wild medicinal species comprise about 70 per cent of the medicinal plants mentioned by our informants. In former times, both wild and cultivated plants were commonly used to treat illnesses. Moreover, some species were well known crop weeds, such as those from cereal fields that were used to treated injuries that occurred during the grain harvest, when instruments such as sickles or scythes made the task potentially dangerous (e.g., all milky latex parts from *Chondrilla juncea* and *Osyris alba* L.).

There are differences in people's perception of wild foods and medicines, though both a lack of food and a lack of healthcare are perceived as social stigmas. The fact is that the collection of medicinal wild plants is considered neither a children's game nor a grown-up idiosyncrasy. On the contrary, it is reported as an expertise belonging to very skilful individuals who, in the past anyway, were respected and famous beyond their village limits. The informants clearly distinguish different levels of expertise among those said to be knowledgeable, identifying for instance the best gatherer or the best user of a wide range of species and treatments. In addition, the ancient pharmacopoeia associates homemade remedies with a kind of magic power that requires a rare natural ability or special training, attributes belonging to few people. Some plants (e.g., garlic, wild roses, St John's wort, peppermint and feverfew – *Tanacetum parthenium* (L.) Sch.Bip.) are not only used for their healing properties, but also as protective amulets and good omens preventing illness and the misfortune of the evil eye.

Around eighty-five taxa were referred to as having only medicinal effects and forty-five species were mentioned as serving both medicinal and nutritional purposes. All these species are recommended for unspecified (e.g., digestive system) or specific (e.g., cramp, cold) disorders, to produce a special effect (e.g., laxative, diuretic), and can be suitable for a specific process such as menstruation, birth or blood pressure.

Interviewees' use-reports (272) can be classified according to therapeutic activity, with the highest number of uses being digestive-system disorders, followed by skin, respiratory and genitourinary afflictions. These results are similar to those obtained in other surveys (e.g., Bonet and Vallès 2003).

According to our key informants, there are a set of 'worthy plants' which they prefer and highly recommend for human or animal primary care because they treat a range of symptoms and a broad spectrum of infections and inflammations (see Table 7.4). Key informants consider that these worthy plants have all the properties and characteristics required: they are not rare or difficult to find; they are easily gathered, preserved, processed and consumed; and they are suitable for the most common health problems, such as digestive, respiratory and urinary afflictions, pains and skin diseases. Some of them are often cited without specifically

Table 7.4. Medicinal species considered as 'worthy plants', in descending order of most frequently mentioned

Scientific name (English name)	Number of quotations	Number of different uses	Main use
Tuberaria lignosa (Sweet) Samp. (perennial herb similar to spotted rockrose)	64	7	Intestinal and genitourinary anti-inflammatory, diuretic, internal and external disinfectant, vulnerary
Juglans regia L. (walnut tree)	55	9	External antiseptic, anti-inflammatory, vulnerary, dandruff, diabetes
Pterospartum tridentatum (L.) Willk.	52	10	Respiratory and digestive systems, skin condition (spots, acne), heart troubles, hypertension
Foeniculum vulgare Mill (fennel)	52	5	Digestive system
Sambucus nigra L. (elder)	48	5	Respiratory system, pneumonia
Glechoma hederacea L. (ground ivy)	46	4	Painful menstruation, restorative
Hypericum perforatum L. (St John's wort)	45	7	External antiseptic, digestive system and liver disease
Cytisus multiflorus (L' Hér.) Sweet (white broom)	45	6	Rheumatic pains, hypertension, diabetes, cholesterol, headache
Mentha suaveolens Ehrh. (pennyroyal)	44	9	Nosebleed, diarrhoea and vomiting, haemorrhoids, cholesterol, coughs, skin
Chelidonium majus L. (celandine)	42	4	Vulnerary, warts, corns
Malva neglecta Wallr. (dwarf mallow)	40	7	Internal and external disinfectant, skin condition, toothache, respiratory and digestive systems
Origanum vulgare subsp. *virens* Hoffmanns. & Link Bonnier & Layens (oregano)	36	5	Coughs, pneumonia, kidney infection
Rosa corymbifera Borkh. (wild rose)	32	4	Diarrhoea and vomiting, eye inflammation, muscle pain

mentioning or identifying the body part or the process treated, as if they are able to cure any problem. Moreover, these species are well known, have higher rates of quotation, are among those with a great number of different medicinal uses (see Table 7.4), and were and are commonly used by the majority of our respondents and relatives; and there is still a large demand for 95 per cent of them.

Standard literature on pharmacognosy and phytotherapy (e.g., Evans 2002; Heinrich et al. 2004) confirms the properties and pharmacological effects of at least half of the 'worthy plants' listed in Table 7.4, and describes their use since ancient times in many other places. Although there is no reference for six species (*Tuberaria lignosa, Pterospartum tridentatum, Glechoma hederacea, Cytisus multiflorus, Mentha suaveolens* and *Malva neglecta*), there is available in the literature useful pharmacological information on the medicinal uses of the same genus or of similar species from the same botanical family.

Local Homemade Remedies

Leaves, flowers and fruits are the most preferred plant parts for homemade preparations. The vegetal materials can be dried, burned, macerated, heated, fried, melted, milled or chopped, and used plain or with some additives. Olive oil, pork fat, honey, lemon juice, alcohol, brandy, wine and vinegar are all mediums. They are consumed or applied in a variety of ways, including infusions, decoctions, gargling, inhalation, maceration, syrups, poultices, baths, vapour baths, lotions and ointments. Medicinal plants are often mixed in order to 'combine efforts and increase quality', which is based on a belief in synergism between several species. Furthermore, they noted that the pharmaceutical industry also develops mixed products: *'Mistura-se pois! Então e os remédios da farmácia? Se calha, são feitos de uma coisa só!'* ('of course we mix herbs! Think about pharmaceuticals, are they made from a single ingredient?').

Generally, wives (and sometimes husbands) prepare the basic remedies to treat the family or the animals. However, particular recipes made of plant mixtures, some herbal extracts, and special lotions and ointments (such as those for wolf and viper bites and for scorpion or wasp stings) are prepared by specially trained healers who provided them on request.

In the folk phytotherapy there are some important remedies that are regarded as essential contributions to the families' well-being. Informants find it remarkable to have such a group of plants they can use for preventative or curative medicine. Widespread examples, most of them still in use, include a decoction made from *Pterospartum tridentatum* flowers and a syrup made from the leaves of watercress (*Rorippa nasturtium-aquaticum*), both used as anti-catarrhals and expectorants. The infusions of elder (*Sambucus nigra*) and hawthorn (*Crataegus monogyna*) flowers are considered the best bronchopulmonary decongestives. An infusion of white broom flowers (*Cytisus multiflorus*) and fruit from red and black bryonies (*Bryonia dioica* and *Tamus communis*), macerated in alcohol, has a powerful anti-rheumatic effect. The root decoction of wild strawberry (*Fragaria vesca*) is used for female genitourinary system disorders. A perennial rockrose (*Tuberaria lignosa*) or St John's wort (*Hypericum perforatum*) decoctions are diuretics, recommended for digestive and liver diseases, and have external and internal anti-inflammatory properties. Walnut tree (*Juglans regia*), clary (*Salvia sclarea* L.), great mullein (*Verbascum thapsus* L.) and self-heal (*Prunella vulgaris*) are commonly used as vulnerary plants. Macerated oak buds (*Quercus pyrenaica*) in alcohol and the latex from various species, already mentioned, promote wound and sore healing. Dwarf mallow (*Malva neglecta*) decoction is a very useful external disinfectant and an internal anti-inflammatory that may be drunk or used in baths. Round-leaved mint (*Mentha suaveolens*) leaves are haemostatic and applied for nosebleed. Pennywort (*Umbilicus rupestris*) leaves prepared with olive oil are an ointment that treats haemorrhoids. The decoction of *Arrhenatherum elatius* (L.) J. & K. Presl. roots and maidenhair spleenwort leaves (*Asplenium trichomanes* L.) was in former times a popular anti-pyretic for children.

The traditional pharmacopoeia suggests that people once exploited quite a lot of the available natural resources; today, some of these resources are not in use any more. Besides herbs, shrubs and trees provided raw materials for medicines and drugs as well. Some of these woody species, mentioned by a great number of interviewees, are: elder (*Sambucus nigra*), ash (*Fraxinus angustifolia*), quince (*Cydonia oblonga*), oak (*Quercus pyrenaica*), helm oak (*Quercus ilex* L. subsp. *ballota* (Desf.) Samp.), ivy (*Hedera helix*) and wild roses (*Rosa canina* and *Rosa corymbifera*).

Medicinal Food

Generally, our informants believe that consumption of 'good food' (flavoured, tasty, without fertilizers and pesticides, mostly greens) is healthy. However, the concepts of food and medicine seem to be quite unrelated in their minds, as they never refer to a species as medicinal food (i.e., medicinal and nutritional uses have always been mentioned separately), although they admit some connection between those uses, particularly between some beverages drunk after meals and their digestive effects. This link has also been reported in other surveyed areas in Europe (Pieroni 2000; Bonet and Vallès 2003; Pardo-de-Santayana, San Miguel and Morales 2006).

Pieroni and Quave (2006) described several ways of perceiving the correlation between food and medicinal value that can be adapted to this study area. In Montesinho, some of the forty-five reported plant species that have both medicinal and nutrient value, mostly herbs and weeds, are multifunctional, without any kind of link between these two uses; others are ingested to satisfy both nutritional and health needs and are interesting in terms of their role as medicinal food; many are simultaneously used as spices, consumed as soft drinks or as herbal teas and for unspecified health purposes. For instance, wild hop (*Humulus lupulus*) flowers, bladderseed (*Physospermum cornubiense*) and fennel (*Foeniculum vulgare*) fruits are consumed in liqueurs, as condiments for standard recipes and traditional baked goods, and as remedies for irritable bowel syndrome. A soup made with water, salt, olive oil and the leaves of ground ivy, stinging nettle or borage was strongly recommended for labour induction and women's post-partum recovery because of its restorative effects. These ingredients were also said to stimulate lactation and prevent excessive bleeding. Moreover, these plants were said to provide a balanced set of nutrients for nourishing newborn babies whose mothers could not breastfeed and for recovering young children who have been ill. Mothers used to give their young children the fruits of species such as sloes, dog rose and hawthorn because they believed they had high vitamin content. Some women mentioned that plants usually sucked (see Table 7.3) have the same, high vitamin content.

Persistence of Wild Food and Wild Medicinal Plant Knowledge and Uses

At the time of our survey, the collection and consumption of wild plants in Montesinho was virtually nonexistent. People were not seen walking around their villages; villagers are old and remain at home most of the day. Also, there were few signs of plant collecting, such as botanicals drying in back yards. Although recognizing, collecting, classifying, managing and manipulating wild plants were once very important skills (Carvalho, Lousada and Rodrigues 2001), today, this local knowledge and those tasks appear to have been set aside by the majority of residents, though several new countervailing tendencies can also be described for the area.

As in some other European regions (Svanberg and Tunón 2000), people are aware that nowadays their lifestyles are very different from those that once prevailed, but they also think that traditional knowledge should be more highly prized by the younger generations. Some people would like to have a combination of today's material well-being with yesterday's appreciation of nature.

Nostalgic older people can remember that just a few years ago, in 2001, the situation was quite different, as a seventy-year-old woman explained, while showing us some dried herbs: '*Lembra-me que ainda não faz três anos que colhi a última alcária*' ('I remember, I collected my last specimen of *rockrose*

(*Tuberaria lignosa*) less than three years ago'). And she concluded: '*Quando saía com o gado sempre ia apanhando, agora já não o há, não vou ao monte e por isso não colho!*' ('I used to gather plants while taking care of the sheep; I don't have sheep any more, so I am not going out and I don't collect so often!').

Other informants commented that 'gathering means to climb up the hills and walk in the woods'. This represents some difficulties they are not able to overcome because 'their legs aren't what they were...' ('*as pernas já não são o que eram ... já nos custa a subir até ao carvalhal*'). Instead they have brought several plants from the wild and planted them in the garden. They think it was a good idea because these plants 'are growing anyway'. Many of them are for herbal teas and used to be gathered every season ('*Fomos trazendo algumas plantas de lá, pusemo-las na horta e até que "prenderam" bem! Muitas são ervas "pro" chá que costumávamos colher todos os anos*').

Along with these more personal reasons, by 2002 the lack of EU subsidies had begun to affect the area dramatically, with farmers and young people giving up and leaving the area. An abnormally long period of drought between 2003 and 2005 only made conditions worse by destroying much of the vegetation resources that might have been harvested.

Even so, traditional knowledge concerning medicinal and edible plants and plant use still remains in the memories of older and middle-aged residents. Moreover, it is still practised by some older women and middle-aged housewives, and occasionally among some young people who claim they care about healthy and safe food and alternative medicine. For a period of four years (2000–2004) the survey showed that 45 per cent of the medicinal applications reported and 15 per cent of the food uses reported persisted in the studied communities. Although some people are not able to gather or prepare wild plants by themselves, the majority of our informants assumed that they would keep using local remedies for basic healthcare (e.g., common cold, digestive disorders, skin injuries or rheumatic pains), food additives and beverages. Usually neighbours, friends and relatives cooperate in providing and changing the required raw material; occasionally certain medicinal herbs and spices (e.g., bay, oregano, mint, lemon balm, common sage or fennel) are found by chance in local markets and sold during village festivals.

According to our informants, the demand for wild vegetables and fruits is lower than it used to be for a number of reasons, including the availability of cultivated alternatives from local homegardens and markets, the general unavailability of wild plants, especially edible ones, in local markets – except for watercress, water blinks (*Montia fontana*), some mushrooms and a few medicinal plants – the increased time and effort required to harvest wild plants, and the age-old symbolic association of wild food plants with poverty and times of starvation.

However, there is a growing interest among the young (30–40 years old) and some returning migrants (50–60 years old) in plant-based natural therapies, which are seen as healthier alternatives to modern pharmaceutical products. These people appear more interested in consuming traditional medicinal plant products than collecting them, and they lack knowledge of where these plants come from, how they are gathered and how they are prepared. Instead, they rely on their elder relatives and neighbours to supply them with these products: '*tenho sempre ervas colhidas e preparadas para a minha filha levar...*' ('I always gather and prepare some herbs for my daughter...').

Nowadays, the village festivals, during summertime and religious celebrations (Easter, Christmas and All Saints' Day), in which all residents participate and relatives living far away are invited to take part, serve as important occasions for reviving and maintaining the role of traditional foods and wild medicinal-plant uses. Traditional meals and rites are recreated, and when their holidays are over and it is time to return home, the 'visitors' leave with a variety of herbal remedies, homemade products and gifts that remind them of 'home' and that they cannot get elsewhere. Unfortunately, those holiday periods are not the times when wild greens are harvested and consumed, in early spring, so they tend not be consumed by visitors. However, one woman mentioned that she had successfully frozen red bryony sprouts because her grandchildren living in France are fond of omelettes made with them.

Uses of the medicinal flora and wild edible mushrooms are more common today than the use of wild food in general. Some medicinal plants have always been traded or sold in local markets, and interest in those species has only increased over the past few years. There is now a certain influx of middle-class, urban people and specialist dealers, able to recognize a large number of species, who come to harvest mushrooms and medicinal plants for free. Several local residents are naturally upset by this activity, which some see as poaching. Although a minority (i.e., five informants from two communities), some residents claim harvesting by outsiders is unacceptable because these people collect without any rules and do not respect the resources and the sites. However, the majority of our informants do not seem to be aware of any potential risk involved with these activities.

In Montesinho, women retain the most valuable knowledge about plants and are primarily responsible for the maintenance of traditional plant-use practices for the purposes of maintaining the well-being of their families (Carvalho 2005). However, that does not stop them from also using and combining traditional remedies together with pharmaceuticals prescribed by conventional biomedicine (Figure 7.2). Insofar as their physical condition allows, women are still gathering or getting from friends and neighbours the species considered 'worthy plants' (e.g., mallows, rockrose, St John's wort), and preparing homemade remedies, amulets and good-luck charms (e.g., a dried branch of St John's wort behind the front door).

It is assumed that the disappearance of many plant uses is related to the arrival of modern medical facilities. However, the loss of the traditional knowledge system is much more linked to the decline of an agro-sylvi-pastoral lifestyle, since most of the gathering tasks and uses were once associated with other agricultural activities. As farming, herding and forestry work disappear the opportunities to get out on the land and harvest wild

Figure 7.2. Edible and medicinal plants from the Montesinho Natural Park, Portugal. a) *Asplenium trichomanes* (maidenhair spleenwort); b) *Hypericum perforatum* (St John's wort); c) *Umbilicus rupestris* (pennywort); d) Collecting and drying medicinal herbs; e) *Cytinus hypocistis*; f) *Tanacetum parthenium* (feverfew); g) Dried bunches of *Melissa officinalis* (lemon balm); h) Traditional remedies are often used together and simultaneously with pharmaceuticals prescribed by conventional medicine and religious or pagan beliefs

plants also begin to decline. Without the continuity of gathering activities, knowledge loses its practical context, becoming just an empty memory or a story, by which few can recreate the practice needed to correctly identify, harvest and prepare wild plant products. Some informants have recognized this in the following statement, *'o que não se faz, a gente esquece!'* meaning knowledge is preserved as long as it is practised, which is also true for other societies described by other authors (Balée 1993).

Conclusions

Gathering from the wild in the Natural Park of Montesinho used to be an important activity performed mainly by skilful women and children while engaged in their daily farming activities. They used to exploit all the villages' natural surroundings and collect roots, leaves, flowers and fruits from well known herbaceous and woody plants. This vegetal material, consumed raw or preserved for year-round consumption, was the basis of several homemade plant products used for daily meals and for primary healthcare of humans and animals.

In our survey of contemporary wild plant use, 140 specimens from forty-four botanical families were reported and identified as potential food, medicine or medicinal foods, which represents 40 per cent of the ethnobotanical flora found in the Montesinho Natural Park. In general terms, these plants have never been under excessive pressure from exploitation as they are common, according to the IUCN Red List (1994).

Although of great importance in former times, the consumption of wild edible plants tends to be considered outdated, the gathering being regarded as very time-consuming, requiring a lot of work and quite unnecessary because there are plenty of cultivated alternatives. Nevertheless some wild species are still widely used for their taste and aroma, as spices and beverages. Some medicinal species that are most frequently mentioned and still used are locally classified as 'worthy plants' and considered to have the essential characteristics for elementary healthcare: widespread, easily gathered and consumed, with a vast array of medicinal properties and pharmacological effects.

As shown in other European surveys, in Montesinho gathering from the wild and consuming wild food or medicinal plants is strongly correlated with poverty, periods of scarcity, and geographic and social isolation. As a result there is often a reluctance to speak of wild plant use, and thus transmission of valuable information about plants and their uses to younger generations or outsiders is inhibited. Also, collecting plants was closely associated with daily farming routines; it is considered that gathering from the wild will not last if these routines are not performed.

In fact, the decline of farming as a viable way of life and other related changes in rural lifestyles have led to a decline in the gathering and consumption of wild foods and folk medicines. Only older women still keep up these practices, and seem interested in discussing them, though there does appear to be some interest in natural therapies from younger people and outside entrepreneurs. Although a huge amount of traditional knowledge concerning these plant uses is still available in their memories, there are few young people around to learn and use this knowledge. While holidays, family visits and demands for homemade products allow for some maintenance of traditions, the knowledge required to find, harvest and prepare these products is not being transmitted.

People admitted that although they are not able to gather by themselves they continue to consume wild plant products, especially the medicinal ones regarded as the best alternatives to conventional medicines, insofar as they can obtain the vegetal material from neighbours or local markets. As a consequence, today, medicinal uses are more popular than wild food uses and some of the medicinal plants have been transplanted into gardens. A recent interest among young adults and returning emigrants in plant-based natural therapies and organic foods is envisaged by some residents as a hopeful trend for the persistence of the collection and consumption of wild food and medicinal plants in their region.

Acknowledgements

We are grateful for all the kindness and hospitality of our informants and their families during the fieldwork. We owe to Ana Paula Rodrigues and Mariana Fernandes a great debt of gratitude for introducing us to several informants.

Note

1. English names were adopted from the website database *Plants for a Future: Edible, Medicinal and Useful Plants for a Healthier World* (PFAF 2003). Scientific names follow Flora iberica (Castroviejo 1986-2009).

References

Aguiar, C. 2001. *Flora e vegetação da Serra da Nogueira e do Parque natural de Montesinho*, Ph.D. dissertation. Lisbon: Instituto Superior de Agronomia, Universidade Técnica de Lisboa.

——— and A.M. Carvalho. 1994. 'Flora leonesa das serras de Nogueira e Montesinho', *Anuário da Sociedade Broteriana* 60: 1–11.

Alves, F.M. 1985[1934]. *Memórias Arqueológico-Históricas do Distrito de Bragança: Arqueologia, Etnografia e Arte.* Volume 9, 4th edn. Bragança: Museu Abade de Baçal.

Balée, W. 1993. *Footprints of the Forest: Ka'apor Ethnobotany. The Historical Ecology of Plant Utilization by an Amazonian People.* New York: Columbia University Press.

Batista, F.O. 1993. *Agricultura, espaço e sociedade rural.* Coimbra: Fora do Texto.

Bonet, M.A. and J. Vallès. 2002. 'Use of Non-crop Food Vascular Plants in Montseny Biosphere Reserve (Catalonia, Iberian Peninsula)', *International Journal of Food Sciences and Nutrition* 53: 225–248.

Bonet, M.A. and J. Vallès. 2003. 'Pharmaceutical Ethnobotany in the Montseny Biosphere Reserve (Catalonia, Iberian Peninsula): General Results and New or Rarely Reported Medicinal Plants', *Journal of Pharmacy and Pharmacology* 55: 259–270.

Camejo-Rodrigues, J.S., L. Ascensão, M.À. Bonet and J. Vallès. 2003. 'An Ethnobotanical Study of Medicinal and Aromatic Plants in the Natural Park of Serra de S. Mamede (Portugal)', *Journal of Ethnopharmacology* 89: 199–209.

Carvalho, A.M. 2002. 'Etnobotánica de Moimenta da Raia: Las plantas en una aldea transmontana', Post-graduate dissertation (Diploma in Evolutionary Biology and Biodiversity, DEA). Madrid: Universidad Autónoma de Madrid.

——— 2005. 'Etnobotánica del Parque Natural de Montesinho: Plantas, tradición y saber popular en un territorio del Nordeste de Portugal', Ph.D. dissertation. Madrid: Universidad Autónoma de Madrid.

———, J.B. Lousada and A.P. Rodrigues. 2001. 'Etnobotânica da Moimenta da Raia: A importância das plantas numa aldeia transmontana', in *Actas do I Congresso de Estudos Rurais.* Vila Real: Universidade de Trás-os-Montes e Alto Douro.

Castroviejo, S. (coord). 1986-2009. Flora iberica. Vol I-VIII, X, XIII-XV, XVIII, XXI. Madrid: Real Jardín Botánico, CSIC.

Costa, J.C., C. Aguiar, J. Capelo, M. Lousã and C. Neto. 1998. 'Biogeografia de Portugal Continental', *Quercetea* 0: 5–56.

Coutinho, A.X.P. 1877. *A Quinta Districtal de Bragança no Anno Agrícola de 1875 a 1876.* Porto: Typografia do Jornal do Porto.

Dias, J. 1984[1953]. *Rio de Onor: Comunitarismo Agro-pastoril.* 3rd edn. Lisbon: Editorial Presença.

Evans, W.C. (ed.) 2002. *Trease and Evans Farmacognosy.* 15th edn. Edinburgh: W.B. Saunders.

Grivetti, L.E. 2006. 'Edible Wild Plants as Food and as Medicine: Reflections on Thirty Years of Field Work', in A. Pieroni and L. Price (eds), *Eating and Healing: Traditional Food as Medicine.* Binghamton, NY: Howard Press, pp. 11–38.

Heinrich, M., J. Barnes, S. Gibbons and E.M. Williamson. 2004. *Fundamentals of Pharmacognosy and Phytotherapy.* Edinburgh: Churchill Livingstone.

INE (Instituto Nacional de Estatística). 2001. *Recenseamento Geral da População*. Lisbon.

IUCN, The World Conservation Union. 1994. *IUCN Red List Categories*. Gland.

Lima, C. 2000. *A Agricultura de Trás-os-Montes e Alto Douro: Diagnóstico Prospectivo. Série Estatísticas e Estudos Regionais*. Lisbon: Instituto Nacional de Estatística.

Novais, M.H., I. Santos, S. Mendes and C. Pinto-Gomes. 2004. 'Studies on Pharmaceutical Ethnobotany in Arrábida Natural Park (Portugal)', *Journal of Ethnopharmacology* 93: 183–195.

Palhinha, R.T. 1946. 'Plantas aromáticas de Portugal: Lista das plantas aromáticas espontâneas, sub-espontâneas e cultivadas que se encontram em Portugal', *Brotéria* 15: 97–113.

Pardo-de-Santayana, M., E. San Miguel and R. Morales. 2006. 'Digestive Beverages as a Medicinal Food in a Cattle-farming Community in Northern Spain (Campoo, Cantabria)', in A. Pieroni and L. Price (eds), *Eating and Healing: Traditional Food as Medicine*. Binghamton, NY: Howard Press, pp. 131–151.

———, J. Tardío and R. Morales. 2005. 'The Gathering and Consumption of Wild Edible Plants in Campoo (Cantabria, Spain)', *International Journal of Food Sciences and Nutrition* 56(7): 529–542.

PFAF. 2003. *Plants for a Future: Edible, Medicinal and Useful Plants for a Healthier World*. Database, accessed June 2006: http://www.pfaf.org

Pieroni, A. 2000. 'Medicinal Plants and Food Medicines in the Folk Traditions of Upper Lucca Province, Italy', *Journal of Ethnopharmacology* 70(3): 253–273.

——— and C.L. Quave. 2005. 'Traditional Pharmacopoeias and Medicines among Albanians and Italians in Southern Italy: a Comparison', *Journal of Ethnopharmacology* 101(1–3): 258–270.

——— and C.L. Quave. 2006. 'Functional Foods or Medicine Foods? On the Consumption of Wild Plants among Albanians and Southern Italians in Lucania', in A. Pieroni and L. Price (eds), *Eating and Healing: Traditional Food as Medicine*. Binghamton, NY: Howard Press, pp. 101–129.

Rodrigues, O. 2000. *Utilização do território e propriedade fundiária*, Ph.D. dissertation. Lisbon: Instituto Superior de Agronomia, Universidade Técnica de Lisboa.

Svanberg, I. and H. Tunón. 2000. 'Ecological Knowledge in the North', *Studia Ethnobiologica* 9. Uppsala: Swedish Biodiversity Centre.

Tardío, J., H. Pascual and R. Morales. 2005. 'Wild Food Plants Traditionally Used in the Province of Madrid, Central Spain', *Economic Botany* 59(2): 122–136.

Torga, M. 2002[1941]. *Um Reino Maravilhoso*. Lisbon: D. Quixote.

CHAPTER 8

The Use of Wild Edible Plants in the Graecanic Area in Calabria, Southern Italy

SABINE NEBEL AND MICHAEL HEINRICH

Introduction

Local knowledge regarding food use, the basis of many cultural traditions, is under pressure as dietary patterns change rapidly all over the world. Food exemplifies local knowledge or traditional ecological knowledge. It is a very basic need, but – provided people do not suffer from starvation – also a pleasant experience. Food and dishes always reflect a 'vision of the world' and consequently peoples, ethnic groups and communities are proud of their special dishes and the plants or breeds of animal they produce and use. Such knowledge gives them a local identity. While local food may well have its origin outside of the regions of use, the crucial aspect of its definition is its local production, gathering or harvesting and consumption (see Heinrich et al. 2005; Pieroni et al. 2005).

Traditional food knowledge is strongly influenced by socio-economic and cultural parameters, as well as religion and history (Johns et al. 1994; Kuhnlein and Receveur 1996). The term 'traditional' is used in this chapter for defining something that has been an integrated part of a culture for more than one generation (Ogoye-Ndegwa and Aagaard-Hansen 2003). All food is part of our everyday experience and the way it is perceived and classified forms the basis for food use in a culture. Around the Mediterranean, a multitude of cultures, religious beliefs, ecological conditions and historical developments has resulted in a multitude of diets which share many elements but also reflect distinct local or regional traditions (Nestle 1995; Noah and Truswell 2001).

The historical migrations of Greek and other peoples in the Mediterranean is reflected in the many ethnic, linguistic and religious minorities which are still found today across the area (Rother 1989). The diaspora of these people and their independent cultural development in different regions gives us a unique opportunity for comparative and historical analyses of food plant use. This is relevant from a variety of perspectives: such an analysis allows for an understanding of the historical evolution of plant use in a discipline largely relying on synchronic methods, and, as importantly, a better understanding of the diverse and complex relationship between cultures and their environments (Balée 1998: 2; Posey 2002).

This chapter explores the traditional food knowledge of ethnic Greek (Graecanic) communities in Calabria, southern Italy, from an ethnobotanical as well as historical perspective. In this interdisciplinary study, ethnobotanical and historical-linguistic methods were used to better understand the role of wild food plants in local nutrition and cuisine. The aim was to identify plants traditionally consumed in rural Graecanic communities in southern Italy, which have been used continuously for many generations. Furthermore, the antiquity of wild food-plant use is analysed by comparing today's wild-gathered food plants in the Graecanic area in southern Italy with wild food-plant knowledge and practice in Greece today and in antiquity. Written sources of Ancient Greek traditions related to plants include, for example, the *Historia Plantarum* of Theophrastus of Eresos (372–287 BC) (Hort 1916) or *De Materia Medica* from Pedanius Dioscorides (ca. AD 40–90) (Gunther 1934).

The ethnobotanical study discussed here represents the initial phase of a European Union-funded research project, which was ultimately aimed at contributing to the continued use of noncultivated food plants, as well as to the search for new 'nutraceuticals' from noncultivated local resources, which are of potential interest in the prevention of ageing-related diseases (The Local Food-Nutraceuticals Consortium 2005).

The Ethnography of Graecanic Communities in Southern Italy

During the eighth century BC, parts of southern Italy, as we know it today, came under Greek influence and were known as Magna Graecia (Kish 1953). The Greek influence continued over centuries until the late-Byzantine Empire (~ AD 1100) (Cerchiaia, Janelli and Longo 2004). Today, the Greek minorities in southern Italy (Graecanic communities) have retreated to fairly remote areas in the peninsula of Salento in the region of Apulia, and in the province of Reggio di Calabria in the region of Calabria (see Figure 8.1) (Pan and Pfeil 2000).

Figure 8.1. Map of Italy

The six Graecanic communities in Calabria are located in the Aspromonte Mountains, the southern tip of the Apennine Mountains: Bova, Amendolea, Condofuri, Gallicianò, Roccaforte del Greco and Roghudi (Figure 8.2) (Condemi 1999). The population of these villages varies from one hundred to one thousand inhabitants. They are, as compared to the surrounding Italian population, characterized by their own dialect (*Grecanico*), culture and history as an ethnic and linguistic minority. *Grecanico* is now only spoken by older people, whilst the younger generation mainly use the Calabrian dialect of Italian. Due to the geographic isolation of the villages they have retained many aspects of their cultural heritage. However, access roads were built in the 1950s, which has led to an increasing Italian influence.

The ethnic Greek minorities living in southern Italy today exemplify the establishment of independent and permanent colonial settlements of Greeks in history. It is a well known phenomenon that when people have travelled, traded and migrated, they have always moved their crops with them (Prance 2005), and so it is not surprising that the Greeks took

Figure 8.2. Graecanic area in the Province of Reggio di Calabria, Region of Calabria.

common crops, like olives and grapes, with them to cultivate in their new colonies (Sallares 1991: 92; Lombardo 1996; Hirschfelder 2001: 58ff.). Today, pastoralism and subsistence agriculture are still the main traditional livelihood activities in the area. In the past, the cultivation of wheat and other grains was of considerable importance. However, economic changes and emigration, as well as serious floods and earthquakes in 1951 and 1971, have led to the gradual abandoning of tillage and pastoral activities and to a sharp decline of the resident population. In addition, extensions of the mountain communities were built along the coast (*la marina*), to which many have migrated over the years (Kish 1953). Nevertheless, the gathering of wild food plants still plays an important role in the traditional diet and is an integral part of Graecanic culture (Nebel 2005).

Studying Wild Plant Use in Gallicianò

Ethnobotanical data regarding traditional food knowledge was gathered during fieldwork in Gallicianò during the spring of 2002 and 2003 and two months during the autumn of 2002. Traditional knowledge regarding food plants was assessed using standard ethnobotanical tools, including free-listing exercises (Alexiades and Sheldon 1996: 53-94; Quinlan 2005), participant observation (Martin 1995: 96) and interview techniques

(Schensul, Schensul and LeCompte 1999). In the first phase of the field research, eighteen participants were asked to list any noncultivated food plants that are used on a regular basis or were used in the past. Data obtained from the free-listing exercise were analysed using ANTHROPAC 4.72 (Borgatti 1992).

In-depth knowledge about the use of wild food plants was collected in semistructured and structured interviews (Schensul, Schensul and LeCompte 1999) with twenty-five older inhabitants of Gallicianò (average 65 years old; thirteen female, twelve male) and nine younger inhabitants (average 32 years old; five male, four female). The interviews were all conducted in Italian, as all informants in the Graecanic communities are fluent in Calabrian (an Italian dialect). Complementary fieldwork (free listing and semistructured interviews) was carried out in the other Graecanic communities of Bova, Ghorio di Roccaforte, Roghudi, Amendolea and Condofuri in 2003 (total of eleven interviews: three in Bova and two in each of the other communities).

The identification of the reported food-plant species followed Pignatti's *Flora d'Italia* (2002). Two voucher specimens of each food-plant species were deposited at the herbarium of the Centre for Pharmacognosy and Phytotherapy, School of Pharmacy, University of London.

In order to better understand the cultural and historical importance of wild food-plant species in the Graecanic area, vernacular plant names of edible plants were compared with plant names mentioned in Ancient and Modern Greek literature. In historical linguistics, a pair or set of words in related languages that have developed from the same ancestor word are known as *cognates* (Trask 1996). Such cognates are a good indication of the historical importance of species, if one compares two geographically distant but linguistically linked groups (Leonti, Sticher and Heinrich 2003). Cognates of plant names have to be treated with caution, since sometimes one plant species might have more than one folk name (over-differentiation) or two or more scientific species might have the same vernacular name (under-differentiation) (Berlin 1973). In this chapter all vernacular names for plants are in *Grecanico* unless otherwise indicated.

Wild Food Plants Traditionally Consumed in the Graecanic Area

The inhabitants of the Graecanic communities in Calabria gather forty-eight wild food-plant species, including greens, fruits, condiments and some mushrooms (Nebel, Pieroni and Heinrich 2006). The plant family contributing most members to wild-gathered food plants are the Asteraceae, with a total of twenty species (42 per cent). The second most important families are the Apiaceae, Brassicaceae and Rosaceae, with three species each.

Ethnotaxonomically, *edible greens* are an intermediate category in the domain of food plants labelled *ta chòrta* in *Grecanico*, which roughly corresponds to what in bioscientific nutritional studies is called 'green, leafy vegetables'. The same term, *ta chòrta* or *horta* (τα χόρτα[1]), was found in the literature to describe 'wild food plants' in Greece (Forbes 1976; Lambraki 2000; Couplan 2005: 200). Usually, *ta chòrta* refers to noncultivated species. However, because in many cases the distinction between wild forms of a certain species and the cultivated one is almost impossible, it is of limited relevance to the local people. Other categories in the domain of food plants (excluding fruits) are *ártema* (condiments) and *mulitária* (mushrooms).

Elderly women are the main keepers of traditional knowledge in the domain of local food plants. Women are directly involved in work with food for the family and therefore possess knowledge that relates to different aspects of food production and use. Men most notably retain specific knowledge about gathering (e.g., identification, where, when) of wild food plants. They play an important role in gathering plants and fungi that grow far from the village. The analysis of the interviews with elderly and younger inhabitants of Gallicianò showed that only very few of the younger generation are able to identify the culturally most important wild edible-plant species. To revalorize local food traditions today, a small handbook on the use of wild food plants with pictures and plant descriptions has been published for dissemination in local communities (Nebel 2005).

The gathering of wild food plants is seasonal: most are gathered during winter and spring (December–May). Informants remark that these plants are very important to their diets (Nebel, Pieroni and Heinrich 2006). They are very valuable as a vegetable substitute in early spring, as they are available several weeks before the garden varieties. Most of these wild food plants are collected by women in the *ambeli* (vineyards) or *uliveto* (olive groves), which are normally located close to the village. A few species are also collected in hedgerows, or on fallow fields. Generally, the men collect any wild species, including mushrooms, from the *bosco* (afforested areas) on the Aspromonte Mountain. Wild fruits like blackberries or prickly pears are preferably gathered and eaten by children (Nebel 2005; Tardío, Pascual and Morales 2005).

In order to determine the most culturally salient plants of the domain 'wild-gathered food plants', a free-listing exercise was conducted (Table 8.1). A high salience, as reported for *Lactuca viminea* (0.49), *Reseda alba* (0.47) or *Reichardia picroides* (0.43), reflects both a high frequency and a high rank in the informants' lists (i.e., these plants appear more often and earlier in free lists). Frequently mentioned plant species among individuals indicate common knowledge, or consensus, within the same cultural domain.

Table 8.1. The most frequently mentioned wild food species in Gallicianò, Graecanic area, Calabria, southern Italy. Free-listing exercise with eighteen informants (seven male and eleven female; aged between 18 and 88, average age 55). Analysed with ANTHROPAC, 4.72 version

Vernacular name	Scientific name	Frequency*	Average rank	Salience	Cognate Greek
Pricaddhída	*Lactuca viminea* (L.) J. Presl & C. Presl	13	3.46	0.49	yes
Másaro	*Foeniculum vulgare* subsp. *piperitum* (Ucria) Cout.	12	4.67	0.33	yes
Gattinaría	*Reseda alba* L.	11	2.72	0.47	no
Gaddhazzída	*Reichardia picroides* (L.) Roth.	10	3.00	0.43	yes
Sculímbri	*Scolymus hispanicus* L.	8	7.50	0.10	yes
Paparína	*Papaver rhoeas* L.	7	5.43	0.16	**
Źuccho	*Sonchus oleraceus* L., or *S. asper* (L.) Hill, or *Urospermum picroides* (L.) Scop ex F.W. Schmidt	7	6.29	0.16	yes
Maruddhaci	*Hypochoeris achyrophorus* L.	7	4.42	0.23	yes

* Frequency of vernacular name mentioned by the informants in free-listing exercise

** *Paparouna* in Modern Greek; *mékon roias* in Ancient Greek (Gunther 1934). Etymology of plant name unclear, probably Latin or Italian origin

The correspondence between folk-generic names and scientific names is mainly a one-to-one correspondence (Berlin 1992). However, sometimes the same phytonym is used to name more than one similar botanical species (under-differentiation), as for example, *źuccho* in *Grecanico* for *Sonchus asper*, *S. oleraceus* and *Urospermum picroides*. The same was reported in Greece, where the same three plant species are called *tsóchós*, *zochós* or *agrio zochós* (Heldreich 1862; Candargy 1889; Tzanoudakis 1982; Lambraki 2000). Interestingly, the taste, as well as the morphology, of all three *źuccho* species is alike. Furthermore, they are all consumed in the same way, as *chórta vramena* or *chórta tiganimena* (see Table 8.2).

Regional specialities and traditional dishes are important elements of local food culture. Many dishes made of young shoots, green aerial parts or basal leaves are part of the everyday cuisine (Table 8.2) and consist of a broad range of species, reflecting the seasonal aspect of gathering, as well as the aspect of taste (*pricìo* (bitter) versus *glicìo* (sweet)).

More than half of the forty-eight gathered plant species in the Graecanic area (58 per cent) are also used as food plants in Greece (Tzanoudakis 1982; Lambraki 2000; Savvides 2000; Kypriotakis 2003; Zeghichi et al. 2003). This

Table 8.2. Local dishes comprising wild food plants in the Graecanic area in Calabria, southern Italy

Name of the local dish in *Grecanico*	**Description, mode of preparation**	**Wild food-plant species used**
Chòrta vramena	Mixed wild greens (*chòrta mimmena*) boiled and seasoned with olive oil and lemon Always: *chòrta pricìa* mixed with *chòrta glicìa*	*Reichardia picroides* (L.) Roth. *Reseda alba* L. *Lactuca viminea* (L.) J. Presl & C. Presl *Papaver rhoeas* L. *Hypochoeris achyrophorus* L. *H. radicata* L. *Sonchus asper* (L.) Hill *S. oleraceus* L. *Urospermum dalechampii* (L.) Scop ex F.W. Schmidt *U. picroides* (L.) F. W. Schmidt
Chòrta tiganimena	Mixed wild greens (*chòrta mimmena*) boiled and sautéed in frying pan with olive oil, garlic and fresh chilli	Same species as *chòrta vramena*
Fasùli me ta másara	Soup with broad beans, young leaves of wild fennel, potato and pasta	*Foeniculum vulgare* ssp. *piperitum* (Ucria) Cout.
Sculímbri me ta lasagne	Leaf stalks of *Scolymus hispanicus* cooked with homemade pasta	*Scolymus hispanicus* L.
Insalata di spèlendra	Raw or cooked as salad, with spring onions, oil and vinegar	*Apium nodiflorum* (L.) Lag.
Frittata	Young shoots mixed with egg, flour and *pecorino* (sheep's cheese) and fried like an omelette	*Hirschfeldia incana* (L.) Lagr.-Foss. *Asparagus acutifolius* L. *Sinapis arvensis* L.

does not come as a surprise, since the vegetation in the Graecanic area is similar to that of Greece (Eberle 1975: 121) and the two regions have common cultural traditions. Most of these plants are also used in other parts of Italy and the Mediterranean: for example, leaves and seeds of *Foeniculum vulgare* as vegetable or condiment; young leaves of *Papaver rhoeas, Reichardia picroides, Sonchus oleraceus* and *S. asper* either eaten raw as salad or mixed and cooked with other herbs; or the leaf stalks of *Scolymus hispanicus* as vegetable (Corsi and Pagni 1979; Aliotta 1987; Paoletti, Dreon and Lorenzoni 1995; Bonet and Vallès 2002; Pieroni et al. 2002; Guarrera 2003; Zeghichi et al. 2003; Pieroni et al. 2005; Rivera et al.

2005; Scherrer, Motti and Weckerle 2005; Tardío, Pardo-de-Santayana and Morales 2006). On the other hand, there are several species, such as *Chrysanthemum segetum* L., *Hedypnois cretica* Willd. or *Lotus edulis* L., that are predominantly used in Greece and the Graecanic area, but not in other parts of Italy (Picchi and Pieroni 2005). The use of these plants might therefore be a residual element of Ancient Greek culture.

Reseda alba, *gattinaría* in *Grecanico*, is an example of a very local food plant, which is used very frequently and appreciated as a vegetable in the Graecanic area, but not in the surrounding Italian communities, where the plant is considered to be too bitter in taste. In Gallicianò the tops of the shoots are either eaten raw, seasoned with olive oil, or cooked and fried in olive oil with garlic, chilli and pepper mixed with other wild greens. In the literature, only two references to the use of *Reseda alba* as food in the Mediterranean were found: first, young leaves of *Reseda alba* used as a vegetable in Greece (Heldreich 1862: 79), and second, as salad by the villagers of the surrounding area of Larnaca in Cyprus (Arnold Apostolides 1991). Interestingly, both records are from regions of the eastern Mediterranean, which were, in historic times, part of the Greek and Byzantine worlds, as was the Graecanic area in southern Italy.

Taste is a key criterion for perceiving, categorizing and characterizing food plants in general (Grivetti 1981; Johns 1986; Nebel 2001). The inhabitants of the Graecanic area often ascribe a specific taste to singular plant species, particularly to *ta chòrta* species. For example, the taste of *Reseda alba* or *Lactuca viminea* is described as *pricìo* (bitter) and that of *Papaver rhoeas* as *glicìo* (sweet). The prefix of the vernacular name of *Lactuca viminea*, *pricaddhída*, indicates the bitterness of the plant. Local gatherers intentionally collect both bitter and sweet herbs to assure a balanced taste of the dishes to be prepared. Generally, bitterness plays a very important role in the local perception of health. The bitter taste of wild greens is perceived as healthy in the sense of 'blood clearing' and 'good for the liver'. In Roghudi, edible greens are also called *chòrta pricía* (bitter wild greens). Already in 1862, Heldreich reported that many plants from the Asteraceae family were considered to be healthy by the Greeks because of their bitterness (Heldreich 1862: 28). Pieroni et al. (2002) reported similar results among ethnic Albanians in southern Italy.

Food as Medicine

Wild food plants are regarded as healthy by most of the informants, and some are also specifically used as medicine (see Table 8.3). The beneficial health effects were described as depurative (help to remove toxins from the blood or internal organs), anti-hypertensive, digestive, diuretic, anti-arthritic, anti-diabetic, anti-cancer or to make one 'feel better'.

Table 8.3. Food plants used as medicine in Gallicianò, Calabria, southern Italy

Scientific plant name	Vernacular name	Family	Part(s) used	AD	Medicinal use
Asparagus acutifolius L.	*Asparagi*	Liliaceae	Young shoots	i	Diuretic
Foeniculum vulgare Mill.	*Spiro másaro*	Apiaceae	Seeds	i	Against digestive disorders
Hypochoeris radicata L.	*Costardeddhe*	Asteraceae	Leaves	i	Blood-cleansing
Lactuca viminea (L.) J. Presl & C. Presl	*Pricaddhída*	Asteraceae	Whorls	i	Perceived as healthy because of bitterness
Laurus nobilis L.	*Lauro*	Lauraceae	Leaves	i	Against stomach ache, digestive
Mentha sp.	*Menta*	Lamiaceae	Leaves	i	Against colds
Opuntia ficus-indica (L.) Mill.	*Sico tu trucu, ficarazzi*	Cactaceae	Flower	i	Diuretic
Reichardia picroides (L.) Roth	*Gaddhazzída*	Asteraceae	Fresh leaves	e	Tooth ache
Reseda alba L.	*Gattinaría*	Resedaceae	Aerial part	i	Perceived as healthy because of bitterness
Rosa canina L.	*Rosa*	Rosaceae	Young leaves	e	Against herpes, especially for young children
Silene vulgaris (Moench) Garcke	*Cavuráci*	Caryophyllaceae	Leaves	i	Diuretic

AD - Administration: e = external application; i = internal application

The consumption of wild food plants is highly relevant for health, as they often contain higher amounts of bioactive compounds than plants that have been under cultivation for many generations (Lionis et al. 1998; Tumino et al. 2002; Stepp 2004; Leonti et al. 2006). Trichopoulou et al. (2000) showed that eight wild-green species consumed in Crete have a very high flavonoid content when compared with cultivated fresh vegetables, fruits and beverages commonly consumed in Europe.

Historical Importance of Wild Food Plants

The comparison of ethnobotanical data with historical written sources in Ancient Greek (e.g., *Historia Plantarum* from Theophrastus of Eresos, or *de Materia Medica* from Pedanius Dioscorides) suggests that people have used several wild food plants since antiquity. The philosopher Plato, in his *Republic*, gives an example of a healthy diet including olive oil, cereals, fruits and 'country dishes of wild roots and vegetables' (Ferrari 2000: 55). This and other passages permit us to conclude that the Ancient Greek diet included a wide variety of wild greens, vegetables and fruits (Vickery 1936; Brothwell and Brothwell 1969; Fidanza 1979; Sallares 1991; Lombardo 1996; Pearson 1997).

A comparison of the vernacular names of food plants that are used both in the Graecanic area and Greece found fourteen cognates of plant species. *Portulaca oleracea* L. (purslane), for example, is used as salad and known as *andrácla* or *andrákla* in the Graecanic area. To the Ancient Greeks, purslane was most probably known as *andrachní* (*ανδράχνη*) (Hort 1916: 61 VII.I.2; Bois 1927: 66; Gunther 1934, No 150) and in Modern Greek purslane is called *andrakla* or *glystrídha* (Candargy 1889: 54; Lambraki 2000; Lardos 2006). In the past and today, the same culinary use is reported for the stems and fleshy leaves, which are eaten fresh in salads or baked in pies (Waterlow 1989; Brussell 2004).

Another example of a cognate is *Foeniculum vulgare* subsp. *piperitum* (wild fennel), known as *másaro* in Gallicianò and *agriomáratho* (*αγριομάραθο*) in Greek (Tzanoudakis 1982; Lambraki 2000). The word *marathon* (*μάραθον*) was found in a list of condiments in Ancient Greek literature and is the Ancient Greek name for wild fennel (Chantraine, Masson and Lejeune 1968: 666). Presumably, the Greek town Marathon was named after the abundance of fennel in the area (Lambraki 2000).

These results show that phytonyms of several wild plant species, in conjunction with their specific food uses, have been preserved for centuries. This might be taken as an indication that these plants have been continuously used as vegetables, condiment or salads at least since Byzantine times and possibly even since the time Magna Graecia flourished in southern Italy.

Conclusion

The Graecanic villages, and specifically Gallicianò, are examples of rapidly changing communities where local traditions compete with modern ways of life. The present study demonstrates how the traditional consumption of wild food-plant species is still deeply embedded in the local culture, providing a strong link between local people and their natural environment. The habit of collecting and cooking edible, noncultivated plants is still alive among the older generation. Their continued appreciation of these traditions can be ascribed to many factors, such as the connection to their 'sense of local/cultural identity', an appreciation for unusual and bitter tastes, as well as the perceived health benefits of the plants. However, most young people today are unable to identify or gather many of the culturally important wild food plants of the Graecanic area. Nevertheless, younger people still eat wild food plants collected and prepared by their parents, as supplementary vegetables. It remains to be seen how threatened this knowledge really is. Perhaps at some time in their adult lives they may develop a desire to learn about the foods of their youth and then one can only hope that there are those around who can teach them.

Several of these wild-gathered food plants are consumed as food because of attributed health benefits, or used both as medicine and food. Thus our findings also demonstrate the convergence of medicinal and culinary practices, which is increasingly recognized as critical for healthy diets everywhere and represents an important intellectual legacy of our ancestors.

This legacy is amply demonstrated by the incredible continuity in plant names and uses from Ancient Greek cultural traditions down through the centuries and across the seas to southern Italy and beyond, presumably transmitted orally from mother to daughter, generation after generation. Given the dynamic history of Europe during this long time it is remarkable to find such cultural continuity, surely a marvellous testament to the importance of wild plants in human lives both in the past, today, and hopefully well into the future.

Acknowledgements

Thanks to the inhabitants of the Graecanic area, particularly the inhabitants of Gallicianò, for their enthusiasm for this project and willingness to share their knowledge with Sabine Nebel, the field researcher; to all the members of the Cumelca Association; to the Mayor of Condofuri, Nucera Giovanni S. and the town councillor, Manti Natale, as well as to all members of the consortium 'Local Food-Nutraceuticals'. The present field research was only possible thanks to the School of Pharmacy, University of London, the British Federation of Woman Graduates (BFWG) and the European Commission, DG Research, for funding under the programme: Quality of Life and Management of Living Resources, key action 1: Food, Nutrition and Health ('Local Food – Nutraceuticals', contract number: QLK1-CT-2001-00173, coordinator: M. Heinrich).

Note

1. Transliteration of Greek characters into Latin characters according to ISO 843: 1997 (http://transliteration.eki.ee/pdf/Greek.pdf). Accessed 8 August 2005.

References

Alexiades, M.N. and J.W. Sheldon. 1996. *Selected Guidelines for Ethnobotanical Research: a Field Manual.* New York: The New York Botanical Garden.

Aliotta, G. 1987. 'Edible Wild Plants of Italy', *Informatore Botanico Italiano* 19: 17–30.

Arnold Apostolides, N. 1991. *Ethnobotanique et Ethnopharmacologie de la flore de Chypre et de l'Est méditerranéen.* Bailleul: Publications du Centre Régional de Phytosociologie.

Balée, W.L. 1998. *Advances in Historical Ecology.* New York: Columbia University Press.

Berlin, B. 1973. 'Folk Systematics in Relation to Biological Classification and Nomenclature', *Annual Review of Ecology and Systematics* 4: 259–271.

——— 1992. *Ethnobiological Classification: Principles of Categorization of Plants and Animals in Traditional Societies.* New Jersey: Princeton University Press.

Bois, D. 1927. *Les plantes alimentaires chez tous les peuples et à travers les âges: Histoire, utilisation, culture, phanerogames légumières.* Paris: Lechevalier.

Bonet, M.A. and J. Vallès. 2002. 'Use of Non-crop Food Vascular Plants in Montseny Biosphere Reserve (Catalonia, Iberian Peninsula)', *International Journal of Food Sciences and Nutrition* 53: 225–248.

Borgatti, S.P. 1992. *ANTHROPAC 4.0 Methods Guide.* Columbia: Analytic Technologies.

Brothwell, D.R. and P. Brothwell. 1969. *Food in Antiquity: a Survey of the Diet of Early Peoples.* London: Thames & Hudson.

Brussell, D.E. 2004. 'Medicinal Plants of Mt. Pelion, Greece', *Economic Botany* 58: S174–S202.

Candargy, C.A. 1889. *Flore de l'île de Lesbos: Plantes sauvages et cultivees.* Uster-Zürich: Diggelmann.

Cerchiaia, L., L. Jannelli and F. Longo. 2004. *Die Griechen in Süditalien: Auf der Spurensuche zwischen Neapel und Syrakus.* Stuttgart: Konrad Theiss Verlag.

Chantraine, P., O. Masson and M. Lejeune. 1968. *Dictionnaire étymologique de la langue grecque histoire des mots.* Paris: Klincksieck.

Condemi, F. 1999. *Galliciano: Acropoli della Magna Grecia.* Reggio Calabria: Laruffa.

Corsi, G. and A.M. Pagni. 1979. 'Studi sulla flora e vegetazione del Monte Pisano (Toscana Nord-Occidentale). V. Le piante spontanee nella alimentazione popolare', *Atti della Società Toscana di Scienze Naturali, Memorie, Serie B* 86: 79–101.

Couplan, F. 2005. *Ce sont les plantes qui sauvent les hommes: ma botanique gourmande.* Paris: Plon.

Eberle, G. 1975. *Pflanzen am Mittelmeer: Mediterrane Pflanzengemeinschaften Italiens und Griechenlands mit Ausblick auf das ganze Mittelmeergebiet.* Frankfurt a.M.: Kramer.

Ferrari, G.R.F. (ed.). 2000. *Plato: the Republic.* Cambridge: Cambridge University Press.

Fidanza, F. 1979. 'Diets and Dietary Recommendations in Ancient Greece and Rome and the School of Salerno', *Progress in Food and Nutrition Science* 3: 77–99.

Forbes, M.H.C. 1976. 'Gathering in the Argolid: a Subsistence Subsystem in a Greek Agricultural Community', *Annals of the New York Academy of Science* 268: 251–264.

Grivetti, L.E. 1981. 'Cultural Nutrition: Anthropological and Geographical Themes', *Annual Review of Nutrition* 1: 47–68.

Guarrera, P.M. 2003. 'Food Medicine and Minor Nourishment in the Folk Traditions of Central Italy (Marche, Abruzzo and Latium)', *Fitoterapia* 74: 515–544.

Gunther, R.T. (ed.) 1934. *The Greek Herbal of Dioscorides / Dioskurides Pedanios.* Oxford: Oxford University Press.

Heinrich, M., M. Leonti, S. Nebel and W. Peschel. 2005. '"Local food – Nutraceuticals": an Example of a Multidisciplinary Research Project on Local Knowledge', *Journal of Physiology and Pharmacology* 56: 5–22.

Heldreich, T. 1862. *Die Nutzpflanzen Griechenlands: Mit besonderer Berücksichtigung der neugriechischen und pelasgischen Vulgarnamen.* Athen: Karl Wilberg.

Hirschfelder, G. 2001. *Europäische Esskultur eine Geschichte der Ernährung von der Steinzeit bis heute.* Frankfurt and New York: Campus Verlag.

Hort, A. 1916. *Theophrastus – Enquiry into Plants and Minor Works on Odours and Weather Signs.* Vol 1 & Vol 2. London: William Heinemann.

Johns, T. 1986. 'Chemical Selection in Andean Domesticated Tubers as a Model for the Acquisition of Empirical Plant Knowledge', in N. Etkin (ed.), *Plants in Indigenous Medicine and Diet: Behavioural Approaches.* New York: Redgrave, pp. 266–288.

———, H.M. Chan, O. Receveur and H.V. Kuhnlein. 1994. 'Nutrition and the Environment of Indigenous Peoples', *Ecology and Food and Nutrition* 32: 81–87.

Kish, G. 1953. 'The "Marine" of Calabria', *Geographical Review (New York)* 43: 495–505.

Kuhnlein, H.V. and O. Receveur. 1996. 'Dietary Change and Traditional Food Systems of Indigenous Peoples', *Annual Review of Nutrition* 16: 417–442.

Kypriotakis, Z. 2003. Personal communication.

Lambraki, M. 2000. *Ta Xopta.* Athens: Ellinika Grammata [in Greek].

Lardos, A. 2006. 'The Botanical *Materia Medica* of the *Iatrosophikon* – a Collection of Prescriptions from a Monastery in Cyprus', *Journal of Ethnopharmacology* 104: 387–406.

Leonti, M., S. Nebel, D. Rivera and M. Heinrich. 2006. 'Wild Gathered Food Plants in the European Mediterranean: a Comparative Analysis', *Economic Botany* 60(2): 130–142.

———, O. Sticher and M. Heinrich. 2003. 'Antiquity of Medicinal Plant Usage in Two Macro-Mayan Ethnic Groups (Mexico)', *Journal of Ethnopharmacology* 88: 119–124.

Lionis, C., A. Faresjo, M. Skoula, M. Kapsokefalou and T. Faresjo. 1998. 'Antioxidant Effects of Herbs in Crete', *The Lancet* 352: 1987–1988.

Lombardo, M. 1996. 'Food and "Frontier" in the Greek Colonies in Southern Italy', in J. Wilkins (ed.), *Food in Antiquity.* Exeter: University of Exeter Press, pp. 256–272.

Martin, G.J. 1995. *Ethnobotany: a Methods Manual.* London: Chapman & Hall.

Nebel, S. 2001. 'Arbereshe Taste Perception of Wild Food Plants', M.Sc. dissertation. Canterbury: Department of Anthropology, University of Kent.

——— 2005. *Ta chòrta: Piante commestibili tradizionali di Gallicianò.* London: University of London, School of Pharmacy.

———, A. Pieroni and M. Heinrich. 2006. '*Ta Chórta*: Wild Edible Greens Used in the Graecanic Area in Calabria, Southern Italy', *Appetite* 47(3): 333–342.

Nestle, M. 1995. 'Mediterranean Diets: Historical and Research Overview', *American Journal of Clinical Nutrition* 61 (Suppl.): 1313S–1320S.

Noah, A. and A.S. Truswell. 2001. 'There Are Many Mediterranean Diets', *Asia Pacific Journal of Clinical Nutrition* 10: 2–9.

Ogoye-Ndegwa, C. and J. Aagaard-Hansen. 2003. 'Traditional Gathering of Wild Vegetable among the Luo of Western Kenya – a Nutritional Anthropology Project', *Ecology and Food and Nutrition* 42: 69–89.

Pan, C. and B.S. Pfeil. 2000. *Die Volksgruppen in Europa: Ein Handbuch.* Vienna: Braumueller.

Paoletti, M.G., A.L. Dreon and G.G. Lorenzoni. 1995. '*Pistic*, Traditional Food from Western Friuli, N.E. Italy', *Economic Botany* 49: 26–30.

Pearson, K.L. 1997. 'Nutrition and the Early-Medieval Diet', *Speculum* 72: 1–32.

Picchi, G. and A. Pieroni 2005. *Le erbe: Atlante dei prodotti tipici.* Roma: AGRA, Rai-Eri.

Pieroni, A., S. Nebel, C. Quave, H. Munz and M. Heinrich 2002. 'Ethnopharmacology of *Liakra*: Traditional Weedy Vegetables of the Arbereshe of the Vulture Area in Southern Italy', *Journal of Ethnopharmacology* 81: 165–185.

———, S. Nebel, R.F. Santoro and M. Heinrich. 2005. 'Food for Two Seasons: Culinary Uses of Non-cultivated Local Vegetables and Mushrooms in a South Italian Village', *International Journal of Food Sciences and Nutrition* 56: 245–272.

Pignatti, S. 2002. *Flora d'Italia.* Bologna: Edizioni Edagricole.

Posey, D.A. 2002. *Kayapó Ethnoecology and Culture* 6. New York: Routledge.

Prance, G.T. 2005. 'The Migration of Plants', in G.T. Prance and M. Nesbitt (eds), *The Cultural History of Plants.* New York: Routledge, pp. 27–28.

Quinlan, M. 2005. 'Considerations for Collecting Freelists in the Field: Examples from Ethnobotany', *Field Methods* 17: 219–234.

Rivera, D., C. Obón, C. Inocencio, M. Heinrich, A. Verde, J. Fajardo and R. Llorach. 2005. 'The Ethnobotanical Study of Local Mediterranean Food Plants as Medicinal Resources in Southern Spain', *Journal of Physiology and Pharmacology* 56: 97–114.

Rother, K. (ed.). 1989. *Minderheiten im Mittelmeerraum.* Passau: Passauer Mittelmeerstudien.

Sallares, R. 1991. *The Ecology of the Ancient Greek World.* Ithaca, New York: Cornell University Press.

Savvides, L. 2000. *Edible Wild Plants of the Cyprus Flora.* Nicosia: private publication.

Schensul, J.S., J.J. Schensul and D.M. LeCompte. 1999. *Essential Ethnographic Methods: Observation, Interviews, and Questionnaires.* London and New Dehli: AltaMira Press, Walnut Creek.

Scherrer, A.M., R. Motti and C.S. Weckerle. 2005. 'Traditional Plant Use in the Areas of Monte Vesole and Ascea, Cilento National Park (Campania, Southern Italy)', *Journal of Ethnopharmacology* 97: 129–143.

Stepp, J.R. 2004. 'The Role of Weeds as Sources of Pharmaceuticals', *Journal of Ethnopharmacology* 92: 163–166.

Tardío, J., M. Pardo-de-Santayana and R. Morales. 2006. 'Ethnobotanical Review of Wild Edible Plants in Spain', *Botanical Journal of the Linnean Society* 152(1): 27–72.

———, H. Pascual and R. Morales. 2005. 'Wild Food Plants Traditionally Used in the Province of Madrid, Central Spain', *Economic Botany* 59: 122–136.

The Local Food-Nutraceuticals Consortium. 2005. 'Understanding Local Mediterranean Diets: A Multidisciplinary Pharmacological and Ethnobotanical Approach', *Pharmacological Research* 52: 353–366.

Trask, R.L. 1996. *Historical Linguistics.* London: Oxford University Press.

Trichopoulou, A., E. Vasilopoulou, P. Hollman, C. Chamalides, E. Foufa, T. Kaloudis, D. Kromhout, P. Miskaki, I. Petrochilou and E. Poulima (2000) 'Nutritional Composition and Flavonoid Content of Edible Wild Greens and Green Pies: a Potential Rich Source of Antioxidant Nutrients in the Mediterranean Diet', *Food Chemistry* 70: 319–323.

Tumino, R., G. Frasca, M.C. Giurdanella, C. Lauria and V. Krogh. 2002. 'Consumption of Wild Vegetables in the EPIC Cohort of Ragusa (Sicily)', in E. Riboli and R. Lambert (eds), *Nutrition and Lifestyle: Opportunities for Cancer Prevention*. Lyon: International Agency for Research on Cancer, pp. 115–116.

Tzanoudakis, D. 1982. 'Popular Names of Cretan Plants [Greek]', *Kritologia* 14–15: 154–164.

Vickery, K.F. 1936. 'Food in Early Greece', *Illinois Studies in the Social Sciences* 20: 1–97.

Waterlow, J.C. 1989. 'Diet of the Classical Period of Greece and Rome', *European Journal of Clinical Nutrition* 43: 3–12.

Zeghichi, S., S. Kallithraka, A.P. Simopoulos and Z. Kypriotakis. 2003. 'Nutritional Composition of Selected Wild Plants in the Diet of Crete', in A.P. Simopoulos and C. Gopalan (eds), *Plants in Human Health and Nutrition Policy*, Vol. 91. Basel: Karger, pp. 22–40.

CHAPTER 9

The Ecology and Use of Edible Thistles in Évora, Alentejo, Southeastern Portugal

MARIA JOSÉ BARÃO AND ALEXANDRA SOVERAL DIAS

Introduction

The Alentejo region in South Portugal lies between the Tagus River and Algarve, representing about one third of the area and only five per cent of the population of Portugal. It is a semiarid region of undulating plains with a Mediterranean climate softened by the mild Atlantic sea breezes, with mild winters and hot, dry summers. With a long history of scarcity and poverty, it shows very particular cultural traits that made Feio (1983) call it 'a country within a country'. Among its defining traits is a unique culinary tradition, generally considered a consequence of necessity, with a high use of bread-based dishes and wild plants, such as wild thistles.

This chapter describes the results of a small ethnobotanical research project conducted in the region surrounding the city of Évora, in the central part of Alentejo. The objective of our study was to gain an initial understanding of the use of wild thistles as food in a small area near the city of Évora where thistles are commonly consumed and also traded in the more traditional markets. We interviewed a small sample of gatherers and sellers of thistles, visited and described the sites where they collect and observed the procedures followed in the harvest and use of thistles. Given that several thistles occur in the area, one of our main goals was to identify which species were in fact used by the local population. Thistles are valued for their high mineral content, and it could be expected that the nutritional quality would vary across the collecting range according to soil mineral content. Thus, we sampled and tested plant and soil mineral content across the region and through time.

Even though this was a small study and in some ways rather preliminary, our findings are surprising in many ways. Only one species is in fact harvested, and its abundance and nutritional quality do not appear to vary across the region, despite some variation in soil conditions. There appears to be little threat to the sustainability of the wild thistle populations, as there are no harvesting regulations or conservation measures in place, though harvesters do have their collecting territories: they harvest only the aerial shoots leaving behind the underground parts that can send up new shoots in the following season. Besides, we found that an alteration in thistles' edibility constrains its human consumption and thereby reduces pressure on the wild thistle populations. We discuss the implications of these findings and the possibilities for the expansion of markets for edible thistles, one potential pathway for sustainable economic development in the region.

Edible Thistles in Alentejo

Edible thistles, locally known as *cardos*, *cardinhos* and also *cagarrinhas* or *tagarrinhas*, are traditionally gathered and consumed in Alentejo in a number of traditional recipes. However, the local names *cardos* and *cardinhos*, derived from the Latin word *carduus* (thistle), are used in a very broad sense, being applied to a variety of weedy, herbaceous, prickly plants, generally species of Apiaceae, Dipsacaceae and Asteraceae (e.g., Coutinho 1939; Fernandes and Carvalho 2003). At least forty-five species of the Portuguese flora belonging to various genera like *Atractylis*, *Carduus*, *Carlina*, *Carthamus*, *Centaurea*, *Cirsium*, *Cnicus*, *Cynara*, *Dipsacus*, *Echinops*, *Eryngium*, *Galactites*, *Scolymus* and *Silybum* may, actually, be called *cardos* as they are all herbaceous, prickly plants.

Edible thistles were reported in Greco-Roman cuisine and several species were probably used in Alentejo during the Roman occupation, such as *Eryngium campestre* L., whose 'roots were consumed like carrots', or *Scolymus hispanicus* L. and *Scolymus maculatus* L., from which 'the green newborn portions were used like asparagus' (Andrés 1961). In Apicius's *The Market-gardener*, Book III of *The Cookery Book* written in the first century AD, there are recipes for thistles, considered as *holera* or 'edible greens' (Apicius 1997).

The oldest known reference to the use of the cultivated cardoon (*Cynara cardunculus* L.) is from the first century AD. The cardoon leaves and heads were boiled or fried with 'a lot of spices' and sold in the markets, but the heads were less prized than the leaves (Andrés 1961). Both were available only to wealthy people, while the poorer people gathered wild thistles for the same purpose (Apicius 1997).

Between the eighth and thirteenth centuries the Arabs ruled the south of the Iberian Peninsula including the area known today as Alentejo (Mattoso 1992). Two recipes with thistles were mentioned in a Hispano-Arabic manuscript (Miranda 1957) dating from the second half of the twelfth century, or the first quarter of the thirteenth century (Rei 2000). Considering the exuberance of the Arabic urban markets, where a great variety of greens from wild harvesting were available (Rosenberger 1998), it seems probable that the use of thistles in Alentejo increased during the Arabic occupation.

In this part of the country, both in the past and today, the use of thistles, as well as other wild edible plants, is frequently associated with poverty, especially hard in past times when poor rural workers had very low or no income during part of the year (Lúcio 1987; Rodrigues 1996; Alves 1997; Brito 1997; Ramos 1997; Pinheiro 2000). Despite some references to thistles' use in other Portuguese regions (Vasconcellos 1967), in no other place have they attained the importance found in Alentejo. However, their use seems also to be restricted to the interior part of the region. In a study among rural workers in south Alentejo, thistles were found to represent a very small fraction (0.14 per cent) of the foods present in the meals in the interior, but were not consumed at all on the coast (Carvalho and Gomes 1973).

In cookery books the more common recipes involving thistles include bean or chickpea dishes (Picão 1947; Lúcio 1987; Alves 1994; Pinheiro 1995; Ramos 1997; Pinheiro 2000; Pulga 2001), bread soups with beans (Alves 1994; Valente 1994; Pinheiro 1995; Rodrigues 1996; Pinheiro 2000) and bread soups with cod (Fialho 1995; Guedes 2000; Pinheiro 2000; Câmara Municipal de Évora, Confraria Gastronómica do Alentejo and Saramago 2001).[1] More rare is their use with rice or vinegar (Lúcio 1987), or cooked, spiced with coriander and served in small bundles (Saramago 1997).

The use of stems and leaf midribs of various edible thistles growing in the study area is referred to in the literature. The golden thistle, *Scolymus hispanicus*, is one such species often used in Alentejo (Salgueiro 2004) and Eurasia (Feijão 1960; Díaz Robledo 1981; Rosselló 1999; Ertuğ 2004; Tardío, Pascual and Morales 2002), as is the spotted golden thistle, *Scolymus maculatus* (Nuez and Hernández Bermejo 1994; Arcidiacono, Pavone and Salmeri 1996). Cardoon, *Cynara cardunculus*, grown since Antiquity for petioles and fleshy leaf midribs (Couplan 1998), is also eaten in Alentejo (Salgueiro 2004). The young roots and stems and leaf midribs, after being artificially blanched, are highly appreciated snacks (Feijão 1960). In addition, the flowers of cardoon are still commonly used as a fermenting agent for making cheese.

Though an old reference, there is a report that the young receptacles and the fleshy bases of the involucral leaves of the globe artichoke, *Cynara scolymus* L., may be eaten either raw or cooked (Hill 1952). Although the consumption of the immature heads is not common in Portugal, a priori we could not exclude *Cynara scolymus* from the list of species potentially used.

The young leaves of milk thistle, *Silybum marianum* (L.) Gaertn., are also reportedly used in Alentejo (Salgueiro 2004), including in the area of this study (Pinheiro 2004). They are consumed fresh in salads (Pinheiro 2004; Salgueiro 2004) and cooked in soups or other dishes (Salgueiro 2004). Roots and heads are simply boiled in water and eaten (Pinheiro 2004).

The leaves, young stems and roots of *Eryngium campestre*, despite their toxicity when consumed in great quantities, are prepared in diverse ways in Europe and Asia (Feijão 1960; Rosselló 1999; Ertuğ 2004; Tardío, Pardo-de-Santayana and Morales 2006). In Alentejo, the young stems are used in salads or cooked in several ways (Salgueiro 2004).

The young shoots and flower stalks of *Cirsium arvense* (L.) Scop. and *Cirsium vulgare* (Savi) Ten. are known to be edible (Berglund and Bolsby 1971; Launert 1981), and roots and young tender stems of various local species of *Cirsium* were consumed by North American indigenous peoples (Kuhnlein 1992; Couplan 1998). However, we could not find any reports on the use of edible *Cirsium* species in Alentejo, or more generally in Portugal.

Tender young stems and leaves of various species of several genera of thistles such as *Carduus*, *Carlina* and *Onopordum* are also reported as edible, as are the young shoots of *Centaurea calcitrapa* L., the star thistle (Couplan 1998). However, although *Centaurea calcitrapa* and various *Carduus* and *Carlina* species are present in the study area, we could not find any reports of its use in Portugal.

Finally, *Onopordum* species, such as *Onopordum acanthium* L. – whose stems and basal leaf midribs may be used raw or cooked (Bremness 1993) and are consumed cooked in the Madrid area (Tardío, Pascual and Morales 2002) – or *Onopordum illyricum* L. – whose roots and basal leaf midribs are used cooked in the Bodrum area of Turkey (Ertuğ 2004) – are not present or are very rare in Alentejo (Franco 1984).

Studying Wild Thistles in Évora

During 2001, exploratory and semistructured interviews were conducted with people who sell or harvest thistles within a 40 km radius of the city centre, considered as influenced by Évora (Gaspar 1972). Harvest sites were then visited, once in the company of the gatherers and then alone, to collect soil and plant samples for laboratory analyses and voucher specimens to identify all thistle species found in the area.

Informants were recruited in the markets of Évora and from previous acquaintances. Thirteen informants, who either sell or harvest thistles, were interviewed about the practice of harvesting.[2] We wanted to know where, when, how and why thistles are harvested. The informants were also asked to relate how they were introduced to this activity and how their practical knowledge was acquired. The criteria used to recognize harvested plants and to choose the harvesting sites were investigated and basic biographical data such as age, education, professional activities and immediate family composition were also obtained.

We soon realized, however, that we had a problem. More than half of our informants avoided or refused to tell us where they actually harvest thistles! They always found some good excuses to keep their harvest places secret. In fact, all of them were sellers at local markets, and involved in traditional commerce more generally, although they claimed to also harvest thistles for home consumption. Given their business interests, there are several explanations for such reticence: perhaps they do not actually do the harvesting, or harvesting is in fact illegal where they go, or perhaps they are suspicious that the interviewers may use this information and increase the competition for scarce resources or a restricted market.

On the other hand, the remaining six informants, who collected thistles mainly for home consumption and as gifts, were eager to show us the sites where they usually did their harvesting. Considering this constraint, we divided the informants into two classes: sellers and gatherers. Sellers were those that did not show us their harvesting sites and from them we obtained information about thistle preparation and its sale in bundles in the market. Gatherers were those who, in addition to providing similar information, allowed us to visit their harvesting sites.

Wild thistles are shared or sold in bundles made up of several plants, each plant consisting of a basal rosette of leaves. The leaf blade is stripped off and the leaves present in the bundles are reduced to the midrib, which is, together with the petioles, the only part used for human consumption. A raffia or nylon thread is used to bind the plants together. Bundles of thistles were observed in the markets and in the gatherers' hands. We acquired some by purchase or offer during the 'thistle season' of 2001, between February and April, and photographed and examined them in order to investigate the variation in the number of plants per bundle and leaves per plant throughout the season (Figure 9.1).<fig 9.1 near here>

Various visits were made with the informants to seven harvesting sites, but only six were used in this study because in the seventh the gatherer had finished collecting all the thistles currently available to be gathered. He explained that he needed all the plants to offer to neighbours.

The informants were asked about their sites and some observations were made in order to characterize the specifics of harvest, namely the land use, area, slope, vegetation, abundance and distribution of the species collected. Additionally, at each site plant and soil samples were collected. Several plants indicated by the informants as *cardos* ('good to use') were selected and tagged in order to wait for blooming and fruiting to allow for botanical identification. Mature plants were collected in June, observed using a stereomicroscope and identified from local floras (Coutinho 1939; Franco 1984). Pressed voucher specimens were taken to the Herbarium at the Departamento de Biologia, Universidade de Évora (UEVH), where the identification was confirmed.

For the purposes of laboratory analyses, fourteen plant samples were collected at six harvesting sites (between one and four samples per site, depending on the area). A plant sample consisted of the shoots of five plants indicated as 'good to harvest and to eat' by the informants. The shoots were harvested, kept in a plastic bag and taken to the laboratory where they were prepared as much as possible in the same manner as the informants would do. The plants at that stage had a basal rosette of leaves but only the petioles and midribs are taken for human consumption. Therefore, the leaf blades were stripped off to obtain just the central veins, which were then scraped with a knife in the same way they would be prepared for eating.

A standard laboratory procedure was then followed. The veins were washed with an aqueous solution with 0.1 per cent detergent and washed again and again. After being washed finally with distilled water, the veins were cut in segments of about 3 cm long and oven-dried at 65°C for 48 hours. After cooling, 2 g of each plant sample were taken, reduced to powder with a grinder and again oven-dried at 105°C for 6 hours. The dry weight was again determined by weighing with a Mettler H35 AR balance, and all analytical results were expressed relative to this dry weight.

Determinations of the content of phosphorous (P), potassium (K), calcium (Ca), magnesium (Mg), sodium (Na), copper (Cu), zinc (Zn) and boron (B) were made from a solution of ashes prepared as described in Barão (2003b). Mineral content was determined using a UV/V Hitachi U-2000 spectrophotometer, a Jenway PFP7 flamephotometer and a Perkin-Elmer 2380 spectrophotometer of atomic absorption (Martí and Munoz 1957; Instituto Agronômico de S. Paulo 1977; Knudsen 1980). The total nitrogen (N) was determined by the Kjeldahl method (Bremmer and Mulvaney 1982). Cellulose and hemicellulose content were calculated from determinations of NDF (Neutral Detergent Fibre) and ADF (Acid Detergent Fibre) using an adaptation of the method of Robertson and Van Soest (1980 cited in Simões 2000).

For each plant sample a soil sample was obtained by mixing twenty-five soil subsamples. The soil subsamples were collected (five subsamples of soil for each plant included in the plant sample) within a 50 cm radius from the base of the shoot, using an auger of 17.5 mm diameter between 0 and 25 cm deep, as recommended by Ritas and Melida (1985). The granulometry was determined with the help of a Sedigraph 5100 sedimetometer (Micromeritics 1988 and Póvoas and Barral 1992, cited by Alexandre, Marques and Ferreira 2001). Organic matter was determined by Anne's method (Anne 1945) and pH (H_2O) was measured by the potenciometric method in a water suspension (1/2,5, v/v) (Piper 1950; Black 1965) using a Crison micropH 2001.

Available phosphorous and potassium were extracted by the Egner-Riehm method (Riehm 1958) and measured as described above. The ammonium acetate method (Thomas 1982) was used to extract cations (Ca^{2+}, Mg^{2+}, K^+ and Na^+) from the soil. Copper and zinc were determined by the Larkanen and Ervio method (FAO 1980; Cottenie et al. 1982).

The relationships between plants and soil mineral content were investigated using parametric correlation analysis. Comparisons were made between the mineral fractions and mineral content of different soils, and between the mineral content of the leaf midribs of plants from different soils using two-tailed Student-t tests. Significance levels of $P=0.05$ were used throughout. The coefficient of variation (CV) corrected for sample size as proposed by Sokal and Rohlf (1995) was used to characterize the homogeneity of the samples of plants and soils.

Thistle Gatherers and Sellers

From our interviews we found that all six gatherers in this study were middle-aged men, born and still resident in the area, and all had a professional activity (agriculture, husbandry or extractive industries) or hobby (hunting or fishing) connected with the land. They were generally married, with one or more sons, who had been at school for four years or more. Their ages ranged from forty-one to seventy-three years, with a mean age of fifty-eight years old. They harvest thistles for their own consumption and as gifts, and in one case also to sell to a restaurant by order. None of the gatherers could tell us precisely when they began or who taught them to harvest and use thistles. This would indicate a very early familiarity, probably starting before four or five years old.

Three of the seven sellers were women, all living in the study area. The mean age of sellers was estimated at about fifty-six years old. Six sellers were married, and five of them had at least a basic level of education. Only one of the sellers could not remember how he learned this practice. The others told us that they learned to harvest and use thistles from their parents.

Local Knowledge of Thistle Biology

Both sellers and gatherers said that 'the thistles [*cardos, cardinhos, cagarinhos*] are all the same' meaning that from their point of view the thistles that are harvested and sold are morphologically similar plants. However, their knowledge about some aspects of the biology of the plants, for instance the colour of the flowers, was variable, especially among the sellers. None could provide a good description of the inflorescences. Three knew that the flowers were yellow but one seller talked about a white flower, another about a blue-purplish flower and two of them disregarded the flower colour as important. These difficulties suggested that sellers' relation with the plants might be restricted to the harvesting period when no flowers are present.

Even for the gatherers, their knowledge about the biology of the thistles they harvested seemed sometimes obscure. They could not clearly explain how they identified the plants and only one of the gatherers wondered about the origin and propagation of the plants – from seed and from the underground taproot, *talaroco*.[3] The others apparently had never thought about this question. One gatherer noted that the plants need some clay in the ground and yet another one noted that when the upright stem appears,[4] the plants become unsuitable for eating. Four gatherers knew about the yellow flowers; one disregarded the flower colour and another did not even know that thistles produced flowers.

Gathering, Harvesting and Consuming Thistles

The plants pointed out by the gatherers were all identified as *Scolymus hispanicus*, one of the species of edible thistles frequently cited in the literature (Feijão 1960; Andrés 1961; Díaz Robledo 1981; Rosselló 1999; Tardío, Pascual and Morales 2002; Ertuğ 2004). Only this species was collected by the gatherers in our study and apparently it is also the only species sold in the markets in spite of the presence of other edible thistles, not only in the region as previously noted, but also, in the case of *Eryngium campestre*, in the harvesting sites.

Scolymus hispanicus is a biennial (Tutin et al. 1994) or perennial species (Valdés, Talavera and Fernández-Galiano 1987; Tutin et al. 1994) that germinates or sprouts in the autumn (October to November). Until the middle of spring the aerial part of the plant is a basal rosette of prickly leaves that may reach 20 cm long (Franco 1984; Tutin et al. 1994) and have white primary veins (Franco 1984). By the end of April, one stem that emits small branches begins to grow and may reach 15–250 cm (Valdés, Talavera and Fernández-Galiano 1987). Stem leaves are harder and much smaller than the basal ones. Inflorescences (heads) of bright yellow flowers are usually solitary, terminal or lateral in the axil of the stem leaves and may be found from May to July (Valdés, Talavera and Fernández-Galiano 1987).

Thistles can be found in the markets from October to May, but only one seller was precise about this period in our interviews; one said that the thistle-selling season was 'now', another said 'spring', and four referred to Easter as the time when selling ceased. The gatherers also emphasised

Easter as a deadline for harvesting the thistles, with only one exception, who harvests between February and April.

This is somewhat surprising because the date of Easter may shift considerably from year to year. For instance, in the period between 1980 and 2001, the date of the Easter celebration shifted between 26 March and 23 April, which is undoubtedly relevant when considering the life span of the aerial edible parts of thistles. However, this is evidence of the persistence of a rural calendar, marked by celebrations linked to sun and moon cycles (see Brito 1996).

Each gatherer harvests thistles in the same place every year. All the harvesting sites have been known for more than five years to the informants, who noted that the land use has varied over that time. Generally the chosen harvest sites are places where the informants live or used to go to in their youth. The most professional gatherer, however, chooses their places using a more economic approach, based on the abundance of thistles and knowledge about the land management, namely the absence of agrochemical applications.

The harvesting sites are situated mainly in set-aside land, sometimes known as 'seven skirt' lands. In fact, the English expression *set-aside* became common in the region after the entry of Portugal into the European Community (INGA 2002), but it is not always well understood. A Portuguese version, *sete saias*, which sounds like the English *set-aside* but means 'seven skirts', was strangely adopted by one gatherer to characterize an area of ground (Barão 2003a).

The vegetation is generally exclusively herbaceous. The areas used by gatherers are generally less than 2,500 m^2. The ground is mainly plain, not rocky, with a slope of less than 4 per cent and a predominantly eastern exposure. With the exception of two places, where high abundance was found, thistles were scarce and the individual plants very scattered. However, the plants tend to form small or mediumsized groups of two to five individuals.

A standard harvesting procedure was found among the gatherers. A little weeding hoe is used to cut the plants off at the soil level, leaving the root intact below the ground to resprout. The plants are usually allowed to dry for a while and then prepared by *ripagem* or 'stripping' off the prickly leaf blade manually from the base to the apex, leaving only the leafstalk and midrib,[5] the only parts which are eaten.

The plants prepared in this manner are tied together in a bundle (see Figure 9.1), which is offered as a gift to friends and neighbours or taken home, where the hard outer skin is said to be scraped off with a knife before the thistles are cooked. These bundles prepared for home and neighbours are quite different from those prepared for market, which are neater and much more carefully composed.

Figure 9.1. Mr Manuel stripping the leaf blades in the back yard of his house; and one of his bundles, ready to offer to neighbours

Thistles are frequently consumed by all of our informants except one seller. Those sellers who did eat thistles often had them in *cozido*, a type of stew with meat, salted pork fat, sausages, potatoes, carrots, cabbages and other vegetables. Half of them also cook thistles with beans. Two references were made by sellers to thistles' use with chickpeas, or in bread soups with *cação* (a small, common, coastal shark), and one seller mentioned their use with fried meat. All the gatherers consume thistles in the meat-based dish *cozido* with chickpeas, called *cozido de grãos*. Five gatherers made thistles *limados* (cooked with flour and vinegar) in bread soups, generally with *cação*, four with beans and two in the meat-based *cozido à portuguesa*.[6]

Marketing Thistles

The prices per bundle of thistles observed during 2001 were very variable and ranged between fifty cents to three euros, being apparently only determined by sellers, taking into account the prices of other sellers. Prices are fixed; the sellers never bargain with the customers.

According to the sellers, the quantities traded are variable and we could not arrive to an estimation of the annual volume traded. They also said that the amount sold decreased recently, with thistles being more important in the past. Thistles are purchased mainly by adults and older people of the lower-middle or working class. Some curiosity is manifested by some urban clients who do not recognize the plants but generally they do not buy thistles or ask for recipes or advice on how to prepare them.

The number of bundles from the market examined (eight) is not enough to draw strong conclusions but some trends are indicated. A priori we would expect that the number of plants in a bundle should decrease over the season as the number of midribs would increase with the progress of the season. However, this was not the case. The number of plants per bundle (between 4 and 17, with a mean of 12 ± 1) and the number of midribs per plant (between 3 and 19, with a mean of 8 ± 1) and per bundle (between 44 and 175, with a mean of 98 ± 14) were in fact extremely variable, but these variations seemed completely random and not related to the progression of the season. We should note, however, that the number of midribs present in the bundle did not correspond exactly to the number of leaves originally present in the plants because the collectors tend to discard those leaves with blemishes or that do not look fresh.

Although the bundles were not actually measured, from the photographs (Figure 9.2) it can be seen that the length increased considerably from February to April. The examination of the bundles, and in particular of the variation of the number of leaves per plant over time, indicates that once the basal rosette is formed, the number of basal leaves tends to stay stable and the growth of the aerial parts is more focused on leaf elongation.

As previously noted, the presentation of the sellers' bundles (Figure 9.2) was more careful than that of the gatherers' bundles (Figure 9.1). The midribs were more completely cleaned of leaf-blade remains and the plants were more carefully arranged, in a way that seemed to be related to making the plants appear more uniform and ordered, less 'wild', to the consumer.

Figure 9.2. Bundles of thistles from an Évora market purchased on 28 February (top) and 14 April (centre and bottom) 2001, from three different sellers. The bundles from the last date are clearly longer and the neglected bundle at the centre was purchased from a less careful (and needy) seller

To have a better-looking bundle, the sellers use a little trick that consists of binding the plants in the bundle more loosely, leaving some space between the thread and the veins. Then, the bases of the bundles are placed in a vessel with water for a while with the objective of expanding the veins and making them appear larger and more firm, before they are put out on display for sale.

The Ecology of Thistles

In this section we discuss our findings from an extensive set of laboratory tests of the soils and the nutritional content of harvested plants. To repeat, we expected that the mineral content, and thus the nutritional quality, of plant individuals would vary depending on the quality of the soils.

Based on texture, two different groups of soils were identified in the study sites. One was of sandy loam soils (eight samples), with (fine and coarse) sand content between 69.5 and 81.5 per cent, lime content between 7.8 and 11.8 per cent and clay content between 10.8 and 19.5 per cent. The other, more rich in clay (six samples), consisted of sandy clay soils, sandy clay loam soils and clay loam soils with sand content between 46.9 and 58.2 per cent, lime content between 10.9 and 24.1 per cent and clay content between 28.4 and 36.3 per cent. The characteristics of these groups were analysed and compared. When no significant differences were found among them, soil samples were pooled.

For pH and C/N ratio as well, no significant differences were found. The pH was relatively homogeneous, from slightly acid to slightly basic (between 5.99 and 7.88, with a medium value of 6.87 and a CV of 9.2 per cent), and the C/N ratio was low (between 6.9 and 11.7, with a mean value of 8.9 and a CV of 19.9 per cent).

The soil organic-matter content may be considered as intermediate (Santos 1991) and differed significantly between the two groups, with the sandy loam soils, poorer in organic matter, ranging between 1.3 and 4.1 per cent, with a mean value of 2.5 per cent and a CV of 20.1 per cent. In the more clayish soils organic matter ranged between 2.0 and 2.3 per cent, with a mean value of 1.6 per cent and a CV of 28.5 per cent.

The soil mineral content was very heterogeneous but only the potassium content was significantly different between the two groups, being higher for the sandy loam soils. This difference was also reflected in the plants, which also showed higher potassium content when growing in the sandy loam soils. This difference is probably due to the higher sand content of these soils, which are richer in mica-releasing potassium (see Table 9.1). For the boron content huge differences were found between the soil types; boron content was higher in the sandy loam soils and nondetectable quantities were present in the majority of the more clayish soils. In this case, however, no differences were found in boron content of the leaf midrib from the two soil groups. However, we found significant differences in the content of sodium and copper in the plants that were not found in their respective soils. In both cases the mineral content was higher in the plants from the more clayish soils.

As shown by the coefficients of variation (Tables 9.1 and 9.2), a high level of heterogeneity was found in the mineral content of the soils contrasting with a relatively stable mineral content of the midribs. Thus it is not surprising that no significant correlation could be found between the variation in mineral content of soil and plants.

Nutritional and Fibrous Content of Thistles

When compared with other vegetables, thistles show higher values of cellulose, protein and mineral nutrients (Table 9.3). These higher nutrient values for thistles can be explained by the differences in the plant tissues

Table 9.1. Mean values (expressed in milligrams per kilogram of dry matter) and coefficient of variation corrected (expressed in per cent) of the potassium (exchangeable and assimilable), sodium, boron and copper contents of soil and thistle leaf midribs. Significant differences were found between the plants growing in the different groups of soils, and between the two groups of soils, when compared by t-Student tests

Group of soil	K		Na		B		Cu	
	Soil Ex./Ass.	Plant	Soil	Plant	Soil	Plant	Soil	Plant
Sandy loamy soils	172/209	76355	68	12101	0.30	26.9	7.6	9.6
CV	63.2/73.3	41.8	79.1	113.7	158.1	17.3	362.5	60.1
Clayish soils	70/66	47206	68	28979	–	26.9	7.6	25.5
CV	26.9/26.8	23.7	79.1	28.0	–	17.3	362.5	65.3

Table 9.2. Mean values (expressed in milligrams per kilogram of dry matter) and coefficient of variation corrected (expressed in percentage) of the nitrogen, phosphorous, calcium, magnesium and zinc contents of soil and thistle leaf midribs

	N		P		Ca		Mg		Zn	
	Soil	Plant	Soil	Plant	Soil	Plant	Soil	Plant	Soil	Plant
Mean	1400	12500	80	3290	2185	14104	359	2843	–	32.9
CV	47.0	29.3	219.3	43.4	56.8	30.8	89.2	40.8	–	42.7

Table 9.3. Cellulose, protein and mineral content of six vegetables and of our thistles (*Scolymus hispanicus*), in 100 g of edible matter. The values for copper and zinc in the vegetables are from Ferreira (1994) and the remainder from the Instituto Nacional de Saúde Dr. Ricardo Jorge (1984). For thistles, the values displayed were obtained in this study

	Cellulose (%)	Protein (%)	P (mg)	K (mg)	Ca (mg)	Mg (mg)	Na (mg)	B (mg)	Cu (mg)	Zn (mg)
Watercress	0.6	3.4	56	-	198	-	-	-	0.6	-
Lettuce	0.6	1.8	46	-	70	-	-	-	0.6	0.4
White cabbage	1.5	5.1	70	-	676	-	-	-	0.7	0.4
Cabbage	1.2	2.9	36	-	234	-	-	-	0.6	-
Spinach	0.6	2.6	45	-	104	-	-	-	0.5	-
Turnip greens	1.0	2.3	64	-	262	-	-	-	0.5	-
Thistles	7.3	7.8	329	6178	1410	284	2054	2.7	1.8	3.3

present in the edible parts of the plant. In the case of thistles, only the leafstalk and midribs, comparatively very rich in vascular tissue, are eaten, while in the other vegetables like spinach or cabbages entire leaf blades are used and concomitantly the proportion of vascular tissue is lower. These high values undoubtedly make thistles valuable as nutritious vegetables.

The midrib cellulose content was relatively stable over time, whereas hemicellulose was not detected until the beginning of April, with a continuous rise thereafter (Figure 9.3). The fibrous quality that the leaf veins acquire before flowering, which reduces the edibility of the plants, may be explained by this rise in the hemicellulose content over the course of the season. Notice that harvesting usually occurs before the plants become very fibrous, preceding the rise in hemicellulose previous to flowering.

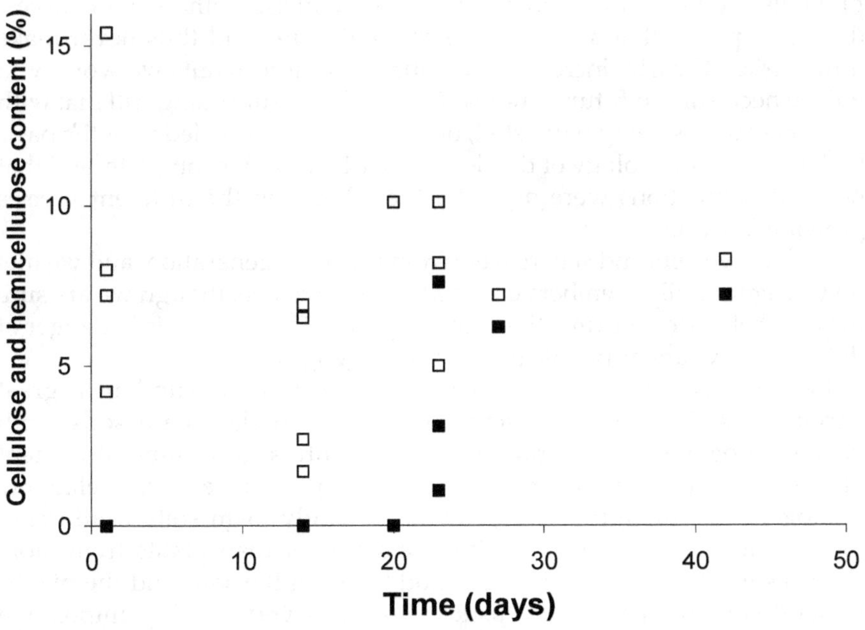

Figure 9.3. Cellulose (white squares) and hemicellulose (black squares) content of the midrib of thistles, expressed as percentage of total dry matter, in samples of five plants each, harvested between 18 March (day 1) and 28 April 2001 at six different sites

Conclusions

In the Évora region, thistles are still consumed and highly appreciated as a local culinary product. Various species of edible thistles can be found in the region and, at the more traditional markets, thistles are available from October to May in the form of bundled leafstalks and leaf midribs. The somewhat skeletal appearance of thistles in these bundles makes identification of the species almost impossible, so the actual species used were unknown when the present study began.

We found that *Scolymus hispanicus* was the only species harvested by the persons we contacted, and apparently it was also the only species present in the markets and used for human consumption. Strangely, other reportedly edible thistle species, such as *Scolymus maculatus*, *Cynara cardunculus*, *Silybum marianum* and *Eryngium campestre*, are present in the region and other authors have reported their use in the area. The latter species was also found by us in one harvesting site. Our informants were thistle gatherers recruited at the markets and among previous acquaintances, so perhaps we have missed others who harvest different species that are only consumed at home and thus never make it to market. Clearly, increasing the number of informants we work with will be necessary in future studies. That said, it is surprising still that with only one species being harvested, our informants' knowledge of the basic biology and morphology of thistles is so variable and incomplete, and that no clear distinctions were made by them between the different species growing in the area.

Both harvesters and sellers were from the older generation, and we met no younger family members engaged in this practice, though we are sure they probably do consume the plants at home. The practice is in danger of dying out if younger people do not become engaged.

In the harvesting sites, *Scolymus hispanicus* was found in a great variety of soil types from sandy loam soils to clay loam soils, with very heterogeneous mineral content. In contrast, the mineral content of the leafstalks and leaf midribs, the only parts eaten, was relatively homogeneous, though plants growing in sandy loam soils were richer in potassium and poorer in sodium and copper than plants from more clayish soils. No correlation was found between the soils and the plants' mineral content. Thus, the thistles showed a very stable composition despite the variety of sites and soils where they grew.

Overall, thistles are a very nutritious edible green vegetable, having higher cellulose, protein and mineral contents than typical cultivated vegetables such as spinach, watercress, cabbage and green turnips.

Despite the thistle's dietary value as a vegetable, its historical and symbolic associations with low status, poverty and hard times make it unlikely that its popularity will increase in the region. Clearly, attempts to open the market for thistles will have to overcome this cultural bias. However, traditional recipes using thistles are present in some local restaurants, where they are appreciated as a luxury, or at least as a unique traditional delicacy, mainly by tourists coming outside the region and the country. It seems likely that a well directed marketing campaign, addressed to niche markets outside the region, might lead to a viable export market for what was once a 'poor man's food'.

However, even if that negative perception is overcome, there is still one characteristic which hinders the spread of thistles as a food source: thier tough, fibrous constitution, which gives them a reputation as hardy, prickly and, for those who do not know them, inedible! In this study we found that people are cleverly harvesting the plant just before the build-up of fibrous tissues necessary for stem formation and flowering, when it is nutritious but also still tender and edible. After this and round about Easter, the increased hemicellulose content in the leaf midribs makes the plant essentially inedible. Even then, there is still a very laborious and time-consuming process that is required after the harvest to make the leafstalks and midribs suitable for cooking. If the thistles must be scraped with a knife before cooking in order to strip the prickly hairs off the outer skin, then this coould represent the biggest stumbling block to maintaining or even spreading the tradition, as people want to spend less and less time preparing and cooking food. Dishes that require long preparation times are thus being abandoned. Overcoming this might require increased management, eventual cultivation, improved postharvest treatments and perhaps even genetic engineering.[7] For now, though, thistle-gathering/eating remains a special tradition. Thistles are collected and eaten by those who know and appreciate the benefits of a prickly but nutritious and tasty wild plant.

Acknowledgements

The authors are grateful to Professor Gonçalves Ferreira, Director of the Laboratório de Química Agrícola (Universidade de Évora/ICAM), where plant and soil analyses were done, for the material and personal support; to Dr Ricardo Freixial for making possible the cellulose and hemicellulose analyses in the Laboratório de Pastagens (Universidade de Évora/ICAM); to Dr Paula Simões of the Herbarium (UEVH) staff for help in the species identification. Thanks also to the reviewers for comments and suggestions on earlier drafts of the manuscript. A special thanks is extended to all informants and people who shared their knowledge with us and made this work possible.

Notes

1. Bread soups are the most typical dishes of Alentejo. They consist of a hot *caldo* (stock) in which slices of (preferably stale) bread are immersed. The *caldo* may be obtained by cooking various foods (e.g., sausages or meat), or more simply by boiling garlic, coriander and salt in water.
2. Thirteen were all the informants we could find. Two more women traded thistles at the traditional markets, but they choose not to cooperate.
3. *Talaroco* is an untranslatable word that means a kind of talus.
4. The upright stem is improperly called the *talo* (talus).
5. The midrib is also improperly called the *talo.*
6. *Cozido à portuguesa* is the same as *cozido*, described above.
7. Sometime after the present study was made, we found other thistle gatherers from Évora who told us that they did not use this time consuming procedure. Our own experiments confirm that cooking the thistles as they are sold in the markets produces tasty results.

References

Andrés, J. 1961. *L'Alimentation et la cuisine à Rome.* Études et commentaires 38. Paris: C. Klincksieck.

Anne, P. 1945. 'Sur le dosage rapide du carbone organique des sols', *Annales Agronomiques* 15: 161–172.

Alexandre, C.J.R., J.R. Marques and A.G. Ferreira. 2001. 'Comparação de dois métodos de determinação da textura do solo: sedimentometria por raios X vs. Método da pipeta'. Encontro Anual da Sociedade Portuguesa de Ciências do Solo – Uso do Solo e da Água. *Revista de Ciências Agrárias* 3–4: 73–81.

Alves, A.F. 1994. *Os comeres dos ganhões: 20 receitas da cozinha alentejana.* Porto: Campo das Letras.

——— 1997. *Os comeres dos ganhões: Memória de outros sabores,* 3rd edn. Porto: Campo das Letras.

Apicius. 1997. *O Livro de cozinha de Apício.* Translated by I.O. Castro. Sintra: Colares Editora.

Arcidiacono, S., P. Pavone and C. Salmeri. 1996. *The Erbe Commestibili dell'Etna.* Universitat de Catania. Retrieved 23 March 2006 from http://www.unict.it/dipartamenti/biologia_animale/webnature/pavone/erbecomm.htm

Barão, M.J. 2003a. 'Foram cardos, foram prosas…', in A.S. Dias (ed.), *Etnobotânica: Perspectivas, história e utilizações.* Évora: Publicações Universidade de Évora, pp. 19–29.

——— 2003b. *O uso de cardos na Alimentação no Alentejo: Contribuição para o seu estudo numa perspectiva etnobotânica e ecológica,* M.Sc. dissertation, Évora: Universidade de Évora.

Black, C.A. 1965. *Methods of Soil Analysis, Part 2: Chemical and Microbiological Properties.* Madison, Wis.: American Society of Agronomy.

Berglund, B. and C.E. Bolsby. 1971. *The Edible Wild: Complete Cookbook and Guide to Edible Wild Plants in Canada and North America.* New York: Charles Scribner's Sons.

Bremmer, J.M. and C.S. Mulvaney. 1982. 'Total Nitrogen', in A.L. Page, R.H. Miller and D.R. Keeney (eds), *Methods in Soil Analysis, Part 2: Chemical and Microbiological Properties.* Agronomy monograph 9, 2nd edn. Madison, Wis.: Soil Science Society of America, pp. 595–624.

Bremness, L. 1993. *Guia prático de plantas aromáticas: Culinárias, medicinais e cosméticas.* Translated by M. Nóvoa. Lisbon: Editora Civilização.

Brito, J.P. 1996. 'Coerência, incerteza e ritual no calendário agrícola', in J.P. Brito, F.O. Baptista and B. Pereira (eds), *O voo do arado.* Lisbon: Museu Nacional de Etnologia, pp. 217–229.

Brito, M.C. 1997. 'Sabores e tradições do Baixo Alentejo', *Arquivo de Beja* 4(3): 109–112.

Câmara Municipal de Évora, Confraria Gastronómica do Alentejo and A. Saramago. 2001. *Concurso de cozinha alentejana.* Évora: Câmara Municipal de Évora.

Carvalho, A. and M.L. Gomes. 1973. *Alimentação e condições de vida de famílias de trabalhadores rurais do Baixo Alentejo.* Oeiras: Fundação Calouste Gulbenkian.

Cottenie, A., M. Veroo, L. Jiekens, G. Velgha and R. Camerlynch. 1982. *Chemical Analysis of Plant and Soils.* Ghent: I.W.O.N.

Couplan, F. 1998. *The Encyclopedia of Edible Plants of North America.* New Canaan, Connecticut: Keats Publishing.

Coutinho, A.X.P. 1939. *Flora de Portugal.* 2nd edn. Lisbon: Bertrand (Irmãos) Lda.

Díaz Robledo, J. 1981. *Atlas de las frutas y hortalizas.* Madrid: Ministerio de Agricultura.

Ertuğ, F. 2004. 'Wild Edible Plants of the Bodrum Area (Muğla, Turkey)', *Turkish Journal of Botany* 28: 161–174.

FAO. 1980. *Working Document No. 3.* European Research Network on Trace Elements, Coordination Centre.

Feijão, R.O. 1960. *Elucidário fitológico: Plantas vulgares de Portugal Continental, Insular e Ultramarino (Classificação, nomes vernáculos e aplicações),* Volume 1, A–H. Lisbon: Instituto Botânico de Lisboa, Artigo de divulgação No. 8.

Feio, M. 1983. *Le Bas Alentejo et l'Algarve.* Évora: Instituto Nacional de Investigação Científica, Centro de Ecologia Aplicada.

Fernandes, F.M. and L.M. Carvalho. 2003. *Portugal Botânico de A a Z: Plantas Portuguesas e Exóticas.* Lisbon: Lidel.

Ferreira, F.A.G. 1994. *Nutrição humana.* 2nd edn. Lisbon: Fundação Calouste Gulbenkian.

Fialho, M. 1995. *Cozinha regional do Alentejo.* 2nd edn. Mem Martins: Publicações Europa-América.

Franco, J.A. 1984. *Nova flora de Portugal (Continente e Açores), Volume II, Clethaceae-Compositae.* Lisbon: Author edition.

Gaspar, J. 1972. *A área de influência de Évora*. Lisbon: Centro de Estudos Geográficos.

Guedes, F. 2000. *Receitas portuguesas: os pratos típicos das regiões*. Lisbon: Dom Quixote.

Hill, A.F. 1952. *Economic Botany*. New York: McGraw-Hill Book Company.

INGA. 2002. *Prémios e ajudas anuais*. Lisbon: Instituto Nacional de Intervenção e Garantia Agrícola.

Instituto Agronômico de S. Paulo. 1977. 'Métodos de análise de solo', Circular No. 63.

Instituto Nacional de Saúde Dr. Ricardo Jorge. 1984. *Tabela de composição dos alimentos portugueses*. Lisbon: Divisão de Documentação e Informação do IQA.

Knudsen, D. 1980. 'Recommended Phosphorous Test for North Central Region', *North Central Regional Publication* 221: 14–16.

Kuhnlein, H.V. 1992. *Traditional Plant Foods of Canadian Indigenous Peoples: Nutrition, Botany and Use*. Oxford: Routledge.

Launert, E. 1981. *The Hamlyn Guide to Edible and Medicinal Plants of Britain and Northern Europe*. London: Hamlyn.

Lúcio, M.C. 1987. *Cozinha regional do Baixo Alentejo*. Lisbon: Editorial Presença.

Martí, F.B. and J.R. Munoz. 1957. *Flame Photometry*. Amsterdam: Elsevier.

Mattoso, J. (ed.) 1992. *História de Portugal*. Volume 1. Lisbon: Círculo de Leitores.

Micromeritics Instrument Corporation. 1988. *Sedigraph 5100, Particle Size Analysis System, Operators Manual*. Norcross.

Miranda, A.H. 1957. 'La cocina hispano-magrebí durante la época Almohade', *Revista del Instituto de Estudios Islámicos de Madrid* 5(1–2): 137–155.

Nuez, F. and J.E. Hernández Bermejo. 1994. 'Neglected Horticultural Crops', in J.E. Hernández Bermejo and J. León (eds), *Neglected Crops: 1492 from a Different Perspective. Plant Production and Protection Series No. 26*. Rome: FAO, pp. 303–332. Retrieved 23 March 2006 from http://www.port.purdue.edu/newcrop/1492/neglected:html#thistles

Picão, J.S. 1947. *Através dos campos*. 2nd edn. Lisbon: Neogravura Lda.

Pinheiro, J.M.M. 1995. *Tradição mediterrânea na alimentação e cozinha alentejana*. Paper given at the 6th International Congress of the International Committee of Anthropology of Food (ICAF), Évora.

——— 2000. *Terra de grandes barrigas onde só há gente gorda*. 2nd edn. Évora: Alentejana.

——— 2004. *As ervas da pobreza e da riqueza (Ensaio etnográfico sobre ervas ou plantas bravas comestíveis e aromáticas do Alentejo)*. Évora: Confraria de Moagem.

Piper, C.S. 1950. *Soil and Plant Analysis*. Adeleide: The University of Adelaide.

Póvoas, I. and M.F. Barral. 1992. *Métodos de Análise de Solos. Comunicações do Investigação Científica Tropical, Série de Ciências Agrárias nº. 10*. Lisboa: Ministério do Planeamento e da Administração do Território.

Pulga, J. 2001. *Alentejanando*. Évora: Casa do Sul Editora.

Ramos, D.L. 1997. 'A arte de comer ervas do campo', *Pública* 27(July): 43–46.

Rei, A. 2000. Escola de Línguas do Instituto Luís de Molina, personal communication.

Riehm, H. 1958. 'Die Ammoniumkaktatessigsäure – Methode zur Bestimmung der leichtlöslichem phosphorsäure in karbonathaltigen Böden', *Agrochimica* 3: 49–55.

Ritas, J.L. and J.L. Melida. 1985. *El diagnóstico de suelos y plantas*, 4th edn. Madrid: Mundi-Prensa.

Robertson, J.B. and P.J. Van Soest. 1980. 'The Detergent System of Analysis and its Application to Human Foods', in W.P.T. James and O. Theander (eds), *The Analisis of Dietary Fibre in Food*. New York: Marcel Dekker, pp. 123–158.

Rodrigues, H.M.M.C. 1996. *Antropologia e alimentação: O papel sócio-cultural da cozinha alentejana como paradigma da transformação social*, M.Sc. dissertation. Lisbon: Universidade de Técnica de Lisboa.

Rosenberger, B. 1998. 'A cozinha árabe e o seu contributo para a cozinha europeia', in J.L. Flandrin and M. Montanari (eds), *História da alimentação, 1: Dos primórdios à Idade Média*. Translated by M.T. Pinhão. Lisbon: Terramar, pp. 305–323.

Rosselló, M.E. (coord.). 1999. *Guía de las plantas del alcornocal*. Mérida: Dpto. Recursos Naturales Renovables, Instituto C.M.C. IPROCOR, Junta de Extremadura.

Salgueiro, J. 2004. *Ervas, usos e saberes: Plantas medicinais no Alentejo e outros produtos naturais*. Lisbon: Marca – Associação de Desenvolvimento Local de Montemor-o-Novo.

Santos, J.Q. dos. 1991. *Fertilização – Fundamentos da Utilização dos Adubos*. Lisbon: Europa-América.

Saramago, A. 1997. *Para uma história da alimentação no Alentejo*. Lisbon: Assírio e Alvim.

Simões, N.M.L.S.S. 2000. *Valor forrageiro de trevos subterrâneos (*Trifolium subterraneum *L.) e luzenas anuais (*Medicago *sp.)*. B.Sc. dissertation. Universidade de Évora.

Sokal, R.R. and F.J. Rohlf. 1995. *Biometry: The Principles and Practice of Statistics in Biological Research*. 3rd edn. New York: W.H. Freeman and Company.

Tardío, J., M. Pardo-de-Santayana and R. Morales. 2006. 'Ethnobotanical Review of Wild Edible Plants in Spain', *Botanical Journal of the Linnean Society* 152(1): 27–71.

———, H. Pascual and R. Morales. 2002. *Alimentos silvestres de Madrid: Guía de plantas y setas de uso alimentario tradicional en la Comunidad de Madrid*. Madrid: Ediciones La Librería.

Thomas, G. 1982. 'Exchangeable Cations', in A.L. Page, R.H. Miller and D.R. Keeney (eds), *Methods in Soil Analysis, Part 2: Chemical and Microbiological Properties*. Agronomy Monograph No. 9, 2nd edn. Madison, Wis.: Soil Science Society of America Inc., pp. 159–165.

Tutin, T.G., V.H. Heywood, N.A. Burges, D.M. Moore, D.H. Valentine, S.M. Walters and D.A. Webb (eds). 1994. *Flora Europaea, Volume 4,*

Plantaginaceae to Compositae (and Rubiaceae). Cambridge: Cambridge University Press.

Valdés, B., S. Talavera and E. Fernández-Galiano (eds). 1987. *Flora vascular de Andalucía occidental*, 3. Barcelona: Ketres Editora, S.A.

Valente, M.O.C. 1994. *Cozinha de Portugal: Alentejo.* Lisbon: Círculo de Leitores.

Vasconcellos, J.L. 1967. *Etnografia Portuguesa.* Volume 5. Lisbon: Imprensa Nacional.

Chapter 10

Spring is Coming
The Gathering and Consumption of Wild Vegetables in Spain

Javier Tardío

Ya viene el mes de los pobres	The poor men's month is just coming.
ya salen a buscar grillos,	They go out looking for crickets,
espárragos y cagarrias,	asparagus and morels,
sombreretes y cardillos	addles and golden thistles
	(Popular Spanish proverb)

Introduction

In Spain, as in other Mediterranean countries, wild vegetables have played an important role in complementing staple agricultural foods, especially during times of shortage, like after the Spanish Civil War at the end of the 1930s. But they have also been very valuable during certain seasons, such as winter and spring, when fresh agricultural products were scarce. Presently, however, with the development of agribusiness and global supply chains, it is easy to find a variety of cultivated vegetables in markets throughout the year. As a result of this, and also because of new socio-economic contexts, the use of noncultivated vegetables has decreased, and nowadays far fewer species are consumed. Furthermore, traditional knowledge of wild food plants is quickly disappearing and, in most cases, survives only with the elderly.

Nevertheless, there is an increasing interest in wild edible plants, especially in relation to their role in maintaining human health. They are considered in a recent, cross-cutting initiative about human nutrition and health by the Convention on Biological Diversity (CBD 2005), and they are possible sources of nutraceuticals, as has been highlighted in recent works (Heinrich et al. 2005; The Local Food-Nutraceutical Consortium 2005).

In this chapter, I analyse and evaluate the gathering and consumption of wild green vegetables traditionally used in Spain, at least during the last hundred years. By traditional use, I mean a history of consumption for more than one generation, what other authors define as the minimum time for using the term 'traditional' (Ogoye-Ndegwa and Aagaard-Hansen 2003; Pieroni et al. 2005). The information used in this chapter was derived from a database of wild edibles traditionally consumed in Spain (Tardío, Pardo-de-Santayana and Morales 2006). Forty-six ethnobotanical and ethnographic sources were exhaustively analysed for that compilation, including our own fieldwork on wild edibles in the province of Madrid (Tardío, Pascual and Morales 2002, 2005) and also our unpublished field data from other Spanish provinces. For the purposes of our database, we established seven categories of food use, to classify wild food plants: vegetables, fruits, beverages, seasoning, preservatives, sweets (including flowers or roots eaten for their sweet flavour) and other food uses, such as oils, flours and pickles. In the category of 'vegetables' (also called 'greens' or 'green vegetables', in Spanish *verduras*) we include those plants whose leaves, stems or even unripe and green fruits or seeds are consumed, either stewed, raw in salads or even raw as a snack.

The analysis of these data addresses a series of ethnobotanical questions in order to characterize the traditional consumption of wild vegetables in Spain. The first question concerns the kinds of plants that people prefer to eat as vegetables, in terms of botanical family, growth habit and life form and their status as weeds or nonweeds. The second question concerns how they were traditionally consumed, in terms of harvesting techniques (including seasonality), plant parts used, processing techniques, the effects of taste, and reasons given for maintaining dishes that use wild vegetables. Finally, wild plants were examined with regard to the social and cultural implications of gathering and their increasing popularity as both food and medicine. The analysis presented here is essentially exploratory: it establishes a descriptive baseline and seeks to uncover interesting or unusual patterns in the data that might be explored in future research.

One caveat: it must be stressed that many of the species registered in our database that were consumed in the past are rarely collected today. Therefore, although there are many species in the compilation, most of them are only a reminder of the consumption of wild vegetables in the past. Perhaps, too, those plants mentioned as still being used today were of even greater importance in the past. For instance, *Anchusa azurea* Mill. (Figure 10.1), predominantly known as *lenguaza* in Spanish, is mentioned in a number of these studies but not by a high number of informants. In the

Figure 10.1. *Anchusa azurea.* **A** Basal leaves; **B** Boiled leaves lightly fried with garlic, egg and ham

province of Madrid this species was mentioned in only two of the sixty-eight surveyed villages. However, the Arabic common name *alcalcuz,* used in those villages, seems to be evidence of a much older tradition of use.

Botanical Characteristics of Spanish Wild Vegetables

This section characterizes the plants selected as wild vegetables, in terms of botanical family, growth habit and life form and their status as weeds or nonweeds.

Botanical Families

Of the roughly 7,000 species listed in the Iberian Flora,[1] 206 species have been recorded in our database as wild vegetable species. Of these about one-third (n=61) are from the Asteraceae family (see Figure 10.2). Other important families are Polygonaceae, Liliaceae, Brassicaceae, Fabaceae, Rosaceae and Apiaceae, with between ten and fourteen species each. However, it is necessary to distinguish if the greater number of species selected in a family is only the result of being a diverse family, or if there are other characteristics that make their species more likely to be used as vegetables. Following Moerman (1991, 1994), a regression analysis between the number of wild vegetables species (VEGSP) and the total species per family in the Iberian Flora (FISP) has been carried out (Figure 10.2). The regression line, defined by the equation VEGSP = 0.02235 + 0.04565 * FISP, shows the predicted number of species used as vegetables from a given family that might be expected if they were chosen randomly. If so, greater families would have a bigger number of green vegetables. The residuals column in Figure 10.2, that is, the distance from each point and the regression line, indicates how far above or below from the predicted values the actual number is. These results suggest that Poaceae (with a residual of -17), Fabaceae (-13), and Caryophyllaceae (-8)

Family	FISP	VEGSP	Res.
Asteraceae (AST)	719	65	32
Polygonaceae (POL)	54	14	12
Liliaceae (LIL)	124	13	7
Brassicaceae (BRA)	287	12	–1
Fabaceae (FAB)	515	11	–13
Rosaceae (ROS)	233	11	0
Apiaceae (API)	117	10	5
Caryophyllaceae (CAR)	300	6	–8
Poaceae (POA)	439	3	–17
Other families*	1705	61	
Total	7000	206	

*Other families that include species used as vegetables

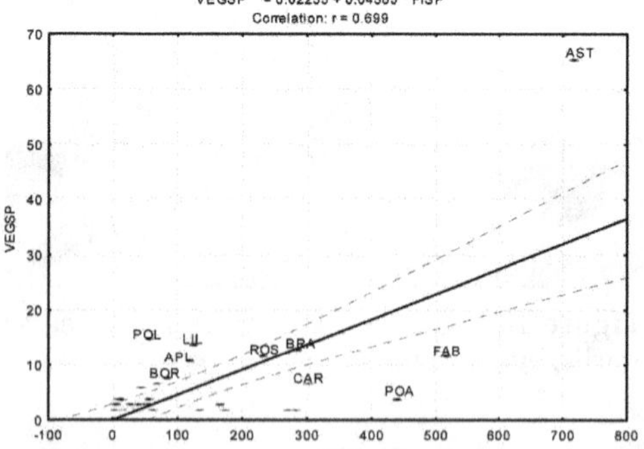

Figure 10.2. Importance of different families of wild species used as green vegetables. Regression analysis of number of wild vegetables species (VEGSP) on total species per family in the Iberian Flora (FISP). Discontinuous lines mark the 95 per cent confidence interval for the predicted values

are greatly underrepresented in spite of the big size of their families. By contrast, the families situated over the regression line, with a positive residual value, have a number of species higher than predicted. So, Asteraceae (with a residual of 32), Polygonaceae (12) and Liliaceae (7) are overrepresented families, whose species are clearly preferred to be eaten as vegetables. What is it about these families that make them preferred as wild vegetables? It appears that these families have species that are widely distributed across Spain; many have desired biological forms, such as perennials with a rosette of leaves in Asteraceae and Polygonaceae, or bulbs and asparagus in Liliaceae; and many species are linked with good nutritional and gastronomic characteristics.

Growth Habit and Life Form

For analysing habit and life form, four simplified categories have been considered: annual herbs, nonannual herbs (both perennial and biennial), shrubs and trees. Figure 10.3 shows that more than half (57 per cent) of the wild vegetables are perennial or biennial plants and that, if taking into account the growth habit, herbs are clearly predominant (88 per cent) over shrubs and trees. The most important group is nonannual herbs (45 per cent), followed by annual herbs (43 per cent), shrubs (11 per cent), and finally trees (1 per cent), with only three species listed: *Ulmus minor* Mill. and *Pinus pinea* L. (immature fruits and seeds respectively), and *Chamaerops humilis* L. (young shoots). The predominance of perennials (or biennials) over annuals seems quite logical. People have selected over millennia those plants that were easy to find again the next season in the same place. Most of these species are perennial herbs, usually with

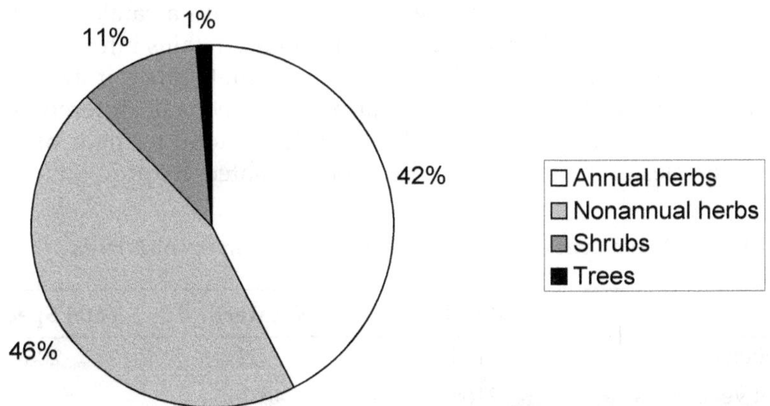

Figure 10.3. Growth habit and life form proportions of wild vegetables traditionally consumed in Spain

a deep root and a rosette of leaves, which withers in the summer and sprouts again in the next wet season. They also have remarkable powers of regeneration from any part of the rootstock that may be severed – a significant feature that enables them to be used in a sustainable way. This is the case, for instance, for *Scolymus hispanicus* L., *Chondrilla juncea* L., *Taraxacum officinale* Weber and *Anchusa azurea* Mill.

Annual herbs are also frequently used as vegetables. They are species that produce a very high number of seeds in favourable environmental circumstances, which have great longevity and easily germinate under many situations, although sometimes this germination is discontinuous. Some of the annuals consumed in Spain are *Borago officinalis* L., *Portulaca oleracea* L., *Silybum marianum* (L.) Gaertn., *Sonchus oleraceus* L. and *Papaver rhoeas* L.

Weeds as a Source of Vegetables

It is well known that a high proportion of wild food plants, especially those used as vegetables, fall inside the cross-cutting category of 'weeds' – *malas hierbas* in Spanish. Whether one is referring to weedy vegetables (Tanji and Nassif 1995; Díaz-Betancourt et al. 1999; Vieira-Odilon and Vibrans 2001; Rivera et al. 2004; Pieroni et al. 2005) or weedy medicinal plants (Stepp and Moerman 2001; Stepp 2004), these are plants whose populations grow entirely or predominantly in markedly human-disturbed habitats, excluding, of course, cultivated plants (Baker 1974). Although the category of 'weeds' is a cultural and ecological concept, these plants have interesting features in common, especially their high reproductive capacity, rapid growth and ability for adapting to different environmental conditions.

Our list of wild vegetables was compared with a catalogue of weeds that grow in Spain (Carretero 2004). This book contains information about 2,341 species, which is perhaps an overestimate since it also includes all those species that occasionally appear as weeds in different kinds of cultivated fields. As shown in Table 10.1, the total number of vascular plant species of the Iberian Flora can be estimated at approximately 7,000

Table 10.1. Weed species as a source for vegetables in Spanish Flora

	Weed	Not weed	Total Species
Vegetables	173	33	206
Not vegetables	2168	4626	6794
Total	2341	4659	7000

$\chi^2 = 243$; $P<0.0001$

(Ramón Morales, personal communication). Of these roughly one-third are considered weeds; and of the 206 wild vegetables in the database, a remarkable 84 per cent can be considered weeds! A Chi Square test ($\chi^2 = 243$; $P<0.0001$) confirms that there is a significant preference for weeds over other plant forms.

In fact, there are many references in Spanish ethnobotanical sources to the exploitation of some weeds while hand-weeding crops, either for personal consumption or for animal fodder. A high number of species, such as *Scolymus hispanicus* L., *Chondrilla juncea* L., *Cichorium intybus* L., *Crepis vesicaria* L. and *Raphanus raphanistrum* L., have been mentioned as gathered that way in the past. Moreover, informants frequently mentioned that plants collected amongst crops are of better quality than those from other environments. They declared that plants are bigger, whiter and more succulent because of growing in tilled soils. Two typical cases were *Chondrilla juncea* and *Cichorium intybus*, whose blanched shoots and leaves are eaten in salads. A modern version of the second species is the 'witloof' or 'Belgian endive', whose compact, elongated head is the result of being forced to sprout in complete darkness to keep new leaves tender and pale.

Another example is *Asparagus acutifolius* L., known widely as *espárrago triguero* ('asparagus from the wheat fields'), because in past times it could be found within the cereal crops growing as a weed. Nowadays, the name seems to make no sense because the species only grows in uncultivated lands. Some scientific names also point out the weedy origins of certain species; for example, the specific name for *Allium ampeloprasum* L. means 'a leek found in the vineyards'. Mesa (1996) mentions that in the province of Jaén it could be frequently found in olive groves. Some people liked it so much as an edible food plant, that its growth was encouraged. Much the same story applies to *Borago officinalis* L., which grows as a 'weed' in the homegardens of Jaén.

Consuming Wild Vegetable Plants

This section, describes and analyses some characteristics of the traditional consumption of wild vegetables in Spain. The preferred plant parts and the mode of consumption in relation to the quality of the species are evaluated, and finally the meaning of some curious recipes made of mixed wild plants is examined.

Part of the Plant Consumed

Among the different species that have been considered as vegetables, the utilized part or parts varies from vegetative organs (leaves, stems, bulbs) to immature reproductive structures (inflorescences, fruits and seeds).

In some species only a part of the plant is used, such as *Silene vulgaris* (Moench) Garcke, whose tender leaves and stems are generally consumed. However, in other plants several different parts are eaten. For example, the midrib of the basal leaves of *Silybum marianum* (L.) Gaertn. is stewed, while the peeled young shoots and the tender parts of the inflorescence are eaten raw as a snack. In other cases, there is a part of the plant that is widely consumed and appears in most of the references, but there is also a different part much less used. This is the case of *Scolymus hispanicus*, a thistle whose basal-leaf midribs are broadly consumed in most parts of Spain (as in many countries of the Mediterranean basin, see Chapter 9 this volume), though some people eat the peeled young shoots raw as a snack.

For analysis, we have established nine categories of the different parts consumed:

1. Tender **leaves**. This category is formed chiefly by plants with a basal rosette of leaves, most of them of the Asteraceae family. Some other species whose leaves are eaten, such as *Malva sylvestris* L. and *Crataegus monogyna* Jacq., are also included here.
2. **Midribs** of basal leaves (*penca*, in Spanish). This is a special case of the former category, as most of the plants are thistles and their prickly leaves need to be peeled. It includes plants of the genus *Scolymus, Silybum, Cynara, Onopordum* and *Arctium*.
3. **Leaves and stems**. This category includes those species whose life form is not a basal rosette of leaves, as above, but includes tender parts of the stems together with the leaves. Typical species are *Silene vulgaris, Montia fontana* L. or *Rorippa nasturtium-aquaticum* (L.) Hayek.
4. Young **shoots**, that sprout from roots or subterranean stems, such as *Asparagus, Tamus, Ruscus*; the tips of some climbing plants, like *Bryonia, Humulus, Clematis*; the peeled sucker of some bushes, as in *Rubus, Rosa*; or leafy shoots such as *Cichorium intybus, Foeniculum vulgare* Mill., *Chondrilla juncea, Sonchus crassifolius* Pourr. ex Willd., *Atractylis gummifera* L.; and sometimes including floral buds as in *Sisymbrium crassifolium* Cav.
5. **Stems**. Includes the stems of *Foeniculum vulgare* or *Allium ampeloprasum*; the basal part of the stems of *Scirpus holoschoenus* L.; *Stipa gigantea* Link and the peduncles of flowers or inflorescences of *Taraxacum* sp.pl., *Silybum marianum, Oxalis pes-caprae* L., *Armeria arenaria* (Pers.) Schult. and *Aphyllantes monspeliensis* L.
6. **Inflorescences**. There are several species of the family Asteraceae, such as those from the genus *Silybum, Onopordum, Cynara* or *Scorzonera*, whose basal part of their tender inflorescences are consumed raw as a snack.
7. Immature **fruits** or **seeds**. For example from *Pinus pinea* L. and from several species of *Malva, Vicia* and *Lathyrus*.

8. **Bulbs**, from *Allium sp.pl.*, and **roots**, from *Campanula rapunculus* L.
9. **Galls**, including the tender and fleshy galls of *Hypochoeris glabra* L. and also those of *Quercus coccifera* L.

The relative importance of the different plant parts consumed can be expressed in terms of the number of species, and also by the number of use-reports for each category (i.e., the number of references that mention the edible use of the plant part), as in Figure 10.4. As was pointed out before, some species have more than one edible part and hence they are in several categories. It can be seen that the most frequently used plant part in wild vegetables are the leaves, which are mentioned for half of the species, followed by leaves and stems (29 per cent) and young shoots (22 per cent). Midribs are also an important plant part, but only if we consider the high number of use-reports for the species *Scolymus hispanicus*, which is one of the most widely consumed thistles in Spain.

Mode of Consumption

Another interesting subject is how the different species were/are traditionally consumed. They could be eaten raw, sometimes directly in the field as a snack, or used for preparing salads. In many other cases the

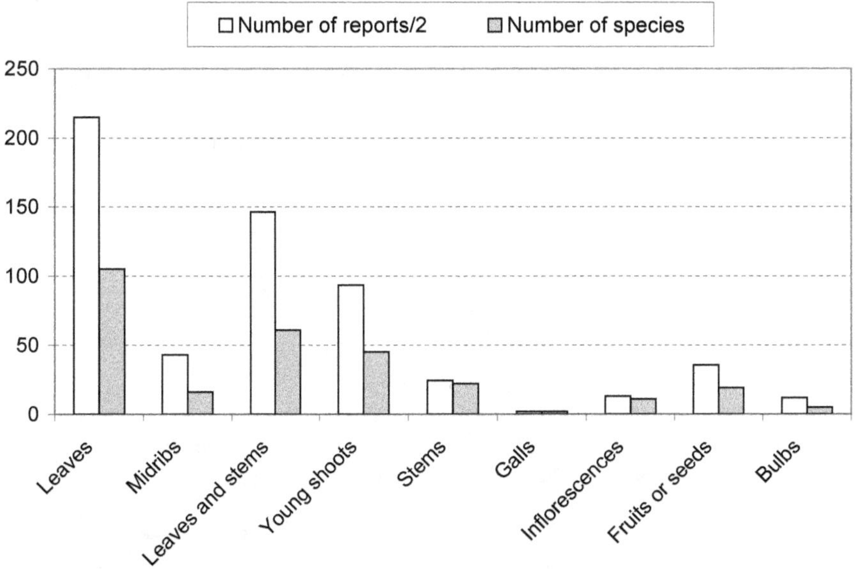

Figure 10.4. Importance of the different plant part consumed, expressed by the number of use-reports and by the number of species mentioned in each category. To maintain the scale, the number of use-reports has been halved in each case

plants are stewed in a variety of ways. With regard to raw consumption, I distinguish between plants eaten raw as a snack and those used to prepare salads, because I think that the latter represents perhaps a greater level of appreciation for those species that are worth taking home to be prepared in a salad. Figure 10.5 shows the number of species consumed in each of the three main categories, and the overlap among them. Note that in some cases (e.g., species of *Malva*), a different part of the plant is used when consumed raw (e.g., immature fruits) or stewed (e.g., young leaves and stems).

Figure 10.5 shows that 141 species (68 per cent) are consumed stewed (see dishes in Figure 10.6), roughly half of these exclusively, while the other half are also consumed raw, either as a snack or in salads, the latter being a somewhat larger category. Are there any common characteristics of the species in these categories? For instance, are any of the species that are only cooked toxic if eaten raw? Do the species that are eaten raw have valuable minerals or vitamins that might be lost if cooked? Where are the species with a reputed higher gastronomic quality?

This last question refers to the species that, perhaps because of their taste or due to their nutritive characteristics, are broadly consumed. I developed an indicator based on the number of use-reports of each kind of use for a particular species; that is, the number of references for one plant used in a particular way. We can suppose that if a widely distributed

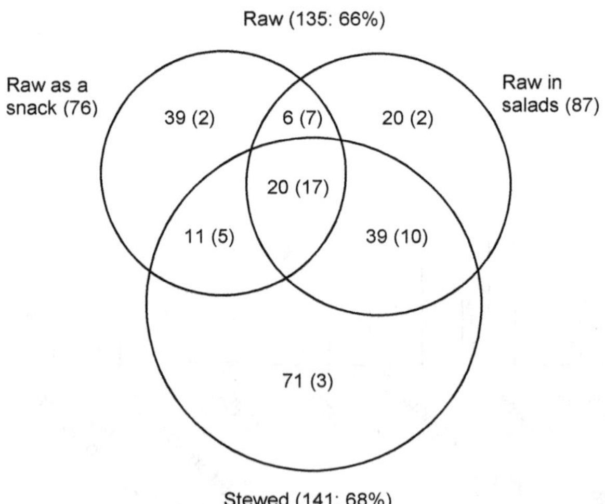

Figure 10.5. Number of species consumed in each category, showing the overlapping among them. In parentheses is the mean number of use-reports for all the species of that category

Figure 10.6. Dishes prepared with *Scolymus hispanicus, Rorippa nasturtium-aquaticum, Tamus communis* and *Silybum marianum.* **A** Midribs of *Scolymus hispanicus* fritters; **B** *Rorippa nasturtium-aquaticum* soup; **C** Soup of *Tamus communis* young shoots; **D** Boiled midribs of *Silybum marianum* lightly fried with garlic and ham

species has been reported as a wild vegetable in many ethnobotanical sources, then it is because of its gastronomic quality. That is the case, for instance, of *Scolymus hispanicus,* a plant that can be consumed in all three ways and was mentioned thirty-four times as being eaten mainly stewed; *Silene vulgaris,* from the group of plants that are consumed stewed and raw in salads, with fifty-one use-reports of its consumption, generally stewed; and finally, *Asparagus acutifolius,* only stewed, which has thirty-five use-reports. Taking into account the use-reports of all the plants that belong to each group we can calculate the mean value for the group. This coefficient reveals whether there are many species broadly reported inside this group. The higher the coefficient, the higher proportion of species with a high number of use-reports will be in this category.

Figure 10.3 also shows, in parentheses, the mean number of use-reports of each category. Therefore, we can see that the groups with the higher mean numbers of reports are in the intersection of plants consumed stewed and plants consumed raw in salads, especially in the intersection of the three modes of consumption. Almost all the species with a greater number of use-reports are located in this area, such as: *Foeniculum vulgare, Scolymus hispanicus* (Figure 10.6A), *Cichorium intybus, Malva sylvestris, Silybum marianum* (Figure 10.6D), and *Taraxacum officinale* in the 'three modes of consumption' group; and *Silene vulgaris, Rorippa nasturtium-aquaticum, Portulaca oleracea, Sonchus oleraceus, Chondrilla juncea* and *Papaver rhoeas,* in the group of plants that are consumed stewed and raw in salads. All of these species are appreciated for their quality; they can be eaten stewed or raw, because of their good flavour. From within these two groups, three of the most extensively consumed plants can be identified. The tender leaves and stems of *Silene vulgaris,* known as *colleja,* are eaten mostly stewed in many parts of Spain, in omelettes, with scrambled eggs, and also as a garnish for *potaje,* a typical Spanish dish often consumed during Lent. The peeled basal leaves of *Scolymus hispanicus* are traditionally boiled and then fried lightly in olive oil with garlic and cured ham and sometimes with hard-boiled or scrambled eggs, often also served as a garnish for *cocido,* another traditional Spanish dish. Finally, the tender leaves and stems of *Rorippa nasturtium-aquaticum* (watercress) are consumed mainly raw in salads but sometimes also in stews and soups (Figure 10.6B).

Another interesting group is the one formed by the vegetables that are always consumed stewed. Most of these species are only consumed in this way because of their taste, which is not so pleasant when eaten raw. That is the case of *Asparagus acutifolius* or *Beta maritima* L. However, some bitter or strongly flavoured plants can be sweetened through repeated boiling, such as *Ruscus aculeatus* L., *Rumex pulcher* L. and *Asparagus albus* L. Other species are unpleasant when raw because of their rough external texture, for instance *Anchusa azurea, Echium plantagineum* L. or *Picris echioides* L., though that can be mitigated when boiled, too.

As stated before, the raw species considered of better quality because of a pleasant flavour are mostly used in salads. The small leaves and stems of *Montia fontana,* the young shoots of *Chamaerops humilis* L., the basal leaves of *Reichardia picroides* (L.) Roth and the leaves and roots of *Campanula rapunculus* are some of the species most widely consumed raw in salads.

Finally, some examples of parts of green vegetables only eaten raw as a snack can be mentioned, such as: the basal part of the stems of *Scirpus holoschoenus*; the leaves of *Oxalis acetosella* L. and other species of the same genus; the peeled young shoots of *Rosa canina* L.; the unripe fruits of several species of *Erodium*; and the immature seeds of *Vicia villosa* Roth and *V. lutea* L.

Mixed Wild Vegetable Recipes

With regard to other curious recipes for cooking wild vegetables, in four provinces of southeastern and eastern Spain (Almería, Murcia, Valencia and Albacete) people eat a blend of some nine to fifteen different wild species in a vegetable stew. The recipes are known by different names, such as *las hierbas* (Martínez Lirola, González-Tejero and Molero Mesa 1997), *ensalada de matas* or *hervido* (Rabal 2000) or *herbes bullides* (Oltra 1998), and with the generic name *collejas* (Verde, Rivera and Obón 1998). In the provinces of Castellón, Valencia and Alicante, also in the east of Spain, a blend of wild vegetables is added to different kinds of vegetable pies known as *pastissets, minxos* or *coquetes* (Mulet 1991; Oltra 1998; Pellicer 2001; Pellicer 2004a). Table 10.2 shows the species used in some of those recipes.

The recipe called *las hierbas* comes from the Natural Park of Sierra del Cabo de Gata-Níjar in the province of Almería. The tender leaves of the different species are boiled, then drained and minced. Subsequently, they are added to a mixture of tomato, onion, pork fat and meat slightly fried in olive oil, along with boiled chickpeas and potatoes, and seasoned with dried red pepper, garlic, cumin, cinnamon, aniseed and salt. The whole mixture is then covered with water and boiled together till reduced.

In Torrepacheco, a town of the province of Murcia, the recipe of *hervido* or *ensalada de matas* is much easier to prepare. All the plants are boiled together, drained, and finally dressed with olive oil, vinegar and salt. Sometimes people prepare the recipe with fewer species, often only with *Beta maritima,* the most common species in the surroundings of the town. The dish *herbes bullides,* from the village of Quatretonda in the province of Valencia, is prepared in a very similar way. Only the less appreciated species are used in this dish. They are boiled and dressed with olive oil and vinegar, and sometimes with some tender young shoots of fennel (*Foeniculum vulgare*). Frequently, the boiled vegetables are slightly fried in olive oil with a bit of garlic.

In the southeast of the province of Albacete, in some of the villages studied by Verde, Rivera and Obón (1998), the blend of vegetables used in stews was called *collejas.* This is the name given in a great part of Spain to one of the most valued species, *Silene vulgaris,* known locally as *colleja fina* (see Table 10.2). *Collejas* vegetables are boiled, drained, made into balls and finally fried lightly in olive oil with a bit of garlic and vinegar, sometimes also with scrambled eggs. A similar recipe, *caldo verde,* is prepared in the same region with fewer species, for instance *Capsella bursa-pastoris, Silene vulgaris, Foeniculum vulgare, Cichorium intybus* and *Roemeria hybrida.*

According to Mulet (1991), the *pastissets de brosses* or *pastissets de verdura* are typical vegetable dishes of the Spanish region of Comunidad Valenciana. This author reported them to be prepared with *Sonchus asper, S. oleraceus, S. tenerrimus* and *Taraxacum officinale.* In the *pastissets de verdura,* mentioned by Pellicer (2001), at least two species are employed: *Silene vulgaris* and *Papaver rhoeas.* Oltra (1998) describes a recipe of *pastissets d'herbes* from Quatretonda (Valencia) with up to ten plant species. He said that this recipe is the favourite for people who like cooking with herbs and they use only the finest species. For making the *pastissets,* a dough of flour, olive oil, salt, lemonade or beer or wine and a pinch of sugar (some people add baking powder as well) is prepared. Circular pieces of dough are covered with the stewed blend of vegetables, codfish or tuna fish and

Table 10.2. Species used in different blends of wild vegetables mentioned in Spain. For comparative purposes, their presence in two Italian recipes is included. The symbol '*' represent the presence of a species and '+' the use of other species of the same genera.

Family / Species	**HIER**	**ENSA**	**COLL**	**HERB**	**PAST**	**PREB**	**MINE**
Total number of species	11	15	15	9	10	38	20-40
Apiaceae							
Eryngium campestre L.		*					
Foeniculum vulgare Mill.		*		*		*	*
Asteraceae							
Anacyclus clavatus (Desf.) Pers.	*	*	*				
Calendula arvensis L.	*						
Carthamus lanatus L.			*	*			
Chondrilla juncea L.			*				
Chrysanthemum coronarium L.	*	*					
Cichorium intybus L.	*	*		*		*	*
Crepis vesicaria L.		*				*	+
Hypochoeris radicata L.				*		*	*
Lactuca serriola L.					*		
Launaea nudicaulis (L.) Hook. fil.		*					
Leontodon taraxacoides (Vill.) Mérat					*	*	
Mantisalca salmantica (L.) Briq. & Cavill.		*	*				
Picris comosa (Boiss.) B.D. Jacks.			*			+	+
Reichardia intermedia (Sch. Bip.) Cout.					*	+	+
Reichardia tingitana (L.) Roth					*	+	+
Rhagadiolus stellatus (L.) Gaertn.			*				
Scolymus hispanicus L.	*						
Sonchus asper (L.) Hill				*			*
Sonchus oleraceus L.		*	*	*		*	*
Sonchus tenerrimus L.		*	*	*			

Table 10.2. *continued*

Urospermum picroides (L.) Scop. ex F.W. Schmidt				*	*	+	
Boraginaceae							
Anchusa azurea Mill.			*				
Borago officinalis L.			*			*	*
Brassicaceae							
Capsella bursa-pastoris (L.) Medik.			*				
Eruca vesicaria (L.) Cav.			*				
Caryophyllaceae							
Silene diversifolia Otth					*		+
Silene vulgaris (Moench) Garcke	*	*	*		*	*	*
Chenopodiaceae							
Beta maritima L.	*	*			*	+	*
Fabaceae							
Scorpiurus muricatus L.	*	*					
Papaveraceae							
Papaper rhoeas L.	*		*		*	*	
Roemeria hybrida (L.) DC.			*				
Plantaginaceae							
Plantago coronopus L.		*		*	*	+	
Plumbaginaceae							
Limonium sinuatum (L.) Mill.	*						
Polygonaceae							
Emex spinosa (L.) Campd.		*					
Rumex acetosa L.	*					*	

HIER: 'Las hierbas', Martínez-Lirola, González-Tejero and Molero Mesa 1997; ENSA: 'Ensalada de matas', Rabal 2000; COLL: 'Collejas', Verde, Rivera and Obón 1998; HERB: 'Herbes bullides', Oltra 1998; PAST: 'Pastissets d'herbes', Oltra 1998; PREB: 'Prebuggiun', Bissio and Minuto 1998; MINE: 'Minestrella', Pieroni 1998.

bits of hard-boiled egg and finally baked in the oven. In a variation of the recipe, the *pastissets* are fried in olive oil.

Minxos, from the province of Alicante, are little pancakes, usually made with corn flour kneaded with a bit of olive oil, covered or stuffed with fried green vegetables. According to Pellicer (2004a), in Castell de la Serrella (Alicante) people use at least *Centaurea calcitrapa*, *Sonchus oleraceus*, *Papaver rhoeas* and *Foeniculum vulgare*. Finally, *coquetes* are rolls stuffed with vegetables, such as *Silene vulgaris* and *Papaver rhoeas* (Pellicer 2001).

Similar examples of recipes with a varied mix of wild vegetables have been noted in Italy, although sometimes with a much higher number of species being used. One of them, from the northwest of Italy (Liguria), is the *prebuggiun*, a blend of plants gathered in the fields and then added to soups, omelettes or stuffing for more elaborate dishes (Bisio and Minuto 1998). Another similar ethnobotanical reference to mixed wild vegetables has been reported by Pieroni (1999) from north Tuscany in central Italy. In that region, the most common culinary preparation of vegetables is in the form of vegetable soups, usually mixing wild and cultivated species. In Gallicano, there is a recipe, called *minestrella*, prepared from twenty to forty different species. Finally, in the south of Italy, in the region of Basilicata, a kind of soup called *mënestra maritate* is prepared with more than ten wild herbs (Pieroni 2003) and in Calabria there are also recipes that include a blend of at least eight wild vegetables (Nebel 2005).

The fact that similar blends of vegetables have been consumed in different parts of southern Europe suggests a common past, perhaps the remains of an ancient Mediterranean custom. But there are enough reasons to consider that these blends of leafy vegetables might in fact be a more general custom, typical of all human groups and acquired during the evolution of our species. According to Johns (1994), these types of wild vegetable mixtures, also consumed in Africa, have been maintained for two main reasons. On one hand, the mixture of bitter and nonbitter vegetables, sometimes even with cultivated ones, can be a way of making the bitter ones more palatable. The same idea is expressed by Verde, Rivera and Obón (1998) and Rabal (2000), who report that the mix of different species is made for covering bad flavours; for example, the use of fennel (*Foeniculum vulgare*) hides the bitterness of *Sonchus oleraceus* and *S. tenerrimus*. On the other hand, following Johns (1994), this behaviour of mixing wild vegetables maintains the knowledge of the usefulness of some plants of lesser quality. So, when the preferred wild vegetables become scarce for climatic reasons, people simply increase the proportion of available edible plants in the mix. This explanation supports findings by Oltra (1998), who states that the dish of *herbes bullides*, although now quite rare, was surely an important part of the diet of their ancestors, and even during the hard times of the last century, was a complement to the daily diet during cold winter days.

The Seasonality of Harvesting and Consumption of Wild Vegetables

Most green vegetables are consumed fresh, therefore harvesting and consumption times coincide. They are primarily collected in spring and hence some popular proverbs connect them with its arrival, as in this chapter's opening quotation. There are several species, such as *Silene vulgaris* and *Rumex pulcher* (Figure 10.7), whose harvest is often for dishes traditionally consumed during Lent, the period from Ash Wednesday to Easter, covering the end of winter and the first part of spring.

There is great climatic diversity across Spain's various regions resulting from the range of latitudes and altitudes. Therefore, the exact time of gathering vegetables varies depending on the region and sometimes, due to the irregularity of the Mediterranean climate, on the year. In warmer areas, especially in the lowlands of the south and east of Spain, wild vegetables can be harvested in winter and in some cases even in autumn, whereas in colder and mountainous areas, harvesting can last until the beginning of summer. Some species, such as *Rorippa nasturtium-aquaticum, Silene vulgaris* or *Allium ampeloprasum*, are collected for a second time in autumn, because of the mild wet weather of the east and centre of the country. In Extremadura, a western region with mild winters, there are several species typically collected from December to March, such as *Montia fontana, Crepis vesicaria, Tolpis barbata* (L.) Gaertn., *Hypochoeris radicata* L., *Leontodon taraxacoides* (Vill.) Mérat, *Rumex pulcher* and *Scolymus hispanicus* (Blanco and Cuadrado 2000).

But in most of the regions, and for most of the species, the typical gathering season for vegetables is the end of winter and the beginning of spring. This strong seasonality of wild vegetables is precisely, in my opinion, one of the reasons for its maintenance in agricultural societies, since it complements the harvesting period for the majority of cultivated plants. Nevertheless, at present many kinds of cultivated vegetables can be found in the markets, at almost any time of the year. So, wild vegetables are no longer necessary and their consumption today is much less than in the past.

Social and Cultural Aspects of Gathering Vegetables

According to Etkin (1994), one of the reasons for neglecting wild plants in diet surveys is the fact that women and children are involved in the collection of these foods. She also suggests that while recent studies have included women-centred activities, they could be expanded by considering children's work, because of the range and quantity of food consumed and also collected by them. Although our focus when compiling the information in the database was on the consumed plant species rather

Figure 10.7. **A** Tender basal leaves of *Rumex pulcher*; **B** Tender leaves of *Silene vulgaris*

than the social and cultural aspects of the gathering, we did also record some relevant data about those subjects.

In Spain, there are many noncultivated vegetables that are typically consumed by children. They are among the species eaten raw as a snack and some of the most frequently mentioned are listed here: the unripe fruits of *Malva sylvestris* and other *Malva* species (called *panecillos* 'little bread'); the basal part of the stems of *Scirpus holoschoenus*; the leaves and stems of several *Rumex* species with an acidic flavour (*R. acetosa*, *R. induratus* Boiss. & Reut., *R. scutatus* L., *R. acetosella* L.); the leaves and stems of *Oxalis acetosella* and other species of the same genus, also with acidic taste; the stem and leaves of *Foeniculum vulgare* and *Scandix australis* L., with their characteristic aniseed taste; the tender parts of the unripe inflorescence, such as the bottom of the inflorescence of *Scorzonera laciniata* L. or the 'artichokes' of *Silybum marianum*; the unripe fruits of several species of *Erodium* (*E. cicutarium* (L.) L'Hér., *E. malacoides* (L.) L'Hér.); or the immature seeds of various species of the Fabaceae family (*Lathyrus cicera* L., *Vicia villosa* and *V. lutea*).

Regarding the role of women across Spain, they have a rich knowledge of wild plant resources because they are usually responsible for the activities of foraging and also for the processing, preservation and preparation of the food. This issue has been reported from different cultures and countries around the world, such us Sudan (Gullick 1999), Nepal (Daniggelis 2003), Canada (Turner 2003) and Italy (Pieroni 2003). In Spanish ethnobotanical sources, some references also state that women usually undertake the gathering and preparing of wild vegetables (Oltra 1998; Verde, Rivera and Obón 1998; Blanco and Cuadrado 2000; Tardío, Pascual and Morales 2002; Pellicer 2004b). They mention that sometimes women go in groups to seek out wild green vegetables and, at times, they collect along the wayside while going about other chores, such as washing clothes in the rivers.

There are some exceptions to this general rule and men are sometimes mentioned as gatherers of wild food plants. That is the case for some of our informants in Madrid who earn extra income selling wild vegetables and mushrooms to restaurants. A more typical case is the consumption of raw vegetables by men while hunting in the field, such as the widely used stems of *Foeniculum vulgare* or the more restricted use of the leaves of *Carduncellus dianius* Webb (Pellicer 2004b). But one of the most important occasions where men collect and prepare wild vegetables is when they are working as shepherds far away from their houses for long periods of time. Typical of transhumance pastoralists, during the winter the animals graze on the lowland pastures of mild-climate regions and at the end of spring the animals are taken up to the mountain pastures where they graze for the summer. During those long periods, shepherds prepare their own food, sometimes using wild plants, such as *Allium ampeloprasum* or *Rumex papillaris* Boiss. & Reut., for making milk soups and other recipes with *Scolymus hispanicus*, *Onopordum macracanthum* Schousb., *Papaver rhoeas* and *Chenopodium bonus-henricus* L. There is even a species of the Asteraceae family, *Mantisalca salmantica* (L.) Briq. & Cavill., whose Spanish name is *pan de pastor* ('shepherd's bread').

Most wild green vegetables are not appreciated by the general public, and are often considered as symbols of poverty (González Turmo 1997). They are frequently associated with scarcity periods, such as 'the years of famine' in the 1940s after the Spanish Civil War. This is particularly the case for species with a bitter taste, such as *Arctium minus* Bernh. and *Cichorium intybus*. For other species, the local name captures this connotation; for instance, one of the Spanish names used for *Bryonia dioica* Jacq. is *espárragos de pobre* (poor people's asparagus). This negative reference can also be found in popular literature, such as the popular Spanish proverb mentioned at the beginning of this chapter.

However, this is not always the case. Wealthy people often consume wild vegetables too, presumably because of their taste. For instance, *Scolymus hispanicus* and *Silene vulgaris* were collected by poor people and sold to the richer citizens of the villages (Sánchez López et al. 1994; Fernández Ocaña 2000). These two species, together with *Asparagus acutifolius* and in some regions *Montia fontana*, *Cichorium intybus*, *Cynara humilis* L. and *Silybum marianum*, have been sold in markets or door-to-door. Nowadays most of them are rarely found in markets, except for the midribs of *Scolymus hispanicus*, which, although not easily available, are possible to obtain prepared and preserved in bottles.

Wild Vegetables: Food or Medicine?

Over the last 150 years or so, biomedicine has increasingly focused attention on the specificity of both disease and treatment and by doing so, has positioned food outside the domain of therapeutics. Science has regarded food as chemically mundane, of no relevance to the disease process (Etkin 1996). But this was not the case either during the history of biomedicine or in Spanish popular culture. Food and health have always been clearly interrelated, as is shown in the seventeenth-century literary classic *Don Quixote*, where Cervantes (1615: XLII, 161v) claims 'the health of the whole body is forged in the workshop of the stomach'. In recent decades, among the scientific and medical elites, the idea of the multifunctionality of foods and their influence on health has been renewed and there is a growing literature and commercial interest in 'functional foods', 'pharmafoods' and 'nutraceuticals'.

Recent ethnobotanical studies have shown that many wild food plants are also used for medicinal purposes throughout the world (Coupland 1989; Etkin 1994; Pieroni 1999; Bonet and Vallès 2002; Pieroni et al. 2002b; Guarrera 2003; Verde et al. 2003; Pieroni and Price 2006). In our compilation of wild vegetables traditionally used in Spain, at least 47 of the 206 species recorded are also said to be used as medicinal plants for internal use. The young stems and leaves of *Malva sylvestris* are considered useful as an anti-catarrhal, either in decoctions (sometimes with the flowers, and sweetened with honey) (Villar et al. 1987; Gil Pinilla 1995), or even eaten boiled as a vegetable (Galán 1993); this last preparation has also been mentioned to cure stomach ache (Pellicer 2001). Both Theophrastus (third century BC) and Dioscorides (first century AD) recorded similar medicinal uses for this species (Teofrasto 1988; Laguna 1555).

There are many other examples of plants used for both purposes, such as *Apium nodiflorum* (L.) Lag., reported as a digestive (Verde et al. 2003) or as an intestinal anti-inflammatory (Bonet and Vallès 2002); *Mantisalca salmantica* (Triano et al. 1998; Pellicer 2004b) and *Reichardia picroides* (Bonet and Vallès 2002; Parada et al. 2002), reported as hypoglycemiants, or *Taraxacum officinale* and other species of the same genus, reported as hepatoprotectors (Bonet and Vallès 2002; Verde et al. 2003).

Other wild vegetables have lost their function as a medicine, which they had in ancient times, and are just considered foods today. *Bryonia dioica* and *Tamus communis* L. (Figure 10.6C), whose young shoots are widely eaten as asparagus, were considered by Dioscorides in the first century AD to be diuretics and have other medicinal qualities (Laguna 1555). In those cases, the current food use could be a reminder of their ancient medicinal use. This seems to agree with the model of 'nonfood first' or 'medicinal first' that was pointed out by Etkin and Ross (1994), studying a Hausa population in northern Nigeria. They suggested that many of the wild edible plants used by Hausa were first discovered and used as medicines, instead of the conventional model of 'food first' that states that people learn about medicines only secondarily to their search for food. In my opinion, local knowledge of food and that of medicine are so interconnected, and evidence for origin of uses is so slim, that knowledge acquisition and transformation could easily be explained by both models. Nevertheless, how foods became medicines or vice-versa is surely less important than the fact of their multifunctionality.

The medicinal value of some foods is not just for healing diseases but also for providing health-enhancing substances other than calories and proteins, such as vitamins and minerals. These micronutrient supplements are critical in a diet based on staple agricultural and livestock products. *Rorippa nasturtium-aquaticum* (watercress), one of the most commonly harvested wild vegetables in Spain and other parts of the world, is a very wholesome and nutritious plant with a well known reputation as a spring tonic because of its exceptional richness in vitamins and minerals, especially iron, found in its leaves (PFAF 2003). That said, recent studies seem to show that the beneficial effects of fruits and vegetables on the risk of cardiovascular disease and cancer may not rely on the effect of the well characterized antioxidants, such as vitamin E and C and β-carotene, but rather on other antioxidants or nonantioxidant phytochemicals, or on an additive action of different compounds present in foods such as α-linolenic acid, various phenolic compounds and fibre (Simopoulos 2003).

Although still quite scarce, there are some recent studies on the nutritional and functional composition of some wild species used in the Mediterranean area. Bianco and Santamaria (2003) investigated the nutritional value of some edible species used in southern Italy and found high levels of minerals often greater than several cultivated vegetables. Zeghichi et al. (2003) described the nutritional composition of twenty-five wild edible plants of Crete and pointed out that all plants contained considerable amounts of antioxidants and minerals. They found, for example, a high concentration of phenols in *Crepis vesicaria*, one of the vegetables used in Spain. This species, and also *Papaver rhoeas*, showed in vitro antioxidant activity (Pieroni et al. 2002a). Schaffer et al. (2005) found promising antioxidant activity in *Cichorium intybus, Sonchus oleraceus* and *Papaver rhoeas*, though they found important differences in plants of different origin and an unclear correlation between the antioxidant activity and the polyphenol content. The works of Guil-Guerrero and collaborators

(Guil-Guerrero, Rodríguez-García and Torija-Isasa 1997; Guil-Guerrero et al. 1998; Guil-Guerrero, Giménez-Martínez and Torija-Isasa 1998; Guil-Guerrero and Rodríguez-García 1999; Guil-Guerrero, Rebolloso-Fuentes and Torija-Isasa 2003) discovered, for example, high concentrations of ascorbic acid in *Chenopodium album* and *Sonchus oleraceus*, high levels of carotenoids in *Urtica dioica* L. and *Sonchus oleraceus* and a high content of α-linolenic in the leaves of *Urtica dioica*. Recent studies of *Silene vulgaris*, one of the main Spanish wild vegetables, have found good nutritional potential with a remarkably high level of essential fatty α-linolenic (18:3ω-3) and linoleic (18:2ω-6) acids (Alarcón, Ortiz and García 2006). Hamazaki and Okuyama (2001) emphasise the great importance of ω-3 fatty acids for normal growth and development. They add that these compounds may play an important role in the prevention and treatment of coronary artery disease, hypertension, arthritis, other inflammatory and autoimmune disorders and cancer.

Finally, medicinal plants are often toxic when consumed in large quantities or if not processed by leaching or cooking. Such toxic plants used as vegetables in Spain include *Tamus communis*, *Bryonia dioica*, *Clematis vitalba* L., *Atractylis gummifera*, *Papaver rhoeas* and *Lathyrus cicera* (Tardío, Pardo-de-Santayana and Morales 2006). Fortunately, however, in all cases the part consumed is either free of toxic compounds or contains only low levels.

Conclusions and Future Possibilities

This chapter has summarized and discussed our current knowledge of the traditional uses of wild vegetables in Spain, but there is scope for plenty of future research, both on the ecology, physiology and nutritional and medicinal properties of the plants, and on the economics and cultural aspects of plant use.

Although wild green vegetables have been a very important supplement to the diet and even used as medicines in the past, nowadays the tradition of using most of them is all but disappearing, mainly because they are thought not to be necessary. The Spanish ethnobotanical surveys carried out in the last decades, and perhaps those that could be urgently completed in the next years, will contribute, at least, to a register of this traditional knowledge and avoid losing it forever. However, no written record of plants and uses can capture all the cultural knowledge embedded in the practices associated with harvesting, processing and consuming wild plants. Thus it may be necessary in future survey work to use video and audio documentation to fully capture the local knowledge needed for an effective knowledge bank, which will make it much easier, more natural and more human to transmit that information to future generations.

Fortunately, the future for wild plant collecting appears more promising than ever, as there appears to be a renewed interest in wild edibles in Spain and elsewhere, especially in mushrooms. There are several reasons for this new interest. One of them may be a higher awareness of environmental issues and the desire of urban dwellers to reconnect with rural areas and lifestyles. Urban people like to look for plants or mushrooms as they walk through the countryside. The rediscovery of the positive influence of the consumption of fruits and vegetables in human health; the popularity of herbal medicine; and also the rising literature on the nutritional potential of noncultivated vegetables of the Mediterranean area may also have contributed to this increasing interest in wild plants. Finally, from a culinary point of view, the gastronomic interest of urban chefs and diners in discovering 'new' flavours from what are really traditional dishes has transformed some wild plants from former poverty foods to speciality products and gourmet ingredients.

For all these reasons, traditional knowledge about wild vegetables could be an attractive and profitable resource for the rural development of many regions, especially the less affluent ones. The exploitation of these natural and cultural resources in small-scale enterprises could represent a significant source of income for the inhabitants of those regions. These activities could be included either in the tourism industry or in the agrofood sector. Local rural or ecotourism enterprises could offer courses to teach urban people how to recognize, gather and cook wild vegetables. Local restaurants, small inns and hotels could include traditional dishes of noncultivated vegetables on their menus.

On a more commercial scale, sustainable harvesting, organic production, processing and marketing of natural plant products for sale locally or overseas is also a future possibility. Spain already exports huge amounts of cultivated vegetables, so adding noncultivated or semicultivated products may not require substantial investments to establish commodity chains. However, many of these products might only appeal to niche markets and so a substantial marketing campaign might be required. But then, with growing interests in organic foods, slow food, etc., now might be the time for Spain's edible greens to expand into other markets in Europe and elsewhere.

Note

1. The Iberian Flora figures are compiled from the incomplete work *Flora Iberica* (Castroviejo et al. 1986–2009) and *Flora Europaea* (Tutin et al. 1964–1980), except for the advanced data from the ongoing project *Flora Iberica* for the families *Lamiaceae* and *Geraniaceae* (Ramón Morales personal communication).

References

Alarcón, R., L.T. Ortiz and P. García. 2006. 'Nutrient and Fatty Acid Composition of Wild Edible Bladder Campion Populations (*Silene vulgaris* [Moench.] Garcke)', *International Journal of Food Sciences and Technology* 41: 1239–1242.

Baker, H.G. 1974. 'The Evolution of Weeds', *Annual Review of Ecology and Systematics* 5: 1–24.

Bianco, V.V. and P. Santamaria. 2003. 'Nutritional Value and Nitrate Content in Edible Wild Species Used in Southern Italy', *Acta Horticulturae (ISHS)* 467: 71–87.

Bisio, A. and L. Minuto. 1998. 'The Prebuggiun', in A. Pieroni (ed.), *Erbi, erbi degli streghi / Good Weeds, Witches' Weeds*. Köln: Experiences Verlag, pp. 34–46.

Blanco, E. and C. Cuadrado. 2000. *Etnobotánica en Extremadura: Estudio de La Calabria y La Siberia Extremeñas.* Madrid: Emilio Blanco, CEP Alcoba de los Montes.

Bonet, M.A. and J. Vallès. 2002. 'Use of Non-crop Food Vascular Plants in Montseny Biosphere Reserve (Catalonia, Iberian Peninsula)', *International Journal of Food Sciences and Nutrition* 53: 225–248.

Carretero, J.L. 2004. *Flora arvense española: Las Malas hierbas de los cultivos españoles.* Valencia: PHYTOMA-España.

Castroviejo S., C. Aedo, J.J. Aldasoro, C. Benedí, F. Cabezas, S. Cirujano, J.A. Devesa, A. Galán, C. Gómez Campo, R. Gonzalo, J. Güemes, I.C. Hedge, A. Herrero, P. Jiménez Mejías, S. Jury, M. Laínz, G. López González, M. Luceño, L. Medina, P. Montserrat, R. Morales, F. Muñoz Garmendia, C. Navarro, G. Nieto Feliner, J. Paiva, E. Rico, C. Romero Zarco, L. Sáez, F. Sales, F.J. Salgueiro, C. Soriano, S. Talavera, M. Velayos and L. Villar (eds). 1986–2009. Flora iberica: *Plantas vasculares de la Península Ibérica e Islas Baleares.* Vols 1–8, 10, 13-15, 18 and 21. Madrid: Real Jardín Botánico-CSIC.

CBD (Convention on Biological Diversity). 2005. *Report of the Consultation on the Cross-cutting Initiative on Biodiversity for Food and Nutrition.* Brasilia, 12–13 March. Retrieved February 2006 from http://www.biodiv.org/doc/meetings/agr/ibfn-01/official/ibfn-01-03-en.doc

Cervantes, M. 1615. *Segunda Parte del ingenioso caballero don Quijote de la Mancha.* Madrid: Juan de la Cuesta. Retrieved February 2006 from http://www.cervantesvirtual.com/servlet/SirveObras/579713971051377285000080/index.htm

Coupland, F. 1989. *Le regal vegetal: Plantes sauvages comestibles, Encyclopedie des plantes comestibles de l'Europe.* Volume 1. Flers: Equilibres Aujourd'hui.

Daniggelis, E. 2003. 'Women and "Wild" Foods: Nutrition and Household Security among Rai and Sherpa Forager-farmers in Eastern Nepal', in P.L. Howard (ed.), *Women and Plants: Gender Relations in Biodiversity, Management and Conservation.* London: Zed Books, pp. 83–97.

Díaz-Betancourt, M., L. Ghermandi, A.H. Ladio, I.R. López-Moreno, E. Raffaele and E.H. Rapoport. 1999. 'Weeds as a Source for Human Consumption: a Comparison between Tropical and Temperate Latin America', *Revista de Biología Tropical* 47(3): 329–338.

Etkin, N.L. 1994. 'The Cull of the Wild', in N.L. Etkin (ed.), *Eating on the Wild Side*. Tucson, Ariz.: The University of Arizona Press, pp. 1–21.

——— 1996. 'Medicinal Cuisines: Diet and Ethnopharmacology', *International Journal of Pharmacognosy* 34(5): 313–326.

——— and P.J. Ross. 1994. 'Pharmacologic Implications of "Wild" Plants in Hausa Diet', in N.L. Etkin (ed.), *Eating on the Wild Side*, Tucson, Ariz.: The University of Arizona Press, pp. 85–103.

Fernández Ocaña, A.M. 2000. *Estudio etnobotánico en el Parque Natural de las Sierras de Cazorla, Segura y Las Villas: Investigación química de un grupo de especies interesantes*, Ph.D. dissertation. Jaén: Facultad de Ciencias Experimentales, Universidad de Jaén.

Galán, R. 1993, *Patrimonio etnobotánico de la provincia de Córdoba: Pedroches, Sierra Norte y Vega del Guadalquivir*, Ph.D. dissertation. Córdoba: E.T.S. de Ingenieros Agrónomos y Montes, Universidad de Córdoba.

Gil Pinilla, M. 1995. *Estudio etnobotánico de la Flora aromática y medicinal del término municipal de Cantalojas (Guadalajara)*, Ph.D. dissertation. Madrid: Universidad Complutense de Madrid.

González Turmo, I. 1997. *Comida de rico, comida de pobre: Evolución de los hábitos alimentarios en el occidente andaluz (Siglo XX)*. 2nd edn. Seville: Universidad de Sevilla.

Guarrera, P.M. 2003. 'Food Medicine and Minor Nourishment in the Folk Traditions of Central Italy (Marche, Abruzzo and Latium)', *Fitoterapia* 74: 515–544.

Guil-Guerrero, J.L., A. Giménez-Giménez, I. Rodríguez-García and M.E. Torija-Isasa. 1998. 'Nutritional Composition of *Sonchus* Species (*S. asper* L., *S. oleraceus* L. and *S. tenerrimus* L.)', *Journal of the Science of Food and Agriculture* 76(4): 628–632.

———, J.J. Giménez-Martínez and M.E. Torija-Isasa. 1998. 'Mineral Nutrient Composition of Edible Wild Plants'. *Journal of Food Composition and Analysis* 11: 322–328.

———, M.M. Rebolloso-Fuentes and M.E. Torija-Isasa. 2003. 'Fatty Acids and Carotenoids from Stinging Nettle (*Urtica dioica* L.)', *Journal of Food Composition and Analysis* 16(2): 111–119.

——— and I. Rodríguez-García. 1999. 'Lipids Classes, Fatty Acids and Carotenes of the Leaves of Six Edible Wild Plants', *European Food Research and Technology* 209(5): 313–316.

———, I. Rodríguez-García and M.E. Torija-Isasa. 1997. 'Nutritional and Toxic Factors in Selected Wild Edible Plants', *Plant Foods for Human Nutrition* 51(2): 99–107.

Gullick, C. 1999. 'Wild Foods: Blessing or Burden?', *ENN, Field Exchange*, February 99: 16–17.

Hamazaki, T. and H. Okuyama. 2001. 'Fatty Acids and Lipids – New Findings', *World Review of Nutrition and Dietetics* 88: 1–260.

Heinrich, M., M. Leonti, S. Nebel and W. Peschel. 2005. '"Local Food-nutraceuticals": an Example of a Multidisciplinary Research Project on Local Knowledge', *Journal of Physiology and Pharmacology* 56(Suppl. 1): 5–22.

Johns, T. 1994. 'Ambivalence to the Palatability Factors in Wild Plants', in N.L. Etkin (ed.), *Eating on the Wild Side*. Tucson, Ariz.: The University of Arizona Press, pp. 46–61.

Laguna, A. 1555[1991]. *Pedacio Dioscorides Anazarbeo, Acerca de la materia médica medicinal y de los venenos mortíferos*. Facsimile edn 1991. Madrid: Comunidad de Madrid.

Martínez Lirola, M.J., M.R. González-Tejero and J. Molero Mesa. 1997. *Investigaciones etnobotánicas en el Parque Natural de Cabo de Gata-Níjar (Almería)*. Almería: Sociedad Almeriense de Historia Natural.

Mesa, S. 1996. *Estudio Etnobotánico y Agroecológico de la comarca de la Sierra de Mágina (Jaén)*, Ph.D. dissertation. Madrid: Universidad Complutense de Madrid.

Moerman, D.E. 1991. 'The Medicinal Flora of Native North American: an Analysis', *Journal of Ethnopharmacology* 31: 1–42.

——— 1994. 'North American Food and Drug Plants', in N.L. Etkin (ed.), *Eating on the Wild Side*. Tucson, Ariz.: The University of Arizona Press, pp. 1–21.

Mulet, L. 1991. *Estudio etnobotánico de la Provincia de Castellón*. Castellón: Diputación de Castellón.

Nebel, S. 2005. *Ta chòrta: Piante commestibili tradizionali di Gallicianò*. London: School of Pharmacy, University of London.

Ogoye-Ndegwa, C. and J. Aagaard-Hansen. 2003. 'Traditional Gathering of Wild Vegetables Among the Luo of Western Kenya – a Nutritional Anthropology Project', *Ecology of Food Nutrition* 42: 69–89.

Oltra, J.E. 1998. *Fer herbes a Quatretonda*. [Quatretonda]: Col·lectiu Cultural Dorresment.

Parada, M., A. Selga, M.A. Bonet and J. Vallès. 2002. *Etnobotànica de les terres gironines: natura i cultura popular a la plana interior de l'Alt Empordà i de les Guilleries*. Girona: Diputació de Girona.

Pellicer, J. 2001. *Customari botànic: Recerques etnobotàniques a les comarques centrals valencianes*. 2nd edn. Picanya: Edicions del Bullent.

——— 2004a. *Customari botànic (2): Recerques etnobotàniques a les comarques centrals valencianes*. 2nd edn. Picanya: Edicions del Bullent.

——— 2004b. *Customari botànic (3): Recerques etnobotaniques a les comarques centrals valencianes*. Picanya: Edicions del Bullent.

PFAF. 2006. Plants for a Future: Edible, Medicinal and Useful Plants for a Healthier World. Database, accessed February 2006 at http://www.pfaf.org

Pieroni, A. 1999. 'Gathered Wild Food Plants in the Upper Valley of the Serchio River (Garfagnana), Central Italy', *Economic Botany* 53(3): 327–341.

——— 2003. 'Wild Food Plants and Arbëresh Women in Lucania, Southern Italy', in P.L. Howard (ed.), *Women and Plants: Gender Relations in Biodiversity, Management and Conservation*. London: Zed Books, pp. 66–82.

———, V. Janiak, C.M. Dürr, S. Lüdeke, E. Trachsel and M. Heinrich. 2002a. '*In vitro* Antioxidant Activity of Non-cultivated Vegetables of Ethnic Albanians in Southern Italy', *Phytotherapy Research* 16: 467–473.

———, S. Nebel, C. Quave, H. Münz and M. Heinrich. 2002b. 'Ethnopharmacology of Liakra: Traditional Weedy Vegetables of the Arbëreshë of the Vulture Area in Southern Italy', *Journal of Ethnopharmacology* 81: 165–185.

———, S. Nebel, R.F. Santoro and M. Heinrich. 2005. 'Food for Two Seasons: Culinary Uses of Non-cultivated Local Vegetables and Mushrooms in a South Italian Village', *International Journal of Food Sciences and Nutrition* 56(4): 245–272.

——— and L.L. Price (eds). 2006. *Eating and Healing: Traditional Food as Medicine*. New York: Food Product Press, Haworth Press.

Rabal, G. 2000. '"Cuando la chicoria echa la flor..." Etnobotánica en Torre Pacheco', *Revista Murciana de Antropología* 6: 1–240.

Rivera, D., J. Fajardo, A. Verde, C. Obón and C. Inocencio. 2004. 'Las plantas y las setas (silvestres y sinantrópicas) recolectadas en la alimentación tradicional de la provincia de Albacete', in A. Verde and J. De Mora (eds), *Actas de las II Jornadas del Medio Natural Albacetense*. Albacete: Instituto de Estudios Albacetenses, pp. 150–162.

Sánchez López, M.D., J.A. García Sanz, A. Gómez Merino and S. Zon Blanco. 1994. *Plantas útiles de la comarca de la Manchuela*. Albacete: Colectivo de Escuelas Rurales de la Manchuela.

Schaffer, S., S. Schmitt-Schillig, W.E. Müller and G.P. Eckert. 2005. 'Antioxidant Properties of Mediterranean Food Plants Extracts: Geographical Differences', *Journal of Physiology and Pharmacology* 56(Suppl. 1): 115–124.

Simopoulos, A.P. 2003. 'Preface', in A.P. Simopoulos and C. Gopalan (eds), *Plants in Human Health and Nutrition Policy*. Basel: Karger, pp. vii–xiii.

Stepp, J.R. 2004. 'The Role of Weeds as Sources of Pharmaceuticals', *Journal of Ethnopharmacology* 92: 163–166.

——— and D.E. Moerman. 2001. 'The Importance of Weeds in Ethnopharmacology', *Journal of Ethnopharmacology* 75: 19–23.

Tanji, A. and F. Nassif. 1995. 'Edible Weeds in Morocco', *Weed Technology* 9: 617–620.

Tardío, J., M. Pardo-de-Santayana and R. Morales. 2006. 'Ethnobotanical Review of Wild Edible Plants in Spain', *Botanical Journal of the Linnean Society* 152(1): 27–72.

———, H. Pascual and R. Morales. 2002. *Alimentos Silvestres de Madrid: Guía de Plantas y Setas de Uso Alimentario Tradicional en la Comunidad de Madrid*. Madrid: Ediciones La Librería.

———, H. Pascual and R. Morales. 2005. 'Wild Food Plants Traditionally Used in the Province of Madrid', *Economic Botany* 59(2): 122–136.

Teofrasto. 1988. [Theophrastus, third century BC]. *Historia de las Plantas*. Introduction, translation and notes by J.M. Díaz-Regañón. Madrid: Gredos.

The Local Food-Nutraceutical Consortium. 2005. 'Understanding Local Mediterranean Diets: a Multidisciplinary Pharmacological and Ethnobotanical Approach', *Pharmacological Research* 52: 353–366.

Triano, E.C., E. Ruiz Cabello, A. Fernández Luque, A. Gómez Miranda, A. Jiménez Conejo, J.A. Gutiérrez Campaña, J.A. Postigo, J. Castro Montes, J.F. Sánchez Najarro, J.R. Marín Osuna, M. Martos, M.D. Mérida Moral, M.J. Mérida Ramírez, R. Moral and R. Hinijosa. 1998. *Recupera tus tradiciones: Etnobotánica del Subbético Cordobés*. Carcabuey, Córdoba: Ayuntamiento de Carcabuey.

Turner, N.J. 2003. 'Passing on the News: Women's Work, Traditional Knowledge and Plant Resource Management in Indigenous Societies of North-western North America', in P.L. Howard (ed.),*Women and Plants: Gender Relations in Biodiversity, Management and Conservation*. London: Zed Books, pp. 133–149.

Tutin, T.G., V.H. Heywood, D.M. Burges, D.H. Moore, S.M. Valentine, S.M. Walters and D.A. Webb (eds). 1964–1980. *Flora Europaea*. Vols 1–5. Cambridge: Cambridge University Press.

Verde, A., D. Rivera, M. Heinrich, J. Fajardo, C. Inocencio, R. Llorach and C. Obón. 2003. 'Plantas alimenticias recolectadas tradicionalmente en la provincia de Albacete y zonas próximas, su uso tradicional en la medicina popular y su potencial como nutracéuticos', *Sabuco. Revista de Estudios Albacetenses* 4: 35–72.

———, D. Rivera and C. Obón. 1998. *Etnobotánica en la Sierras de Segura y Alcaraz: las plantas y el hombre*. Albacete: Instituto de Estudios Albacetenses.

Vieira-Odilon, L. and H. Vibrans. 2001. 'Weeds as Crops: the Value of Maize Field Weeds in the Valley of Toluca, Mexico', *Economic Botany* 55(3): 426–443.

Villar, L., J.M. Palacín, C. Calvo, D. Gómez García and G. Montserrat. 1987. *Plantas medicinales del Pirineo Aragonés y demás tierras oscenses*. Huesca: CSIC, Diputación de Huesca.

Zeghichi, S., S. Kallithraka, A.P. Simopoulos and Z. Kipriotakis. 2003. 'Nutritional Composition of Selected Wild Plants in the Diet of Crete', in A.P. Simopoulos and C. Gopalan (eds), *Plants in Human Health and Nutrition Policy*. Basel: Karger, pp. 22–40.

CHAPTER 11

Plants as Symbols in Scotland Today

VEERLE VAN DEN EYNDEN

Introduction

The thistle is the plant that has come to symbolize Scotland and the Scottish identity, and there exist various theories as to why the thistle has become such a ubiquitous Scottish symbol (Mabey 1996: 455; Milliken and Bridgewater 2004: 143). Initially used as a personal emblem by the Stuart kings in the fifteenth century, the thistle has been a national emblem since the sixteenth century. Better known, however, is the legend that the thistle was adopted as a Scottish emblem after the cries of a tenth-century Norse invader, who had stepped on thistles, alerted the Scottish to an imminent attack.

The thistle is truly used everywhere in Scottish society nowadays, as a logo for businesses and national institutions (Scottish Natural Heritage, National Trust for Scotland, the Scottish rugby team), as well as on stamps, Scottish pound coins, jewellery, biscuits, in architecture, etc. The Scottish Tourism Board awards not stars, but thistles, to value tourism facilities. The thistle represents Scottish identity, both to Scottish people themselves, and to visitors.

This is a typical example of a contemporary symbolic use of plants. During years of lecturing Scottish students on people and plant interactions, and collecting information on plant uses in Scotland, my experience is that most people think that when talking about plant symbolism, we refer to traditional symbolic uses or beliefs that belong to the past. People remember a rich Scottish plant lore, which has been widely described, even in contemporary books such as those by Beith (1995), Darwin (1996), Mabey (1996) and Milliken and Bridgewater (2004). Indeed, many of the plant symbols, beliefs and perceptions described in these books belong to the past, but this does not mean that certain old uses may not persist today, or that novel meanings of plants may have come into existence recently.

There is a need to update the status of plant symbolism in present times and to assess how relevant it may be nowadays. It is worth knowing how people today view historic uses and beliefs, and which traditional and contemporary uses are practised at present. The information presented here is by no means the result of a systematic review of all cultural aspects of plants in Scotland today. It is an initial collection of examples and themes that were observed by the author and about which informants were questioned, or that were mentioned by informants.

Information presented is based on personal observations made throughout the Scottish Highlands and Islands, informal talks with over eighty Scottish people from a variety of backgrounds and areas, and information elicited from undergraduate students of the BA Culture Studies at the University of the Highlands and Islands Millennium Institute (UHI). The thirty-two female and six male students, aged between 18 and 65, that have studied an ethnobotany module over the last three years, represent a cross-section of the population of the Scottish Highlands and Islands. Living and studying in Shetland, Orkney, Lewis, Mallaig, Inverness, Argyll and Moray, they have provided information from a variety of often remote areas and their communities. Between 2001 and 2005, students and informants were questioned about their perceptions of certain existing plant uses that were observed by the author, about plant-related beliefs they may have and about plant uses they may practise themselves.

This paper, rather than covering all ethnobotanical aspects of plant use, focuses on noneconomic types of plant use, which are typically symbolic in nature and found in art, ornamentation, literature, song, belief, ritual and everyday discourse. In this sense, then, I am interested in the *meanings* that plants have for a group of people.

Plant Symbols Today

Returning to the Scottish thistle, people give a variety of explanations when asked why they believe the thistle to be such an omnipresent Scottish symbol. Approximately one third of people questioned refer to the legend of a Scottish victory over Vikings (Mabey 1996: 455). The remainder see the thistle as embodying the Scots character and personality (resilient, prickly, tall and proud) or ascribe the symbolism to the thistle's common habitat on waste ground. The Scottish thistle does not refer to any particular species of thistle, although botanists have tried, without success, to establish a botanical link. For symbols, however, botanical identification is usually not important. When asked which plants people consider as typically Scottish, only half mention the thistle. The other half mention plants such as heather (*Calluna vulgaris* (L.) Hull), Scots pine (*Pinus sylvestris* L.), harebell (*Campanula rotundifolia* L.), rowan (*Sorbus aucuparia* L.), primrose (*Primula vulgaris* Huds.) and other species that are all common, abundant or linked with childhood memories.

A different contemporary plant symbol is the poppy (*Papaver rhoeas* L.). Each year in the weeks leading up to Remembrance Day (11 November), people throughout many Commonwealth countries wear paper poppy badges. On Remembrance Day itself paper poppy wreaths are laid on war memorials. The poem 'In Flanders fields the poppies blow / Between the crosses, row on row…', written by John McCrae at the war front in 1915, made the poppy a symbol for all victims of the First World War (Goody 1993: 306). This symbolism remains very strong today. Interestingly, the poppy became a symbol for Remembrance Day in Britain, but not in Flanders itself, nor anywhere else in the world.

Ornamental plants in general are being used as symbols of beauty and civic pride in villages and towns in certain parts of Scotland. Many communities participate in *Scotland in Bloom* competitions. Here, plants and flowers are used to give the village or town a certain image of neatness, prettiness and attractiveness. Some towns go beyond the traditional floral garden and park displays and combine plants with stereotypical Scottish symbols by arranging them in three-dimensional figures or statues that symbolize the town or area to the outside world (particularly to visiting tourists). Statues of golf players, dolphins, salmon and the Loch Ness monster, filled with living flowers and plants, decorate parts of northeastern Scotland all year round. Local residents, when asked, express mixed feelings about such displays and often do not feel that they personally relate to this symbolic plant use or that it represents their identity.

Plants for Celebrations and Life Events

Plants and flowers are frequently used in a variety of celebrations. Often people feel that any flower or plant can be used as a gift or for decorations. For certain occasions or purposes particular plant species are used. Red roses are the typical example, mentioned by most informants, symbolizing love. This is not a typical Scottish use, as roses are used for this purpose in many cultures. Use of plants or flowers for celebrations thus also has a certain symbolic character.

Holly (*Ilex aquifolium* L.), mistletoe (*Viscum album* L.), ivy (*Hedera helix* L.) and pine trees are frequently mentioned by informants as plants used for celebrations, in this case for Christmas. These evergreens have been used for this purpose for centuries. Informants, however, also associate other plants typically with Christmas celebrations: sage (*Salvia officinalis*

L.), an important ingredient of turkey stuffing; spices like cinnamon, clove, orange peel and nutmeg used in mulled wine; the local vegetable kale (*Brassica oleracea* var. *acephala* DC.); and poinsettia plants (*Euphorbia pulcherrima* Willd. ex Klotzsch).

Some people mention carnations (*Dianthus caryophyllus* L.) being used for buttonholes at weddings. Carnations are used for this purpose throughout Europe and further afield (Goody 1993). At Scottish weddings, white or purple heather (*Calluna vulgaris*) buttonholes are often worn. Elder informants report that heather would only be used if the groom were wearing highland dress (kilt and associated clothing) and that in the past it was not traditional to wear highland dress for weddings throughout Scotland. They say that the associated use of heather buttonholes is fairly recent. Nowadays, also at weddings where no highland dress is worn, heather buttonholes are frequently used (personal observation). This use of heather thus seems to be partly symbolic of a Scottish identity. Occasionally thistles are used as buttonholes, in a similar way.

Some informants associate lilacs (*Syringa vulgaris* L.) and lilies (*Lilium* spp.), in particular white lilies, with funerals and death and report that they would therefore not put these flowers in the house or even the garden. Flowers in general are extensively used at funerals and graveyards, with no apparent special preference for certain flower species.

A more recent plant use involves laying flowers on accident sites. This use has spread throughout Britain in the last decades, probably influenced by the media (Walter 1996). It is fairly common today, especially in the case of child victims (personal observation). As a consequence, local councils in Scotland have since 2004 started introducing guidelines for the removal of such displays after a certain time, with Aberdeenshire Council being the first to do so, not always to the liking of the public (Argo 2004; Gadher 2004).

Surviving Historical Plant Uses

Certain plant uses that were more common in the past survive to date, though their meanings may have changed or been lost. In some places in Scotland, *clootie* trees and wishing trees are still found. *Clootie* trees (*clootie* is Scots for a piece of cloth) are rag trees, from a variety of species, usually found near *clootie* wells. Wells and water were important in the past for healing purposes, and certain wells were used for healing very specific illnesses (Beith 1995). Some wells (not all) served as *clootie* wells. Traditionally, water from the well was drunk and a piece of clothing dipped in the well and hung in a nearby tree, to rid oneself of a disease or to make a wish. The piece of clothing would have been linked to the body part causing a problem.

Nowadays, the healing powers of the *clootie* wells may be dismissed, but rags, clothes and even plastic bags are still frequently hung in *clootie* trees. The Munlochy *clootie* well (or Saint Boniface well) on the Black Isle is the best known in Scotland. In fact, this *clootie* well inspired Ian Rankin in his novel *The Naming of the Dead* (Rankin 2008). Many trees of various species that stand alongside the well are covered in rags. Other *clootie* trees still in use today are those found near Saint Mary's well in Culloden Wood near Inverness, those near Saint Anthony's well in Lothian and a tree on Doon Hill in Aberfoyle. *Clootie* trees form part of tourist attractions in Ballachulish (Glencoe) and Traquair House (near Peebles). *Clootie* trees have also gained a different meaning, as can be seen in family activities like 'making a *clootie* tree' that sometimes feature during community events. A *clootie* tree carrying peace messages written on paper was made in Kelvingrove Park in Glasgow in 2003 (Friends of Kelvingrove Park, personal communication).

Similarly, coins may be pressed in the trunk of a wishing tree, whilst making a wish. The wishing oak on Innes Maree, a small island in Loch Maree in the western Highlands, has not survived this use and died of metal poisoning decades ago. The dead tree trunk is still there, covered in old and new coins, with euro coins indicating the contemporary use by tourists. Various nearby living trees on the island are now substituting the original wishing oak and have ever increasing numbers of coins pushed into their trunk (personal observation 2004). This wishing tree is also linked with a well, considered in the past to have medicinal properties for curing mental illnesses (Beith 1995: 138–141). Loch Maree and its islands now form a National Nature Reserve and Ramsar site, managed by Scottish Natural Heritage since 1994. Another wishing tree that only recently died is the hawthorn money tree near Oban.

About half of Scottish people spoken to had never heard of or seen *clootie* or coin trees. Most did not consider the hanging of rags or pressing of coins in trees to be a Scottish custom, and said they could not relate in any way to such custom. Most people see the continuation of this custom as an act of curiosity by foreign visitors, or as the result of a herd effect. Margaret Bennett (personal communication) and Beith (1995) have both reported local people still practising this custom. *Clootie* trees are indeed not typically Scottish. Rag trees can be found throughout the world, though the meaning of hanging rags in trees may vary from group to group (Dafni 2003; De Cleene and Lejeune 2003). Only in Scotland do they seem associated with wells. They certainly form a tourist attraction nowadays, as can be seen from the frequency with which the Munlochy trees are photographed and visited, and the presence of foreign coins in the Innes Maree wishing tree.

Another example of a traditional plant use that survives today is the *burry* man parade that takes place each August in South Queensferry near

Edinburgh. The *burry* man goes dressed in a costume covered from top to toe with *burrs*, the hooked fruit-heads of burdock (*Arctium minus* Bernh). The meaning of this plant use seems lost. Theories are that the *burry* man symbolizes either a spirit of vegetation and fertility or a scapegoat, whereby the clinging *burrs* symbolize the taking away of the town's evils (Darwin 1996: 75–76; Milliken and Bridgewater 2004: 166–168).

A last example of a traditional plant use that has drastically altered in time is in the making of Halloween lanterns. Although pumpkins are increasingly used throughout the Western world for this purpose, Scottish children, especially in northeastern Scotland, Orkney and Shetland, still use *neeps* (turnips, *Brassica rapa* L.) to make so-called 'neepy lanterns'. Informants also mention the occasional use of *tatties* (potatoes, *Solanum tuberosum* L.) for this purpose.

Plant Beliefs

Common Scottish plant beliefs mentioned by informants are those of the four-leafed clover (*Trifolium* spp.) and white heather (*Calluna vulgaris*) bringing luck, and rowan trees (*Sorbus aucuparia*) being planted near the entrance to the garden to protect a house from evil spirits. For the same reason rowan trees should not be cut down. Although most people questioned deny personally holding such beliefs nowadays, rowan trees are still frequently found in Scottish gardens, and some informants admit to have planted a rowan tree in their new garden out of tradition. Similarly, some informants admit to kissing under the mistletoe (*Viscum album*) at Christmas time, out of tradition. The traditional belief is that kissing under the mistletoe holds a promise of love, good fortune and marriage (De Cleene and Lejeune 2003: 416). As mentioned earlier, some people would not have plants typically associated with death, like white lilies or lilacs, in their house or garden.

Conclusion

These are selected examples of the mixture of traditional and altered plant uses that exist today in Scotland alongside more contemporary ones, and the views and opinions local people have about them. Certain uses are related to special occasions, celebrations or events, whereas others pertain to everyday activities. Many uses are plant-species specific, as is the case for heather, poppy and burdock. At other times, the use involves general categories of flowers, trees or any plant, as is the case for thistles, *clootie* trees and flowers used as gifts or memorials. Most uses described here are not necessarily typically Scottish, but rather British, pertaining to a western European culture or even more globally widespread. Those that are Scottish are plant symbols like thistle and heather, or uses related to specific local traditions or events. Particular uses may be unimportant to local people or even cause some aversion, but still play a role in society as a symbol of Scottishness or by providing an attraction for visitors.

This review is by no means complete. Cultural values can be found in many daily activities involving plants. Student projects of the BA Culture Studies course have indicated that plants still feature frequently in children's play (Christie P. unpublished work; Garson S. unpublished work). In the popular British pastime of gardening, plants often carry images of personal memories and associations for the gardener, which may be more significant than the more common aesthetic and economic values (Gee C. unpublished work; McCullagh C. unpublished work). With regard to contemporary representations of plants, heated debates are held over what to think of the use of fake pine trees to disguise mobile-phone masts in the landscape, as can be seen near Stirling.

This first overview and update of Scottish plant symbols forms the first phase of ongoing research into present-day cultural aspects of plants in Scotland, in particular the knowledge and perceptions Scots have about them. One thing is certain already: plant use in Scotland is loaded with contemporary meanings and is more than just folklore and past traditions.

References

Argo, A. 2004. 'Parents Condemn Memorials Move', *The Courier* (Dundee), 30 August 2004. Retrieved 30 November 2005 from http://www.thecourier.co.uk/output/2004/08/30/newsstory6279103t0.asp

Beith, M. 1995. *Healing Threads.* Edinburgh: Polygon.

Dafni, A. 2003. 'Why Are Rags Tied to Sacred Trees of the Holy Land?', *Economic Botany* 56(4): 315–327.

Darwin, T. 1996. *The Scots Herbal: the Plant Lore of Scotland.* Glasgow: Mercat Press.

De Cleene, M. and M.C. Lejeune. 2003. *Compendium of Symbolic and Ritual Plants in Europe.* Gent: Mens & Cultuur Uitgevers.

Gadher, D. 2004. 'Councils Ban Shrines to Road Crash Victims', *The Sunday Times.* 24 October 2004. Retrieved 30 November 2005 from http://www.timesonline.co.uk/article/0,,2087-1325531,00.html

Goody, J. 1993. *The Culture of Flowers.* Cambridge: Cambridge University Press.

Mabey, R. 1996. *Flora Britannica.* London: Sinclair-Stevenson.

Milliken, W. and S. Bridgewater. 2004. *Flora Celtica – Plants and People in Scotland.* Edinburgh: Birlinn Limited.

Rankin, I. 2008. Retrieved 23 July 2008 from http://www.ianrankin.net/pages/books/index.asp?PageID=12

Walter, T. 1996. 'Funeral Flowers: a Response to Drury', *Folklore* 107: 106–107.

CHAPTER 12

The Botanical Identity and Cultural Significance of Lithuanian *Jovaras*

An Ethnobotanical Riddle

DAIVA ŠEŠKAUSKAITĖ AND BERND GLIWA

Introduction

In a Lithuanian folk song a riddle is asked: *Kas nežydi vasarėlėj?* (What does not blossom in summertime?). It is answered: *Jaunas berneli, Kas aš do būčiau, Kad aš to nežinočiau? Dievo medelis, Žals jovarėlis – tas nežyd vasarėlėj* (Young lad, who would I be if I would not know the answer? The God tree, green *jovaras,* does not blossom in summertime) (Juška 1954c). The riddle raises an interesting question, what is a god tree? And why should anyone care that a tree does not blossom in the summer? In trying to answer this, it seems obvious that we must at least look to the answer offered, but this only provides us with a new riddle. What is *jovaras*?

The Lithuanian *jovaras,* synonymous with *jievaras,* appears frequently in folk songs and games. This is also true for the Polish *jawor,* Croatian *javor,* Ukrainian *javir,* Russian *javor* and other Slavic cognates. In most cases it is evident that the concept *jovaras* names a tree. In other cases it appears to refer to something made of timber, boughs or bast fibre. But which tree, and why is it so prominent in folklore?

This chapter presents the results of our hunt for the botanical identity and cultural significance of this concept and the underlying meaning of the riddle of the nonblossoming 'god tree'. In order to determine its identity, we examined dictionaries and the scientific literature. This research suggested several taxa for which the term *jovaras* has been or still is in use. We then looked at its mention in folklore, such as songs, which are often overlooked as a resource in ethnobotanical studies. The final step of our analysis required synthesising the evidence to produce etymologies of the Baltic and Slavic words, explaining both their formal derivations and their semantic development. From this analysis, we think we now understand the importance of the riddle above for Lithuanians and others in eastern Europe in the past and even today.

The Plant Names in Dictionaries and Botanical Literature

The first set of evidence we investigated indicates how difficult the task actually is, as the Lithuanian Botanical Dictionary (Dagys 1938), compiled from a great amount of botanic literature, plant name collections and other manuscripts lists *jovaras* as one of several names referring to the species *Acer pseudoplatanus* L. (sycamore maple), *Carpinus betulus* L. (European hornbeam), *Populus nigra* L. (black poplar) and the genera *Platanus* (sycamores) and *Crataegus* (hawthorns). The synonym *jievaras* shares similar referents, *Acer pseudoplatanus*, *Populus alba* L. (white poplar), and the genera *Platanus*. In fact, various names appear for the same plants: *javorėlis, ovaras, ievaras, ievarys, jovarklevas.* The twenty-volume Lithuanian Academic Dictionary (LKŽ 1956–2002), which absorbed the Lithuanian Botanical Dictionary, contains additional data collected during the latter half of the twentieth century, but offers no new evidence for *jovaras*'s botanical identity, and in some cases only confuses the whole matter. The LKŽ also contains material from folk texts. However, instead of using the texts to illustrate the use of the Lithuanian term, these examples are listed under the assumed botanical identity of the local name. To give an example, the following line from a folk song, '*mano žalias vainikėlis jovarų žiedelių*' (my green wreath [made from] blossoms of *jovaras*) is not listed under the entry for *jovaras*, but instead is given as an example of the meaning of *Populus nigra* (Figure 12.1). However, the catkins of the poplar would not be called *žiedelių* (blossoms) in Lithuanian. More likely are such terms as: *žirgutis* (catkin of birch, alder, hazel), *kačiukas* (catkin of willow) or *spurgana* (hop strobile, alder catkin or strobile), rather than *žiedelių*. So, it is unlikely that this entry is correct, as by this description the folk term *jovaras* does not refer to a poplar.

If we look at dictionary entries for cognate terms of *jovaras* in Slavic languages, we find *Acer pseudoplatanus* as a referent for *jawor* in Polish (Podbielkowski and Sudnik-Wójcikowska 2003), *javor* in Russian, *javor* and *avor* in Bulgarian, and *javor* in Macedonian (Trubačev et al. 1974). Different meanings appear in certain dialects of these languages. *Javor* in Slovincian (an extinct dialect of the Prussian province of Pomerania) probably refers to *Acer saccharinum* L. (silver maple) or *A. platanoides* L. (Norway maple). *Jawor* in Upper Sorbian (a Slavic language spoken today in German Saxony) refers to *Acer platanoides*. In Russian *javor* refers to

Figure 12.1. Wreath made of oak (*Quercus robur* L., Fagaceae, Lith. *ąžuolas*) leaves weaved into lime boughs (*Tilia cordata* Mill., Tiliaceae, Lith. *mažalapė liepa*) and forking larkspur (*Consolida regalis* L., Ranunculaceae, Lith. *dirvinis raguolis, dirvinis pentinis*). These wreaths are given on the occasion of a naming-day or anniversary. They are hung on the door or round the neck, or put on the head of the celebrating person

Acer saccharinum and *Platanus orientalis* L. (oriental plane). In Czech the meaning of *javor* is restricted to the genus *Acer* (Trubačev et al. 1974), as is the Polish cognate *jabór* (Karłowicz 1901). All of these trees are morphologically quite similar, so the sharing of generic names between them is not surprising.

Besides trees, there are several plant names probably derived from Slavic *javor*. These are Croatian *javornica* (*Laurus nobilis* L., laurel), Ukrainian *javirnice* (*Ribes rubrum* L., red currant), and Slovenian *javornik* (a cultivar of *Vitis vinifera* L., wine grape) (Trubačev et al. 1974). Remarkably, the Istro-Romanian *iåvorica* (*Vaccinium myrtillus* L., blueberry) (Pieroni et al. 2003) is likely to be a loan from a Slavic language.

But how are these plants related to the maples, sycamores and poplars? Do any of them not have summer blossoms? How are these terms actually used in Lithuania today? Are any of these trees and plants still referred to as a 'god tree'?

Current Use of *Jovaras*

Due to the Lithuanian national movement and recent political independence, people and authorities responsible for standardizing Lithuanian do not like to use words of Slavic origin if there is another possibility. Since linguists claimed *jovaras* is a loan-word from a Slavic language, it is currently not in use in Lithuanian botany for any of the plants mentioned above (Jankevičienė 1998). However, a few botanists (Navys 2004) would like to reestablish *jovaras* as a common or vernacular name for *Acer pseudoplatanus*, out of respect for the history of Lithuanian botany. *Acer pseudoplatanus* and *Platanus* spp. are rare plants in Lithuania, found only in parks (Navasaitis 2005), so only botanists are likely to need to speak of them, and they usually use Latin. However, in some west-Lithuanian dialects *jovaras* is used today to refer to *Populus nigra*, and *javorelis* remains in common use for *Crataegus* spp. in north Lithuania (Genelytė 2005, personal communication).

Not influenced by political thought, *jawor* is common in Slavic languages today, referring to *Acer pseudoplatanus*. It is widely accepted in the official botanical nomenclature (e.g., Machek 1954; Podbielkowski and Sudnik-Wójcikowska 2003).

Occurrence of *Jovaras* in Folk Songs

Given the difficulties of identifying the plant referred to by *jovaras*, both in the literature and in public discourse today, we have investigated those folk songs that mention the plant to try and get clues as to its identity and cultural significance.

Jovaras and *jievaras*[1] appear synonymously in a remarkable number of folk songs. Due to similarities, most can be grouped into a few semantic categories: *'auga ievaras laukuose'* (ievaras grows in the field), *'linko ievaras vartuosna'* (ievaras tilts to the gate), and *'šalia kelio ievaras stovėjo'* (ievaras stands beside the road) (Misevičienė and Puteikienė 1993). In all of these songs, no other tree is mentioned.

With regard to the first type of song, the refrain *'Ei jovar jovar žaliasai'* (Hey, green jovar jovar!) appears frequently. These songs have been reported to be sung while harvesting crops, particularly rye, and may in fact function to thank the goddess of the harvest or the spirit of the grain for the bountiful crop. Singers comment that they have been sung while harvesters have been making a harvesting wreath (Figure 12.2). The harvesting wreath may also be called *ievaras* or *jovaras* (Misevičienė and Puteikienė 1993; LKŽ).

Figure 12.2. Wreath of rye (*Secale cereale* L., Poaceae, Lith. *rugiai*) weaved during the end of the harvest celebration, from the last sheaf of rye. Occasionally, various flowers may be woven into it, such as cornflowers (*Centaurea cyanus* L., Asteraceae, Lith. *rugiagėlė*) or corncockles (*Agrostemma githago* L., Caryophyllaceae, Lith. *raugė*). When the wreath is brought home, the harvesters are doused with water. Then the wreath is handed over to the hosts, who treat the harvesters to dinner

The second category of song relates to the association of *jovaras* with *vartu* (gate) (Figure 12.3). A frequent opening stanza in these songs goes like this:

> *Augo jovaras terp vartų, O jo šakelės an lango, Jo viršūnėlė in dangų, Oi kas mūs dvaro gražumas!* (*Jovaras* grows in the middle of the gate, its boughs on the window, its top in heaven. How nice is our farmstead!) (Misevičienė and Puteikienė 1993).

Another version suggests that *jievaras* branches lie on the gate and should be raised to allow the harvesters to pass through:

> *Augo ievaras Pas vartus, O jo šakelės, O jo šakelės Ant vartų. Išeik, močiutė, Ant dvaro, Atkelk šakeles Nuo vartų* (Ievaras grows beside the gate, its branches are on the gate. Come on mother in the yard, raise the branches from the gate!) (Misevičienė and Puteikienė 1993).

The third use of *jovaras* in songs is illustrated by the following:

> *Šalia kelio vieškelėlio Žalias jovarėlis, Jovarėlio pašaknėlės Skambančios kanklelės, Jovarėlio vidurėlyj Gaudžiančios bitelės, Jovarėlio viršūnėlėj Sakalėlio lizdas*

Figure 12.3. Wooden gate decorated with greens of various plants and tree boughs on the occasion of a village or family celebration. Plants are tied to the structure with strips of bark

(Green *jovaras* grows beside the road, beneath the roots-jingling zithers, in the middle-buzzing bees, on the top-a hawk's nest) (Misevičienė and Puteikienė 1993).

The symbolism here points to the mythical 'world tree', uniting the underworld (roots) with the present (trunk) and the heavens above (canopy leaves), which is common in Baltic and Slavic folklore (Katičić 1989). Perhaps this reference reflects one interpretation of the concept of a 'god tree' mentioned in the riddle at the start of this chapter?

Another song that alludes to *jovaras* as something mythical is this surrealist verse:

Marių vandenėly, pačioj gilumėlėj, o ir išdygo žalias jovarėlis. Po tuo jovarėliu, po tuo šimtašakėliu, o ten stovėjo du jaunu broleliu. (In the sea, in the deepest place, green *jovaras* sprouts. Beneath the *jovaras*, beneath the hundred branched, stand two young brethren) (Juška 1954a).

How is it that *jovaras* could sprout beneath the sea? It is clear that we need a much deeper appreciation of the use of natural symbols in Lithuania's past in order to understand the meanings being communicated here.

A rather cryptic reference to the tree is found in songs that appear to comment on matters of courtship, but also offer us some morphological clues to its botanical identity:

> *Tėvulio dvare ievarėlis, Ievarėlis devyniašakis. Kas viena šakelė – aukso spurgelė, In viršūnėlę raiba gegulė. Kukavo rytą ir vakarėlį, Kol iškukavo motkos dukrelę. Išlėkė gegulė iš ievarėlio, Išvarė dukrėlę iš tėvo dvaro* (In father's court – a ievaras with nine boughs. Every bough with a golden catkin, on the top a mottled cuckoo. He cuckooed in the morning and in the evening until he cuckooed out mother's daughter. Cuckoo flew out of the ievaras, banished the daughter) (Čiurlionytė 1999).

This nuptial song tells of a cuckoo visiting the bride and the groom. They want to have the bird caught, brought into the granary,[2] clothed in gold and saturated with wine:

> *O kad jį pagaučiau į svirną įleisčiau. Su aukseliu aprėdyčiau, Vynu pagirdyčiau.'* The bird answers: *'Lazdynėlio tankūs krūmai, – Tai mano svirnelis, Jovarėlio viršūnėlė, – Tai mano aukselis, Nuo rūtelių šalta rasa, – Tai mano vynelis* (Hazel brushes are my granary, *jovaras*'s tree top is my gold, cold dye of herbs is my wine) (Juška and Juška 1955).

At least here we have some reference to boughs with golden catkins, which may suggest a willow. They may also resemble the 'golden' cones found on hop vines. It is also possible that the cones refer to the red or green catkins found on the poplars, especially *Populus nigra*, black poplar, which has red catkins on the male trees and green ones on the female ones.

Another song offers more botanical details, though this one suggests a completely different set of species:

> *Per mano dvarą Upelė teka – Tai žaliojo vynelio. Po mano langu Ievarų krūmas – Tai su saldžiom uogelėm. Žali lapeliai, Balti žiedeliai, – Vis raudonos uogelės* (A brook flows through my yard. It is green wine. Beneath my window a ievaras bush with sweet berries. Green leaves, white blossoms, always red berries) (Jonynas et al. 1962).

A similar botanical profile is found in a song about a young man courting; he is advised to hobble his horse beneath the *jovaras* tree and not beneath the guelder rose (*Viburnum opulus* L.):

> *Pucinėlio krūmelis – Tai neslaunas medelis. Jo nesaldžios uogelės, Jo nemiklios rykštelės .... Jievaro krūmelis – Tai laimingas medelis. Tai jo saldžios uogelės, Tai jo miklios rykštelės* (The guelder rose is a useless tree. Its berries are not sweet, its twigs are not flexible .... The jievaras bush is a fortunate tree. Its berries are sweet, its twigs are flexible) (Četkauskaitė 1981).

A few songs point to *jovaras* growing in an orchard, e.g., '*O kieno žali sodai, žali jovarėliai?*' (Whose orchards are these, the green *jovaras*?) (Juška 1954b), suggesting the tree is not cultivated for fruit, but perhaps is grown for something else. There are indeed several songs that suggest the tree has useful branches or timber.

Jievaras is always present in the game *Jievaro tiltas* 'Jievaras bridge' and related songs:

> *Gręskime mergos Ievaro tiltą, Aleliuma, rūtela, Ievarėlio tiltą. Iš ko mes gręsim Ievaro tiltą, – Iš beržo šakų, Amalėlio lapų. O ar praleisit Ievaro žmones? Visus praleisim, Tik vieną pasiliksim. Oi, ir mes eisim Ir jūs nebijosim! .... Oi, mes pasilikom Pačią geriausią Pačią geriausią Linų verpėjėlę. Oi, jūs pas'likot Pačią aršiausią Kačių piemenaitę* (Girls, let's make us the ievaras bridge. From what should we build the ievaras bridge? From birch rods and mistletoe leaves. Do you allow the ievaras people to pass through? We let them pass all, only one we will take. We will pass and not be afraid of you .... O, we took the best, the best flax spinner! O, you took the most impetuous cat herdswoman!) (Čiurlionytė 1955).

Two groups play the game and sing the verses: one group forms the bridge, the other members represent those walking over the bridge. The game appears to symbolize the rites of passage that young women go through, having to demonstrate their skills, especially in spinning flax, and a strong work ethic, in order to be allowed to pass over the bridge into a new life as a bride (Skrodenis 1965). Despite being called *jievaro tiltas*, only one text names jievaras as material for building the bridge: *'beržų lapų, jievarėlio šakų'* (birch leaves, jievaras boughs), so the significance of the game and its connection to jievaras remains unknown. One might suggest that *jievaro* in this case may be referring not to a particular species of tree, but to its connotation as the mythic 'world tree', which is often portrayed as a link between worlds, much like the bridge links the worlds of the adolescent and the adult in the game.

A different explanation is suggested by a recent description of *jovaras* from a nearly eighty-year-old singer, A. Kučinskienė from Šlienava, who also played the game *jievaras tiltas* in her youth.

> The tree has been called *jovaras* in our village. On the other side of the Neman River it has been called *jievaras*. It is a tall tree with branches directed steeply upwards. Leaves are similar to bird cherry (*Prunus padus* L.) but more narrow. Blossoms are white and not arranged in umbels. It was nice and honourable to get a blooming twig presented or attached to one's clothes. Its red berries ripen while harvesting rye. The game *jievaras tiltas* was performed during harvest. But the *jovaras* tree has nothing in common with the wreath *jievaras*.

This distinction is clearly important, suggesting that *jovaras* and *jievaras* may not be as semantically related as their lexical similarities might suggest, at least for the villages on this side of the Neman River. In fact it seems more likely that the wreath *jievaras* refers to the annual sacrifice to the deity of harvests, which is made from the braided stalks of the ripe rye grain left uncut during the harvest. It would be a mistake, then, to conclude that this ritual and the accompanying game have anything to do with the tree *jovaras*, except that once in a while the bridge may be said to be constructed from it.

These songs reveal several different meanings and uses of *jievaras*, and several botanical descriptions and uses of *jovaras*. Their connection remains tenuous, either they are synonyms of the same tree, or *jievaras* actually is not a tree and represents a deity, spirit or a mythical 'world tree'. Our understanding of the identity and meaning of *jovaras* seems no greater. From the evidence of the folk songs taken as a whole, it appears then that different trees could fit the general archetypical picture of *jovaras*, a tall tree with white blossoms and red or blackberries. Perhaps an investigation of folklore in Slavic countries may shed some light on the mystery.

The game and song *jievaro tiltas* has a counterpart in Polish and Belarusian folklore:

> *Jaworowi ludzie, – czego tam stojicie? Jawor, jaworowi ludzie. Stojemy, stojemy, - mosty budujemy, Jawor, jaworowi ludzie. Z czego budujecie, – z czego i plecicie? Jawor, jaworowi ludzie. Z dębowego liścia, – z brzozowego kiścia, Jawor, jaworowi ludzie. Dajcie nam tam dajcie, – stado koni przegnać, Jawor, karetą przejechać. Damy, damy, damy, - jedno otrzymamy. Jawor, jaworowi ludzie* (Jawor people, why are you standing there? Jawor, jawor people. We stand and build a bridge, jawor, jawor people. From what do you build, what do you plait, jawor, jawor people? From oak leaves, from birch cudgels, jawor, jawor people. Give us there a herd of horses being driven, jawor, a carriage being brought. We give, we get one. Jawor, jawor people.) (Skrodenis 1965).

It is remarkable that Karłowicz (1901) describes the game *Jaworze drzewo* 'Jawor tree' as a war game, while a year later the same author (Karłowicz, Kryński and Niedźwiedzki 1902) talks about a children's game with this example: *'Jaworowi ludzie, co wy to robicie? ... Budujemy mosty dla pana starosty'* (Jawor people, what work do you do there? ... We build a bridge for the mayor).

There is also a Belarusian example: *'Явар, явар, яварове людзі, Чаго вы стаіце За мастом маленкім'* (Jawor, jawor, jawor people, why are you standing behind the little bridge?) (Grynblat et al. 1983). As in Lithuanian songs, *javar*, or sometimes *jagor*, may appear in Belarusian songs as a lucky tree, while guelder rose is unfortunate:

> *Явар – дзерева шчаслівае: У яварным камлі чорныя бабры, У явары ў сярэдзіне райскія пчолкі, У явары ў вярху сівыя сакалы.* (Javar is a lucky tree: there are black beavers at the base of the stem, there are paradise bees in the middle of the javar, there are grey hawks on the javar tree top) (Grynblat et al. 1980).

A few texts mention pine or birch as a fortunate tree and rowan or fir as an unfortunate tree.

In both Polish and Belarusian songs the javar is situated in the courtyard (Grynblat et al. 1980; Katičić 1989: 90) or at the gate: *'А зелёны ягор у варот'*, again with the form *jagor* used instead of *javor* (Grynblat et al. 1980). Ukrainian texts suggest they grow on hills and fields (Katičić 1990). Croatian songs add the edge of water as a habitat:

> *Javore, zelen bore, lijepo ti t`je ukraj vode! Iz stabra ti voda teče, a iz grana čele lete, A u stabru tvoga duba sjedi jato od goluba.* (Javor, green pine, you stand on the edge of the water! Water swells out of your stem, bees fly out of your boughs. Squabs sit on the stem) (Katičić 1989).

Another Croatian song with the same opening stanza, *O javore, zelen bore,* reports that the tree has apples on its twigs (Katičić 1990).

Thus, these Slavic songs appear to offer little in the way of botanical clues or common everyday uses for *jawor,* or its cognates; instead they tend to appear in songs that have ritual contexts and sacred meanings. As with most references to *jievaras* in Lithuanian, *jawor* should be taken to represent a mythical 'world tree' (Katičić 1989, 1990), symbolizing the bridging between worlds, and used in social contexts to symbolize the transition or passage to new social roles, such as that of a married adult. What is the connection between the objective tree and a symbolic, mythical, world tree? Clearly the songs only serve to muddle the search for a clear distinction. Are *jovaras* and *jievaras, jawor* and *jagor* really lexically and semantically linked to a real tree? Or have these terms become detached from some anchor in nature? Perhaps the problem lies in the language itself, and the history of these words and their cognates.

The Linguistics of *Jovaras*

A History of Borrowing?

In this section we challenge the standard approach to explaining Lithuanian *jovaras* and the Slavic *jawor,* and offer an alternative of our own. The evidence is largely derived from a linguistic analysis but we do include ethnographic data when available.

Taking the Slavic *jawor* first, the standard explanation suggests that *jawor* once meant *Acer pseudoplatanus.* It was borrowed from the Old High German *ahorn* referring to *Acer* species in general, but in Proto-Slavic it specifically refers to *Acer pseudoplatanus* (Vasmer 1973; Kortlandt 1975). There was a change from *h* > *w* and the German *ahorn* became the Slavic **aworъ* (Trubačev et al. 1974). Critics of this story point to the lack of comparable cases of this shift in consonants between the two languages, and find the loss of *-n-* also needs special argumentation. Machek (1954) concludes that German *ahorn* and the Slavic **aworъ* arise from an unknown substratum language. This thought seems a rather hasty and convenient explanation; it would make any search for an etymology impossible. One difficulty is that a lot of evidence has never been considered in the discussion.

According to the standard theory, the Lithuanians are believed to have acquired *jovaras* from the Polish *jawor* (Skardžius 1931), probably in the mid-sixteenth century[3]. If the borrowed term originally referred to just *Acer pseudoplatanus,* it was soon confused by a mislabelling of *Populus nigra* (black poplar) as *javaras* in a Lithuanian translation of the Lutheran bible in the late-sixteenth century (Genesis 30:37). Since the black poplar is not native to Lithuania (though it is now found in the country), one can assume no local name for the tree existed at the time of the translation. We guess that Bretke either took the Polish term for poplar, *tapalas,* and sought out a similar, Lithuanian tree name, or he wrongly associated the German *Pappelbaum* with the *Acer pseudoplatanus* (sycamore maple), rather than the poplar.

Whatever the reason, the mistaken translation became orthodox as the printed bible spread across Lithuania and became the authority on matters beyond local folk botany. The Clavis dictionary (Anonymous ca 1680), a dictionary for reading the Bible, uses Genesis 30:37 to define *Jóvaras – Pappelbaum 'Populus* spp.' and gives the Lutheran text *'Stäbe von grünen Pappelbäumen'* ('Staff from green poplar' or 'staff of blooming poplar'), as the example.

One reason many of the botanical referents in Lithuanian derive from Polish terms, such as *jawor,* is that the early works on Lithuanian botany, at the end of the eighteenth and beginning of the nineteenth centuries, were written in or translated into Polish. It was only in 1870 that the first Lithuanian work on botany, a plant list of Lithuanian names and their Latin translations, was published, by Ivinskis (Navys 2004). It is assumed that many of his local names were copied from earlier texts written in Polish. For instance, Ivinskis gives *jowaras* as the local name for the genus *Platanus* (plane tree), a clear derivation from the Polish *jawor,* and not too surprising given its morphological similarities to *Acer pseudoplatanus.*

How can it be that other plants so different from *Acer pseudoplatanus,* such as *Crataegus* spp., *Ribes rubrum* and a cultivar of *Vitis vinifera,* should also be labelled as *jovaras*? It is possible that the label *jovaras* was applied because the leaves of all of these species are similar in shape.

Homonymy

The standard model tacitly supposes that all instances of *jovaras* and *jievaras* have the same origin. Having one sure example where the word is loaned, as in the case Polish *jawor* > Lithuanian *jowaras 'Platanus'*, an overgeneralization is made such that all *jovaras* must be loaned. In this section we challenge this conventional view, and introduce the concept of homonymy as the basis of an alternative explanation for the origins and meanings of *jovaras.*

Homonymy, where two words sound or are spelled the same yet have different meanings (e.g., *to, too, two*), may facilitate the incorporation of a new loan-word into one's own language. On the other hand, a loaned word may be reshaped to resemble a well known, native word – thus creating homonymy. These two processes can help to explain how semantic relationships or shared origins can be wrongly inferred between words that sound alike. For instance, a chainsaw is called *varna* in Lithuanian villages, a term that usually means 'crow'. However this homonymy has nothing to do with some perceived relationship between crows and chainsaws. The term has been derived from the chainsaw manufacturer's name *Husquarna*, and now also applies to saws made by other manufacturers.

This has been the case with Lithuanian *rūta* (twig, herb) and *Rūta*, the Biblical name for Ruth (Šeškauskaitė and Gliwa 2002), and the Biblical name Eve, which became *Ieva* in Lithuanian, and is the common name for *Prunus padus*. Both *Rūta* and *Ieva* are now popular women's names in Lithuania, but have no necessary connection to their botanical homophones. However, it is possible that some names of girls and boys are indeed derived from plant names and parts, as are Rose, Daisy, Marigold and Hazel in English. In the Baltics, historical sources (Einhorn 1649) report pagan Latvians giving bird names to their daughters.

It appears that there are some homonymous words for the Lithuanian and Slavic plant names under discussion, which may help us in discovering their botanical identities. In Old Russian the word *javor* refers to 'a bundle of hay ready for transportation'. It leads to the Proto-Slavic form **aworъ*, which is identical to the Proto-Slavic term for the Polish plant name *jawor* (Trubačev et al. 1974). *Jawor* is usually seen as having no convincing etymology (Smoczyński, personal communication 2005), yet this homonymy points toward an origin as a borrowed term for bundles of hay that then developed a distinct meaning referring to a tree species.

Two more instances of secondary homonymy from Slavic languages should be mentioned. Ukrainian *javor, javir* is also reported for *Acorus calamus* L. beside *ajir, ajer, hajir, hajvir*, etc. (Makowiecki 1936). Slovenian *javora* (*Laurus nobilis*) is seen as a transformation of *laurus* via **lavor* > **javor*; there is also *javorishk* (laurel wreath) (Bezlaj 1976).

In Lithuanian there are several different meanings reported for both *jievaras* and *jovaras*. They both name the wreath made from rye at the harvest festival. In comments concerning songs of the types '*Auga ievaras laukuose*' (ievaras grows on the field), '*Linko ievaras vartuosna*' (ievaras tilts to the gate), and '*Šalia kelio ievaras stovėjo*' (ievaras stood beside the road) (Misevičienė and Puteikienė 1993), singers made clear that these songs are sung while the harvesters twist or plait a ritual wreath, which is then carried in a ritual procession into the court. Ethnographic data from Lithuania and elsewhere in Europe confirm the rye wreath, or a bundle of harvested grain, as being symbolic of the fertility of crops. Grains taken from the wreath will be mixed into the seeds for the next crop.

Another meaning of *jievarys* is 'dense twigs in the crown', probably 'tree crown' (LKŽ), and *jievaras* can also refer to 'two trees with twisted together twigs' (Skrodenis 1965). Both of these homonyms, and the ones for *jovaras* and *jawor*, suggest some relationship between our mystery tree and ritual wreaths made of twisted, plaited grains or twigs.

Thoughts on Semantics and a New Suggestion

In most cases, the semantics of *jovaras* and its various cognates are or may be related to the process of twisting or plaiting. One may suppose that *jovaras*, in the harvesting song, is not a tree at all but a material object like a straw bundle or a maypole, plaited or decorated with straw and, maybe, herbs. This could explain why *jovaras* grows in the middle of the gate or why it has to be lifted to pass through the gate. The song texts, where *jovaras* is situated in the gate, at the window, as well as inside the house, seem to support the assumption that what is discussed is indeed an artefact, which may be easily divided into several parts.

The crown of a tree can be viewed as composed of plaited twigs, certainly when gazing upward into it. This analogy appears in Lithuanian *vainikas* (tree crown, wreath), derived from the Indo-European root (IE) **wieH*$_1$ (to twist, loop). The wreath of newly harvested crops is plaited from stalks of rye, or other harvested grains, and is bound by winding straw around a bundle.

This results in a new view of the phrase *jievaro tiltas* (jievaras bridge), as an object that traverses the ground like a path, rather than spanning a gulf or river as is implied in the English sense of a bridge. Evidence for this can be found in comparing Baltic and Slavic words for 'bridge' or 'path', which appear to have connotations of 'to braid', 'to lift up' or 'to pave, or boulder'. For example, Old Prussian *pintis* (way), Russian *put'* (path) and Lithuanian *pinti* (to braid) are all cognate terms (Mažiulis 1996; Eichner 2004).[4] In Slavic languages, we find Russian *most* (bridge), *mostki* (planked footway) and *mostit'* (to pave, cobble).[5] When looking at the derivation of 'to lift up', one finds Lithuanian *kelias* (road, way), *keltas* (ferry), *kelti* (to lift) and *tiltas* (bridge) < IE **telH*$_2$- (to lift up; to endure) (Smoczyński 2003).

The explanation for this sense of 'bridge' may be found in the landscape itself, as the homeland of Balts and Slavs is famous for its bogs, swamps and marshes. There is often no solid bank on which to build a traditional bridge. Rather, in a marshland a bridge or a path is often a log causeway created by laying twigs, boughs and timber on the ground. This makes it clear why the derivation starts with 'to lift up' or 'to pave, boulder'. The connotation 'to braid' requires an explanation of how these paths are constructed. Before laying logs across the path, a layer of twigs is thrown onto the soggy, wet ground to stabilize the area.[6] The twigs inevitably become enmeshed, or braided, in a kind of net and ensure that the logs stay in position.

In Slavic folk songs the bridge is most often built from *kalina* (guelder rose, *Viburnum opulus*), a shrub or small tree often used for hedges, with white flowers and red berries, that thrives in wet and soggy soils. By the way, the symbolic meaning of guelder rose is mostly related to marriage – at least in Slavic folklore. In Lithuanian folklore, *putinas* is also very popular but has a strong connotation of death. Gold and jewellery are sometimes mentioned as material for the bridge, but only rarely in Lithuanian folklore where the suitor is asked to build such a bridge overnight. Slavic songs refer to bridges of gold and jewellery more often.

Returning to the game of the *jievaro tiltas*, we see that the material for the bridge is noted as consisting of boughs, twigs and leaves. The bridge itself is said to be *grįstas* (paved), resembling the Lithuanian log causeway, *kūlgrinda*, containing the same element, *grind-* (pave). So, we would claim then that *jievaras* in *jievaro tiltas* concerns not the material, but the construction method, as in the 'braided bridge or path'. As already mentioned, the different meanings of *jievaras* and *jovaras* point to a semantic origin as 'to twist, loop' or 'to get caught', as are two trees that have grown into each other. This is also stated clearly in the Polish text given above, '*Z czego budujecie, - z czego i plecicie?*' (From what do you build, what do you plait?).

Thus, we may distinguish two layers of meaning in Lithuanian *jovaras* and Slavic **aworъ*. The older one starts with 'to twist, loop' or 'to get caught'. The folklore material refers to this layer as well as the Old Russian 'hay bundle'. Then the Croatian '*O javore, zelen bore!*' should be translated like this: 'O, tree with dense crown, you green pine!'. The newer connotations associated with the names of trees are likely to have been borrowed, and probably originally only referred to *Acer pseudoplatanus* and *Platanus* spp. The lexical forms of these newer referents have been derived from the older forms, both in Baltic and Slavic. This explains the strange sound changes during the borrowing process.

Solution of the Riddle

If our approach is correct, the riddle's answer is obvious. *Jovaras* in its earliest sense is a material object, an artefact, and only later does it come to refer to natural species. Artefacts do not bloom. The artefact is used ritually – hence it is called *Dievo medelis* (God tree). The poem-riddle is a cultural pun of sorts, as it plays on the layered meanings of the term *jovaras*, as both braided objects and living trees, and captures the overlap and ambiguity of its varied meanings. The riddle itself captures the symbolic history of the term and its transformations, which when uncovered, or unbraided, in fact tells the story of past uses in path construction, spring rites, marriage ceremonies, children's games, handicrafts and botanical nomenclature.

When applying historical linguistics to material of folk texts one has to be careful. The form of words and phrases will quite often be well conserved, but this is not true of their meaning. In this case we have seen major changes resulting in a complete loss of any clear meaning, especially in the refrains of folk songs. Thus, to uncover the origins and meaning of local botanical names, we must not only consult dictionaries but also consider the role and the meaning of the folk text itself. In this way folk texts together with historical linguistics may reveal deep insight into the historical uses and local classifications of plants.

Acknowledgements

We thank Prof. Radoslaw Katičić (Vienna), Prof. Wojciech Smoczyński (Cracow), Prof. Axel Holvoet, Dr Gina Kavaliūnaitė and Dr Aurelija Genelytė (Vilnius) for interesting and constructive discussions and help with literature.

Notes

1. *Jievaras, jevaras* and *ievaras* are only spelling variations of the same word.
2. The *klėtis* or *svirnas* (granary) has also been the building where adult daughters and sons, married or unmarried, lived.
3. The oldest text written in Lithuanian dates to 1547.
4. Smoczyński (2000, 2003 and personal communication 2005) does not agree that Prussian *pintis* is cognate to this word, nor does he agree with the derivation from a verb 'to wind, braid'.
5. Katičić (1989) notes that the phrase *mostiti mosty* (paving a bridge), which is frequently found in Slavic folk songs, is apparently a *figura etymologica* and acts as a sign of sacral poetry; and that the bridge as an important artefact in wedding ceremonies.
6. This is the way Lithuanians create log causeways nowadays, when they have to work with a horse or a tractor in marshland forests during mild winters.

References

Anonymous. ca 1680. *Clavis Germanico-Lithvana,* manuscript. Facsimile provided by V. Drotvinas and A. Ivaškevičius (eds). 1995–1997. *Clavis Germanico-Lithvana,* vols 1–4. Vilnius: Mokslo ir enciklopedijų leidykla.

Bezlaj, F. 1976. *Etimološki slovar slovenskega jezika,* vol. 1 A-J. Ljubljana: Institut za slovenski jezik / Mladinska knjiga.

Četkauskaitė, G. (ed.). 1981. *Dzūkų melodijos.* Vilnius: Vaga.

Čiurlionytė, J. 1955. *Lietuvių liaudies dainos.* Vilnius: Valstybinė grožinės literatūros leidykla.

——— 1999. *Lietuvių liaudies melodijos.* 2nd edn. Vilnius: Lietuvos muzikos akademija.

Dagys, J. (ed.). 1938. *Lietuvių botanikos žodynas.* Kaunas: Varpas.

Eichner, H. 2004. 'Die Etymologie von lat. pontifices'. Paper presented at *12th Congress of the Indogermanische Gesellschaft, Cracow, October 11–16.*

Einhorn, P. 1649. 'Historia Lettica: Das ist Beschreibung der lettischen Nation'. Reprinted in N. Vėlius (ed.). 2003. *Baltų religijos ir mitologijos šaltiniai,* vol. 3. Vilnius: Mokslo ir enciklopedijų leidybos institutas.

Genelytė, A. 2005. Personal communication.

Grynblat, M.J., L.A. Malaš, Z.J. Mažeika and A.S. Fjadosik (eds). 1980. *Vjaselle pesni,* vol. 1. Minsk: Navuka i technika.

———, L.A. Malaš, Z.J. Mažeika and A.S. Fjadosik (eds). 1983. *Vjaselle pesni,* vol. 3. Minsk: Navuka i technika.

Jankevičienė, R. 1998. *Botanikos vardų žodynas.* Vilnius: Botanikos instituto leidykla.

Jonynas, A., V. Barauskienė, B. Kazlauskienė and B. Uginčius (eds). 1962. *Lietuvių tautosaka,* vol. 1: *Dainos.* Vilnius: Valstybinė politinės ir mokslinės literatūros leidykla.

Juška, A. 1954a. *Lietuviškos dainos,* vol. 1, 2nd edn, reprinted and commented by A. Mockus. Vilnius: Valstybinė grožinės literatūros leidykla.

——— 1954b. *Lietuviškos dainos,* vol. 2, 2nd edn. Vilnius: Valstybinė grožinės literatūros leidykla.

——— 1954c. *Lietuviškos dainos,* vol. 3, 2nd edn. Vilnius: Valstybinė grožinės literatūros leidykla.

——— and J. Juška. 1955. Lietuviškos svotbinės dainos, vol. 1, 2nd edn. Vilnius: Valstybinė grožinės literatūros leidykla.

Karłowicz, J. 1901. *Słownik gwar polskich,* vol. 2: F–K. Cracow: Nakł. Akad. Umiejętn.

———, A.A. Kryński and W. Niedźwiedzki. 1902. *Słownik języka polskiego,* vol. 2: H–M. Cracow: Nakł. Akad. Umiejętn.

Katičić, R. 1989. 'Weiteres zur Rekonstruktion der Texte eines urslawischen Fruchtbarkeitsritus', *Wiener Slavistisches Jahrbuch* 35: 57–98.

——— 1990. 'Weiteres zur Rekonstruktion der Texte eines urslawischen Fruchtbarkeitsritus (2)', *Wiener Slavistisches Jahrbuch* 36: 61–93.

Kortlandt, F. 1975. *Slavic Accentuation – a Study in Relative Chronology.* Lisse: The Peter de Ridder Press.

LKŽ. 1956–2002. *Lietuvių kalbas žodynas.* 20 vols. Vilnius: Valstybinė politinės ir mokslinės literatūros leidykla, Mintis, Mokslas, Mokslo ir enciklopedijų leidykla, Mokslo ir enciklopedijų leidybos institutas, Lietuvių kalbos instituto leidykla.

Machek, V. 1954. *Česká a slovenská jména rostlin.* Prague: ČSAV.

Makowiecki, S. 1936. *Słownik botaniczny łacińsko-małoruski.* Cracow: Nakładem polskiej akademii umjętności.

Mažiulis, V. 1996. *Prūsų kalbos etimologijos žodynas*, vol. 3: L–P. Vilnius: Mokslo ir enciklopedijų leidykla.
Misevičienė, V. and Z. Puteikienė (eds). 1993. *Lietuvių liaudies dainynas*. Vol. 6. Vilnius: Vaga.
Navasaitis, M. 2005. *Dendrologija*. Vilnius: Margi raštai.
Navys, E. 2004. 'Kai kurių medžių ir krūmų rūšių senųjų lietuviškų vardų pagrindimas', *Dendrologia lithuaniae* 7: 92–98.
———, M.E. Giusti, H. Münz, C. Lenzarini, G. Turkovic and A. Turkovic. 2003. 'Ethnobotanical Knowledge of the Istro-Romanians of Žejane in Croatia', *Fitoterapia* 74(7–8): 710–719.
Podbielkowski, Z. and B. Sudnik-Wójcikowska. 2003. *Słownik roślin użytkowych*. Warszawa: Wydawnictwo rolnicze i leśne.
Skardžius, P. 1931. 'Die slavischen Lehnwörter im Altlitauischen', *Tauta ir žodis* 7: 3–252.
——— 1965. 'Lietuvių liaudies žaidimas "Jievaro tiltas"', *LMA darbai Serija A* 1(18): 245–258.
Smoczyński, W. 2000. *Das deutsche Lehngut im Altpreussischen*. Cracow: Wyd. UJ.
———. 2003. *Studia bałto-słowiańskie*. Vol. 2. Cracow: Wyd. UJ.
——— 2005. Personal communication.
Šeškauskaitė, D. and B. Gliwa. 2002. 'Rūta, die Nationalblume der Litauer. Zur Kulturgeschichte der Weinraute (*Ruta graveolens* L.) und zur Etymologie von litauisch rūta und deutsch Raute', *Anthropos* 97(2): 455–467.
Trubačev, O.N., V.A. Merkulova, Ž.Ž.Varbot, L.A. Gindin, G.F. Odincov, L.V. Kurkina, I.P. Petleva, T.V. Gorjačeva and V. Mikhailovič. 1974. *Ėtimologičeskij slovar' slavjanskych jazykov*. Vol. 1. Moscow: Nauka.
Vasmer, M. 1973. *Ėtimologičeskij slovar' russkogo jazyka*. Translated and expanded by O. Trubačev. Vol. 1. Moscow: Progress.

CHAPTER 13

Norway's Rosmarin (*Rhododendron tomentosum*) in Past and Present Tradition

TORBJØRN ALM AND MARIANNE IVERSEN

Introduction

This chapter reviews the uses of *Rhododendron tomentosum* (Stokes) Harmaja in Norway, with particular emphasis on North Sami material. By comparing past tradition and present use, it will also assess the extent to which former traditions related to this (and by inference, probably many other species) still survive in living practice or at least in living memory in Norway.

Like many other plant species and materials, *R. tomentosum* has been traded by pharmacies in Norway; it was listed in the *Pharmacopoea Danica*, also valid in Norway, in 1772 (Anonymous 1772) but not in the first Norwegian version (*Pharmacopoea Norvegica*, Anonymous 1854). However, with the sole exception of the Norwegian vernacular name *rosmarin* (see below), there is nothing to suggest that material from scholarly medicine or herbals has passed into folk tradition in Norway. On the contrary, all data on uses and beliefs related to *R. tomentosum* derive from areas where it grows more or less abundantly, and is thus easily available, and nothing at all has been recorded in those parts of the country where the species is absent.

Although closely related to *Rhododendron*, *Ledum* has traditionally been recognized as a separate genus, comprising only a few species, including the widespread Eurosiberian *Ledum palustre* (L.). Kron and Judd (1990) united the two genera, and treated *Ledum* as a subsection within *Rhododendron* subgenus *Rhododendron* section *Rhododendron*. This merger has been supported by DNA studies (Kron 1997). Accordingly, the new combination *R. tomentosum* is now the valid name for the plant formerly known as *Ledum palustre* (Harmaja 1990, 1991).

R. tomentosum is a small bush, usually about 30–50 cm high (to 1 m), with narrow, evergreen leaves (Figure 13.1). Both the twigs and the underside of the leaves carry numerous brown hairs. The flowers are white. The plant is rich in etheric oils (Greve 1938), and has a strong turpentine scent. Its typical habitat is boreal forests, mires and damp heaths.

In Norway, *R. tomentosum* belongs to an element of eastern, Boreal species that reach westwards to the interior parts of both northern and southern Norway; it is much more widespread in adjacent Finland and Sweden (Granlund 1925; Hultén 1971; Hultén and Fries 1986). Within Norway, large stands occur in the interior and eastern parts of Finnmark, the northernmost county. Here, the species is well known in folk tradition, especially among the Sami – not least since its area of distribution in Finnmark overlaps with some of the major Sami settlements.

Ethnobotanical Evidence

Within Norway, *R. tomentosum* has served mainly as a medicinal plant, especially in North Sami tradition. Qvigstad (1932) provided an extensive review of Sami folk medicine, based on literature and other sources, which also incorporated his own substantial material. Steen (1961) reworked and summarized this material, and added some supplements. Alm (1993) provided a brief review of the ethnobotany of *R. tomentosum* in Norway, adding some recent records of vernacular names and uses. Substantial new information has been collected since then.

For the purposes of this chapter, we will consider all but our own material (see below) and the brief notes in Vars (2000) and Rasmussen (2006) as historical, and thus as providing information on past uses. With

Figure 13.1. Norway's rosmarin (*Rhododendron tomentosum*)

these exceptions, Vorren (1971) and Høeg (1974) are the last authors to provide new data. We will compare this past body of information with our own, recent records, as a means of assessing to what extent the previously known uses and other traditions remain extant, either in terms of present use, or at least surviving as knowledge of such use among members of the present population.

Our own material comprises interviews and other records made in Norway by the first author from the late 1970s onwards. From the early 1990s onwards, both of us have collected ethnobotanical material on Sami plant names and uses in Finnmark; for example, as a number of taped interviews in 1996. Transcripts of all these records are stored by the first author; they are referred to below as EBATA + year and record number. Informants are not identified here; such information is available from the original material.

All quotations have been translated here. Unless otherwise stated, all information given here pertains to North Sami tradition recorded in Finnmark. North Sami vernacular names are spelled according to the present (1979 onwards) orthography. Material from other ethnic groups, mainly the Norwegian majority population, is specified as such. In addition, Qvigstad (1901, 1932) provided a few records from the East Sami population of Sør-Varanger, easternmost Finnmark.

Past Uses

General

Rhododendron tomentosum does not resemble any other species found in Norway, and is easily recognized by its smell alone. Thus, one can safely assume a one-to-one relationship or identity between the scientific and folk species. Voucher specimens, for example, in the eighteenth century herbarium of J.E. Gunnerus (in herb. TRH), confirm this identity.

At least where *R. tomentosum* occurs abundantly, it is usually known to the locals by a vernacular name (and quite often several, even within a single area). It has also given rise to some toponyms (Lagerberg, Holmboe and Nordhagen 1956: 101; Rynning 2001). Porsanger, one of the major fjords in Finnmark, is a likely example, probably from Norse *pors* and *angr*, although its origin remains disputed (Granmo 1992). Further south in Norway, the plant name *pors* is mostly restricted to *Myrica gale* L. In general, Norwegian tradition uses the term *pors* for these two species only; that is, for plants with a strong scent that can be used in beer (cf. Alm 2003).

A number of sources from the seventeenth century onwards provide information on how *R. tomentosum* – and not least its strong smell – was conceived. As noted by Alm (1993), there is a peculiar difference between the conception of its odour as described by ethnic Norwegians and by the Sami. At least in Finnmark, the former usually praise *R. tomentosum* as well scented, and refer to it in positive terms. *Wilde-Rosmarin* (in modern Norwegian: *vill rosmarin*, 'wild rosemary') is a vernacular name used by District Governor Hans H. Lilienskiold in his large topographical account of Finnmark, completed about 1698 (Solberg 1942–1943: 230–231). It is also mentioned by Knud Leem in an eighteenth-century manuscript on the plants of Finnmark (published by Dahl 1906), and by Sommerfelt (1799: 118; cf. Alm 1992). Lilienskiold mentioned 'wild rosemary or Finnmark's well-scented *pors*', and noted that it was 'beloved', and Leem commented that 'due to its fine scent, this herb is loved here' (Dahl 1906: 100). Gunnerus (1772) listed *Skov-Rosmarin* ('forest rosemary') as one of several Norwegian names. Høeg (1974: 424) recorded *rosmarin* as a vernacular name in Alta, western Finnmark, and Sør-Varanger, eastern Finnmark.

The name *rosmarin* for *R. tomentosum* has obviously been introduced, as similar usage is widespread in Europe, for example in sixteenth- and seventeenth-century herbals. There is no reason to believe that the name has been transferred directly from the 'real' rosemary (*Rosmarinus officinalis* L.) in Norway, where this mainly Mediterranean species can hardly be grown outdoors, and has been of little importance even as a spice. The manuscripts and publications mentioned above are probably not responsible for its transfer into folk usage in Finnmark, though it is worth noting that all records of *rosmarin* as a vernacular name derive from areas that have served as administrative centres for Finnmark – Alta in the west and Varanger in the east. Presumably as political centres they also served as hubs for trade in materials and information. Whether or not the records in Lilienskiold, Leem and Sommerfelt reflect something that already was an established folk use is impossible to tell.

The only pejorative Norwegian name known seems to be *stinkgras* ('stench grass'), recorded by Gunnerus (1772: 6). North Sami vernacular names are generally less benevolent. The most frequent one is *guohcarássi* ('stinking plant' or 'rotten plant') (Nielsen 1934: 208), a name first recorded from eastern Finnmark by Gunnerus (1772). A number of related or similar names are known (Qvigstad 1901), for example, *jeaggeguohc* ('mire stench'), recorded in eastern Finnmark by C. Weldingh in the eighteenth century (Dahl 1893: 56, voucher specimen in TRH). Vorren (1971: 15) also noted *bahčarássi* ('bitter plant' or 'evil-tasting plant') as a vernacular name in Finnmark. The species is known by about a dozen other North Sami vernacular names, including a more neutral *jeaggerássi* ('mire plant') (Qvigstad 1901: 316; Alm 1993; Svanberg 2004: 44). A couple of East Sami vernacular names noted by Qvigstad (1901: 314) belong to the first group, such as *guohcdaŋas* ('stinking heather').

A single Quain (Finnish) name has been recorded in Norway. According to Gunnerus (1772: 6; see also Dahl 1893: 56, citing a voucher specimen in TRH), C. Weldingh recorded *suunburre* among the Quain minority of Varanger, eastern Finnmark. This is probably an erroneous form of *suopursu* or *suonpursu*; the former is the usual name for the species in Finland. As noted by Vorren (1971: 16; 1978), *pursu* is a Germanic loan-word in Finnish (cfr. Norwegian *pors*).

Food and Drink

A single nineteenth-century source mentions that *R. tomentosum* was used as a substitute for salt by the Sami of Guovdageaidnu/Kautokeino, interior Finnmark: 'this same [plant] they also use instead of salt in their food, when they cannot reach the trading posts, it also gives almost the same taste as salt – and is thus called in Sami: *jægrasi* [= *jeaggerássi* in present-day spelling] or *saltrasi* [= *sálttrássi*], mire- or salt grass, as *jæg* means mire' (Budde 1808: 527).

To some extent, *R. tomentosum* may also have served as a beverage (Gunnerus 1772: 6), although decoctions are usually mentioned in a medicinal context. Again, Budde provided an interesting account: 'The mountain Sami also use a kind of decoction (*Ledum palustre*), which they drink as a tea, it has a pleasant taste and is good for the chest' (Budde 1808: 527).

Gunnerus (1772: 6) recorded *finnete* ('Sami tea') as a Norwegian vernacular name, suggesting Sami use it as a beverage. In the unpublished, original material of O.A. Høeg, a single Norwegian informant mentioned that *'rosmarin (finnmarkspors)* was also used as tea' – in this case probably by the Norwegians, and seemingly as a beverage (NFS O.A. Høeg 90; record dated 1938).

Medicine

At least in the past, the medicinal use of *R. tomentosum* in Norway may largely have been a Sami tradition. It was mainly used to treat cold and related symptoms or diseases, and as a cure for rheumatism, though a variety of other uses are known.

Cold. In Guovdgeaidnu/Kautokeino, two or three cups of *R. tomentosum* 'tea' were sometimes drunk as a cure for coughing. Less detailed records of use for cold and coughing are available from Porsanger (Gunnerus 1772: 6), Deatnu/Tana, Unjárga/Nesseby and Sør-Varanger. At least in the two latter areas, the decoction was made of the leaves. In Unjárga/Nesseby, it was specified that the flower buds had to be avoided, as they were considered poisonous (Qvigstad 1932: 27). In Deatnu/Tana, people considered that 'boiling *finnmarkspors* and drinking the water is good for chest complaints' (NFS Qvigstad 36a: 33). If the nose was affected, they placed *R. tomentosum* leaves there and left them for a while (Qvigstad 1932: 29). In East Sami tradition, a decoction of the dried plant was used for whooping cough (Qvigstad 1932: 31).

A single note on similar Norwegian use of *R. tomentosum* derives from Sør-Varanger, eastern Finnmark. According to Høeg (1974: 424), a decoction of the leaves was considered a good cure for cold and bronchitis; the effect was best if the plant material was collected prior to or just after flowering; that is, in early summer.

Rheumatism. A decoction of *R. tomentosum* was also used to treat rheumatism. It was either rubbed onto the affected part (Qvigstad 1932: 46), or used as a bath (Steen 1961: 18; Vorren 1971). Cures of this kind are known from several areas in Finnmark – Guovdageaidnu/Kautokeino, Kárášjohka/Karasjok, Lebesby, Deatnu/Tana and the Varanger area, and in adjacent Finland. According to a female informant from Kárásjohka/ Karasjok, cited by Vorren (1971: 16), there were still old people who considered such a decoction a good cure for rheumatism. The affected joints were bathed in it, relieving the pain. Some preferred to apply *R. tomentosum* as a kind of poultice, chopped finely and mixed with *Stellaria media* (L.) Vill. and fibres of *Populus tremula* L. This cure is known from Guovdageaidnu/ Kautokeino (Steen 1961: 18) and Deatnu/Tana (Qvigstad 1932: 48).

Frost damage. Nielsen (1926: 122) and Steen (1961: 18) noted that sores caused by frost damage could be treated with a decoction of *R. tomentosum*. If joints affected by frost became swollen, the East Sami of Sør-Varanger drank a three-days-old decoction (Qvigstad 1932: 120).

Other medicinal uses. From Guovdageaidnu/Kautokeino, Steen (1961: 17–18) noted that a decoction or tea of *R. tomentosum* was used for a variety of other complaints. People affected by high blood pressure could be relieved by drinking it while fasting. It was also used as a cure for diphtheria. Furthermore, it was believed to cure bladder catarrh. According to Vorren (1971: 16), people in Kárášjohka/Karasjok used the vapour for the latter complaint – in all probability by sitting with their underbelly above it, a cure otherwise in Norway mainly carried out with *Equisetum arvense* L. (Høeg 1974: 341).

Other Uses

Insect repellent. The only source that mentions a use of *R. tomentosum* in southern Norway is a brief note from Østfold: '*Påsst*, that is, *finnmarkspors*, *Ledum palustre*, which has a strong, spicy smell, was used to get rid of fleas' (NFS Hvidsten). However, *Myrica gale* was much more frequently used as an insecticide in this area (Høeg 1974: 457; EBATA 2006: 29).

Present-day Traditional Knowledge

Smell and Taste

Recent records confirm the past ethnic dualism as to conception of the smell of *R. tomentosum*. When asked, Norwegian inhabitants tend to describe the species as pleasantly scented, as in a comment from Alta, Finnmark: 'It has a very pleasant fragrance' (EBATA 1992: 3). Another emphasised that *R. tomentosum* had fragrance (Norwegian: *duft*), whereas other plants – with *Prunus padus* L. mentioned as an example – only possessed some kind of smell (Norwegian: *lukt*) (EBATA 2005: 77). In accordance with this positive view of *R. tomentosum*, *rosmarin* (rosemary) has survived as a Norwegian vernacular name in Finnmark, although the species is more frequently known as *pors*. *Finnmarkspors* ('pors of Finnmark') is also sometimes used as a vernacular name, at least in Alta, Porsanger, Deatnu/Tana and Sør-Varanger (EBATA 1992: 3, 1992: 5, 2005: 77, 2005: 82, 2006: 2, 2006: 13, 2006: 23, 2006: 32).

The Sami, on the other hand, usually describe the scent of *R. tomentosum* as unpleasant, in accordance with vernacular names describing it as 'rotten', 'evil-smelling' or 'piss-like'. Exceptions may occur. A woman from Porsanger thought the smell was 'quite fresh' (EBATA 1996: 7), and another female informant from Buolbmát in Deatnu/Tana described it as having a 'spicy smell' (EBATA 1996: 8). A third comment, from interior Finnmark, commented more neutrally on an 'extremely strong smell' (EBATA 2005: 66). *Guohcarássi* ('stinking plant') remains the most frequently encountered vernacular name (e.g., EBATA 1992: 4, 1992: 10, 1992: 14, 1996: 6, 2006: 55, 2007: 37), but at least half a dozen other names remain in use.

In Deatnu/Tana, eastern Finnmark, we were also told that the strong smell of *R. tomentosum* may adversely affect the popular and much collected fruits of *Rubus chamaemorus* L., making them more or less inedible: 'It poisoned the cloudberries as well. It gave them a bad taste' (EBATA 1996: 3).

Food and Drink

There are no recent records of any uses of *Rhododendron tomentosum* in food, as a salt substitute or as a beverage – except the widespread use of decoctions for medicinal purposes, described below. A single informant from Deatnu/Tana mentioned that it was sometimes used for smoking fish (EBATA 2006: 32), whereas a woman from Kárášjohka/Karasjok noted that others warned against this practice (EBATA 2006: 60).

Medicinal Uses

Our recent research confirms the extensive use of *R. tomentosum* as a medicinal plant, especially to treat cold and rheumatism.

Cold. As a cure for cold, *R. tomentosum* remains well known in the northernmost part of Norway. This is mainly a Sami tradition, though some Norwegian informants at least know the cure. A couple of recent interviews are quoted here. The first one derives from Válljohka in Kárášjohka/Karasjok: '*Guohcarássi* ... one boiled it and drank it as a cure for cold. It should be almost boiling hot when one drinks it, as hot as one can possibly bear' (EBATA 1992: 14). The same cure was used at Sieiddá in Deatnu/Tana: 'It was used here when we had a cold. Then she boiled tea from it. The taste was awful' (EBATA 1996: 7).

The mode of use was somewhat more varied at Varangerbotn in Unjárga/Nesseby, according to a woman born in 1931:

> Grandma boiled a decoction of *finnmarkspors* when we had a cold. Then, you were covered with a blanket, and should sit over the vapour rising from the kettle. We also had to drink the decoction. The plant was dried and used throughout the winter as a cure for cold (Rasmussen 2006: 232).

Some further details were provided in a recent interview: 'Yes, it was boiled, simmered. [They] had it in a kettle or pan, which was placed on a stool. And the children had to sit over it. And then you got to sweat, and you had to breathe it in.' This method is no longer used, but the decoction certainly is. 'Yes, we do [use it]. I have a female cousin who uses it a lot.' 'I still use it. It is used by a lot of people' – at least in the Unjárga/Nesseby and Sør-Varanger area. Only the leaves are used; the flowers are supposed to be poisonous (EBATA 2007: 37).

In some cases, *R. tomentosum* was included in 'combination' cures, for example, in Guovdageaidnu/Kautokeino: '[It] is good for pneumonia and cold'. A 'good mixture' consisted of *R. tomentosum*, juniper (*Juniperus communis* L.) and *Hoffmansdråper*; the latter a mixture of alcohol and ether, bought in pharmacies. 'The mixture is so strong that it never freezes.' In this case, the best time of year for collecting plant material was said to be 'in autumn and late summer' (EBATA 1992: 4).

Rheumatism. The only other medicinal use of *R. tomentosum* that remains well known is as a cure for rheumatic complaints. At Porsanger, we were offered a fine account of such use: [It] 'was very much used for medicine. As children, we picked and collected a lot. If people were affected by rheumatism, they boiled it and used it as a bath, and as compresses'. The cure was external only, 'it was not drunk' (EBATA 2001: 4). According to a woman from Kárášjohka/Karasjok, her father had made a decoction to cure rheumatism; he placed his aching feet in it (EBATA 2006: 55).

For bathing purposes, one could use the whole shrub. In other cases, the leaves and bark were collected and applied as a compress: 'When one needed such [decoctions] for bathing, the whole bush was boiled in large kettles. At that time, one had those large kettles that were used in the barn [to boil additional fodder for livestock, such as kelp and fish], a hundred litres or so'. For compresses, '[w]e rubbed off those needles [the leaves], and then the bark, and boiled it, and placed it inside towels'. These were placed on the 'aching parts' (EBATA 2001: 4).

A record from Čuovddatmohkki in Kárášjohka/Karasjok adds what is obviously an element of magic. Before the decoction is applied, one should proclaim the purpose. Afterwards, 'when one has used it, one should carry it [the remnants] back to the same tussock' – that is, back to where the plant had been gathered – in order to transfer the disease there (EBATA 1992: 10).

Other informants were less sure of the details, as recorded, for instance, at Vestertana/Deanodat and Buolbmát in Deatnu/Tana: 'I know it was used as a medicine for rheumatism. I am not sure if she boiled it and rubbed it on' (EBATA 1996: 2); 'it was used as medicine for rheumatism. They used to boil it, very much of it, in a cooking pan' (EBATA 1996: 6). Vars (2000: 25), citing two informants from Guovdageaidnu/Kautokeino, also mentions its use to treat rheumatism.

Mouth wash. A single informant from Guovdageaidnu/Kautokeino mentioned that the leaves of *R. tomentosum* served as a mouth wash. 'We used to chew it. The old people used to say that we would get a clean mouth from doing so' (EBATA 1992: 2).

Snowblindness. A recent interview provided a detailed record of how *R. tomentosum* could be used to treat people who had become snowblind: 'The old Sami used it, when they were busy herding reindeer, and their eyes became just like smoke. They boiled it in water and the one who had such eyes had to lie over the vapour covered by a black blanket' (EBATA 2005: 66). This cure was mostly used in early spring (April), when the combination of snow-covered surfaces and strong sunshine could easily take its toll on the eyes. At this time, the reindeer herds are still in the interior of Finnmark, where *R. tomentosum* occurs in abundance. Thus, it was not necessary to keep any provision of the plant. It was easy to get hold of, protruding above the thin snow cover with its evergreen leaves (EBATA 2005: 66).

Other Uses and Beliefs

Insect repellent. At Láhpoluoppal in Guovdageaidnu/Kautokeino, plant material or a decoction is used as an insecticide; for example, to keep the bed free of lice (Vars 2000: 25). In Sør-Varanger, eastern Finnmark, *R. tomentosum* has some reputation as an insect (mosquito) repellent, applied on the skin.

Tobacco substitute. According to an old female Sami from Láhpoluoppal in Guovdageaidnu/Kautokeino, her mother had used the leaves of *R. tomentosum* as a tobacco substitute. Although she hinted at some kind of narcotic effect, she certainly did not believe that this was the main reason for such use; the incentive was straightforward: 'my mother used to smoke it, she used to put it in her pipe. Lighted it ... while we were out collecting cloudberries' (*Rubus chamaemorus*). As to the reason for doing so: 'She had run out of tobacco. Then she took it and smoked it. But people will do anything when there's no tobacco left' (EBATA 2006: 23).

Sauna whips. Whips for this purpose are usually made of birch (*Betula pubescens* Ehrh.) or rowan (*Sorbus aucuparia* L.) branches. According to a female informant, people in Kárášjohka/Karasjok had used *Rhododendron tomentosum* twigs for the same purpose, adding some of it to the steaming hot water. This had led to a hallucinogenic experience, with disturbed vision (EBATA 2006: 57).

Calendar. The woman from Láhpoluoppal, cited above, also informed us of a traditional belief that the flowering of *R. tomentosum* and the rutting season of the brown bear [*Ursus arctos* (L.)] coincide: 'The bear is in heat at the time of the flowering, that is why he is yelling' (EBATA 2006: 23). It appeared from her story that the likelihood of meeting and seeing the bear was greater at this time than during other parts of the summer. The time of year is correct; *R. tomentosum* flowers in early summer (June), and the rutting time of the brown bear in Norway is in May and June (Sørensen 1990: 62). Male bears may range widely at this time looking for female bears.

Trends: Past and Present Traditions in Norway

In general, our data from the late 1970s onwards confirm the main uses of *R. tomentosum* as described by Qvigstad (1932), Steen (1961) and other authors. The species remains well known to people, at least where it occurs abundantly, and a fair number of vernacular names, both Norwegian and Sami, survive to the present day. The most frequent North Sami names recorded by us were also noted by Qvigstad (1901). Other names in our material are previously unrecorded – which only reflects the sparse material previously available on Sami ethnobotany in Norway.

Ethnobotanical traditions related to how people perceive the species' scent also survive, retaining the peculiar ethnic difference between Norwegians, praising its fragrance, and the Sami, detesting its stench. This is, however, easily explained, as the Norwegians generally live in coastal areas where the species occurs in restricted numbers. In the main Sami settlement areas of interior and eastern Finnmark, *R. tomentosum* may form vast stands, sometimes filling the air with its overpowering turpentine smell – to such an extent that it may even cause a headache, as noted by Lagerberg, Holmboe and Nordhagen (1956: 100), Vorren (1971: 15) and some of our informants (EBATA 2006: 55).

Some medicinal cures noted by previous authors have not been rerecorded in our interviews; other uses are new, if only in the sense that no previous records are available from Norway. Our record of using *R. tomentosum* vapour to treat people who have become snowblind is an interesting addition to the array of medical uses. Although recorded in 2005, it is probably obsolete. The springtime sun may still shine brightly, but modern sunglasses afford good protection. In addition, a major reason for such complaints in the past was that reindeer herders had to spend almost all their time outdoors in April, as the flock were becoming restless at this time of year. They started wandering towards the coast, and needed intensive herding (EBATA 2005: 66). Nowadays, snow scooters have made reindeer herding much less time-consuming than before.

It is evident from our material that only the uses of *R. tomentosum* as a cure for cold and related diseases (internally, by drinking a decoction), and to treat rheumatism (externally, by bathing) survive to the present day, in the sense that they may still be resorted to. This is hardly surprising, as both cold and rheumatism are frequent complaints in a cold, northern climate, and both remain difficult or impossible to treat by conventional medicine. Other plant species still used for medicine in Norway also meet the needs of people affected by diseases for which conventional medicine offers but limited relief, for example, the use of *Linnaea borealis* L. to treat shingles (*herpes zoster*) in parts of southern Norway (Alm 2006).

In our experience, the extent to which knowledge of *R. tomentosum* and its uses survives in present-day practices reflects the general state of folk plant knowledge in northern Norway. At least older people, both of Norwegian, Sami and Quain (Finnish) ethnic origin, still know the local names of numerous plants, and can provide details of their uses, as shown by interviews carried out and other records made throughout northern Norway in 2005 to 2008 (EBATA 2005: 1–96, 2006: 1–146, 2007: 1–95, 2008: 1–91), some running into dozens of pages of information.

Although people often remembered a multitude of past plant uses, the old traditions survive mainly when plants are needed for food and other general uses, for instance as fuel, or if they are used in children's games, or as decorations. In addition, some plants are needed for speciality purposes, for which no synthetic or modern alternative exists, or plant material is still considered superior, at least by some users. For instance, sedge leaves, mainly of *Carex aquatilis* Wahlenb. and *C. rostrata* Stokes, are still regularly collected and processed for shoe lining by the Sami, though in much reduced quantities (Alm and Iversen 1998: 16), whereas Norwegians gave up this practice long ago.

In terms of medicinal uses of plants, there is perhaps more information surviving among the Sami than in the Norwegian majority population, for several reasons. The Sami settlement areas are sparsely populated, and most elderly people here grew up in a society where doctors were difficult to access, both in terms of distance and culture. Until recently, there were few doctors fluent in the Sami language, and the language barrier may also have kept people from seeking medical advice. Doctors were probably not resorted to as long as other options – for example, old and trusted plant cures – were considered viable alternatives.

Some Notes on the Uses of *R. tomentosum* outside Norway

Rhododendron tomentosum (syn. *Ledum palustre*), *R. subarcticum* Harmaja (syn. *Ledum decumbens* (Aiton) Lodd. ex. Steud.), and *R. groenlandicum* (Oeder) Kron and Judd (syn. *Ledum groenlandicum* Oeder) are closely related. Hultén and Fries (1986) considered them to be subspecies within the *Ledum palustre* complex. Of these three taxa, *Rhododendron tomentosum* is widespread in northeast Europe and parts of Asia, and *R. subarcticum* is widespread in the northern hemisphere (northern Eurasia, the northern part of North America), whereas *R. groenlandicum* is restricted to Greenland and North America (Hultén and Fries 1986, map 1451). Throughout their range, they have found similar uses in folk tradition, both culinary, mainly as a 'tea', and medicinal.

Whereas *R. tomentosum* in Norway is rarely used as a drink per se, *R. groenlandicum* and *R. subarcticum* (or 'Labrador Tea') have been much used as a beverage among Indian and Inuit tribes in North America (Porsild 1953; Turner et al. 1990: 21; Moerman 1998: 300–301; Small and Catling 2000; Griffin 2001: 106) and Greenland (Egede 1741), including the Algonquin, Anticosti, Bella Coola (Smith 1929: 63), Cree, Gitksan, Salish, Thompson and many other tribes, partly as a substitute for tea and coffee (Turner et al. 1990: 240); recently also to add flavour to ordinary tea (Griffin 2001: 106).

The main medicinal uses of *Rhododendron* subsect. *Ledum* are comparable throughout its range. Among several native tribes of North America, *R. groenlandicum* is used to treat cold and head colds (Moerman 1998: 299), including the Kitasoo, Micmac (Chandler, Freeman and Hooper 1979: 58), Oweekeno and Abenaki (Rousseau 1948: 154); this cure is also known by the Cup'it Inuits of Nunivak Island (Griffin 2001: 114), in the latter case with *R. subarcticum*. The Abenaki included it in a snuff used to treat nasal inflammation. As a cure for pneumonia and whooping cough, *R. groenlandicum* has been used by the Cree Indians (Leighton 1985: 42).

Antirheumatic use of *R. groenlandicum* is also known in North America (Moerman 1998: 299–300). The Cree Indians of Hudson Bay used it externally (Holmes 1884: 303), whereas it was used internally for the same purpose by the Quinault (Gunther 1973: 43).

Moerman's (1998) survey of native American uses of *R. groenlandicum* include a range of other diseases and complaints; for example, for asthma, burns, fever, headache, scurvy or tuberculosis, as a diuretic, for dermatological and gynaecological aid, and to purify the blood. The Cup'it Inuits of Alaska treated diseases of the digestive tract with *R. subarcticum* (Griffin 2001: 114).

Some of these cures are also known from the Eurosiberian area. Chikov (1976), commenting on *R. tomentosum* as an economic plant in the Soviet Union, noted that young shoots were gathered for various medicinal purposes; for instance, as a remedy for coughing. Furthermore, leaves and stems had been used for tanning (see also Holmboe 1911: 33), and could serve as an insecticide. Chikov also noted that *R. tomentosum* honey was toxic.

In Finland, Rautavaara (1980) noted that *R. tomentosum* was still used as a cure for rheumatism, in baths, decoctions and compresses. Use for gout in Finland is mentioned by Tuovinen (1984), in this case with an infusion made from *R. tomentosum*, hops (*Humulus lupulus* L.), juniper 'berries' (*Juniperus communis* L.), and needles or young shoots of spruce (*Picea abies* L.) and pine (*Pinus sylvestris* L.). *R. tomentosum* has also been used to cure asthma, and to drive out tapeworms; in the latter case, one should drink a decoction after fasting for two days (Rautavaara 1980).

In Sweden, *R. tomentosum* was used to drive away mice (Linnaeus 1737: 121–122), and as an insecticide, against bedbugs, as a decoction to treat livestock infected by lice, and as a ground cover for pigs to avoid this problem (Linnaeus 1745: 6, 60; Holmboe 1911: 33; Tunón 2005: 430). An interesting group of Swedish vernacular names (*skvattram* and similar) may be related to the way the plant folds its leaves back and down during the winter (Nordhagen 1948: 11ff; cf. Rydén 2001, 2003).

R. tomentosum is supposed to be somewhat psychoactive (Festi and Samorini 1996; Rätsch 1998), but this property is weakly reflected in ethnobotanical traditions. The Kwakiutl Indians of Vancouver Island and British Columbia considered the leaves of *R. groenlandicum* to be narcotic (Turner and Bell 1973: 283), but this is the only such record mentioned by Moerman (1998: 299–300). In Europe, *R. tomentosum* has found some use in beer, as in Russia (Nordhagen 1946: 63), Germany and adjacent areas, Finland (Rautavaara 1980), Norway and Sweden (Linnaeus 1737;

Locke 1859; Schübeler 1888: 212; Holmboe 1911: 33; Hofsten 1960); it may have contributed to an intoxicating brew (Nordhagen 1946: 63–64; Sandermann 1980; Seidemann 1993; Alm 2003). An alcoholic extract may contain poisonous volatiles, and is hardly a good recipe for a healthy or tasty beer; it induces strong headaches, as noted by Tonning (1773: 105) and Rautavaara (1980). As suggested by the widespread use of *R. groenlandicum* beverages, ordinary decoctions in water are probably much to be preferred, in terms of both taste and potential medicinal effects.

According to Novik (1989: 34), Siberian shamans used *R. tomentosum* and spruce (*Picea abies* L.) during their rites, as a drink intended for their helping spirits, and as incense – in which case an intoxicating effect on the shaman is conceivable.

Extensive use of *Rhododendron* subsect. *Ledum* to treat cold and rheumatism, both in the Old and the New World, strongly suggest a positive medicinal effect. These complaints remain difficult to treat with conventional medicine. Thus, plants that serve as cures – at least according to folk tradition – stand a good chance of surviving in folk use, both at present and into the future. In accordance with this, Crellin (1987: 120–121) noted that no less than twenty-two plant species were still used to treat rheumatism in the southern Appalachian area of the United States. Tyler (1985, 1987), recording home remedies in Indiana, also in the United States, received more submissions of cures for cold and flu (61) than for any other diseases, followed by coughs (36), stomach trouble (36) and rheumatism/arthritis (28).

Whether or not *Rhododendron tomentosum* may be turned into some kind of pharmaceutical product for such complaints remains to be seen; some preliminary studies have been carried out in Tromsø (Aasen et al. 1998). Although *R. tomentosum* is generally considered as somewhat poisonous, with *ledol* as the prime suspect of toxicity (Small and Catling 2000), no negative effects of using a decoction have been recorded in folk tradition in Norway. Some constituents are extremely volatile (Ylipahkala and Jalonen 1992), and the strong turpentine smell of fresh material is absent from decoctions or 'tea'. It is possible that the medicinal effect may be due to nonvolatile components, such as complicated sugars.

Conclusion

Rhododendron tomentosum has a restricted distribution in Norway, but occurs abundantly in the Sami areas of interior Finnmark. Due to its strong turpentine smell, it cannot fail to catch people's attention; large stands may even cause headache. The main medicinal uses on record are for cold and rheumatism, frequent diseases that remain difficult to treat with modern medicine. Thus, *Rhododendron tomentosum* is still sometimes resorted to as a cure, especially for cold. These factors increase the likelihood that at least some ethnobotanical knowledge of the species will survive into the future, including vernacular names for it. In similar fashion, a fair number of other plant species that remain useful to people as food, for utility purposes, or as cures for specific diseases, are still well known, at least in the rural areas of Norway, both among ethnic Norwegians and the Sami and Finnish minorities. Other species, with more or less obsolete uses, are now known only to older people, and are much more likely to be forgotten.

Acknowledgements

We thank Hartvig Birkely, Anna Maria Iversdatter Sara Buljo, Inge Heika Hætta Eickelmann, Marit Sofie Holmestrand, Ellen Inga O. Hætta, Dagny Larsen, Ole Larsen, Kirsten Losoa and many others for providing information on past and present uses of *Rhododendron tomentosum* (and a substantial array of other plant species), mainly in North Sami tradition. Mikko Piirainen, University of Helsinki, provided help with Finnish literature; Dr Ilona Blinova, Kola Science Centre, with Russian material; and Jan Wesenberg, Natural History Museum, University of Oslo, with Russian translations.

Unpublished Archival Sources

EBATA: Interviews, letters and other records on Norwegian, Quain (Finnish) and Sami ethnobotany collected by T. Alm and M. Iversen (Sami only) from the late 1970s onwards; material stored by the first author.

NFS: Norsk folkeminnesamling/Norwegian Folklore Collection: NFS Hvidsten, undated manuscript, 10 pages; NFS Qvigstad 36a; NFS O.A. Høeg.

References

Aasen, A.J., T. Alm, T. Anderssen, W. Stensen and J.S.M. Svendsen. 1998. 'Planter og innholdsstoffer', *Ottar* 220: 49–55.

Alm, T. 1992. 'Amtmann Sommerfelts botaniske opptegnelser fra Finnmark 1799', *Polarflokken* 16: 225–252.

——— 1993. 'Finnmarkspors (*Ledum palustre*) i samisk og nord-norsk folketradisjon', *Polarflokken* 17: 219–228.

——— 2003. 'Ales, Beer and Other Viking Beverages – Some Notes Based on Norwegian Ethnobotany', *Yearbook of the Heather Society* 2003: 37–44.

——— 2006. 'Ethnobotany of *Linnaea borealis* (Linnaeaceae) in Norway', *Botanical Journal of The Linnean Society* 151(3): 437–452.

——— and M. Iversen. 1998. 'Samisk Etnobotanikk', *Ottar* 220: 13–16.

Anonymous. 1772. *Pharmacopoea Danica*. Hauniae: Apud Heineck & Faber.

——— 1854. *Pharmacopoea Norvegica Regia Auctoritate Edita*. Christianiae: Brøgger & Christie.

Budde, S.B. 1808. 'Blandede Efterretninger om Koutokejno's og Afjovarra's Præstegjeld, i Finmarkens Amt og Throndhjems I) Stift i Norge', *Fallelsens Theologiske Maanedskrivt* 12: 502–529.

Chandler, R.F., L. Freeman and S.N. Hooper. 1979. 'Herbal Remedies of the Maritime Indians', *Journal of Ethnopharmacology* 1: 49–68.

Chikov, P. (ed.). 1976. *Atlas arealov i resursov lekarstvennych rastenij SSSR*. Moscow.

Crellin, J.C. 1987. 'Folklore and Medicines – Medical Interfaces: a Kaleidoscope and Challenge', in J. Scarborough (ed.), *Folklore and Folk Medicines*. Madison, Wis.: American Institute for the History of Pharmacy, pp. 110–121.

Dahl, O. 1893. 'Biskop Gunnerus' virksomhed fornemmelig som botaniker tilligemed en oversigt over botanikens tilstand i Danmark og Norge indtil hans Død. II. Johan Ernst Gunnerus. Tillæg I. C. Gunnerus' visitatsreiser i Nordland og Finmarken og der indsamlede planter. D. Planter indsendte til Gunnerus fra Stadsbygden, Aafjorden, Nordland og Finmarken', *Det Kongelige Norske Videnskabers Selskabs Skrifter* 1892, no. 2: 1–61.

——— 1906. 'Biskop Gunnerus' Virksomhed fornemmelig som botaniker tilligemed en oversigt over botanikens tilstand i Danmark og Norge indtil hans død. III. Johan Ernst Gunnerus. Tillæg II: Uddrag af Gunnerus' brevveksling, særlig til belysning af hans videnskabelige sysler. Hefte 8. G. Breve angaaende Seminarium lapponicum, dettes adjunkter og Gunnerus' assistence ved Prof. Knud Leems videnskabelige arbeider', *Det Kongelige Norske Videnskabers Selskabs Skrifter* 1906, no. 4: 1–102.

Egede, H. 1741. *Det gamle Grønlands nye Perlustration, eller Naturel-Historie og Beskrivelse over det gamle Grønlands Situation, Luft, Temperatur og Beskaffenhed*. Kiøbenhavn.

Festi, F. and G. Samorini. 1996. 'Psychoactive Card VI. *Ledum palustre* L.', *Eleusis* 6: 31–37.

Granlund, E. 1925. 'Några växtgeografiska regiongränser. *Betula nana, Erica tetralix* och *Ledum palustre* i Sverige', *Geografiska Annaler* 7: 81–103.

Granmo, A. 1992. 'Pors, finnmarkspors og Porsanger', *Håløygminne* 18: 404–412.

Greve, P. 1938. *Ledum palustre* L. Monographie einer alten Heilpflanze (Eine botanisch-chemisch-pharmazeutische Bearbeitung), Ph.D. Dissertation. Hamburg: Hansischen Universität zu Hamburg.

Griffin, D. 2001. 'Contributions to the Ethnobotany of the *Cup'it* Eskimo, Nunivak Island, Alaska', *Journal of Ethnobiology* 21(2): 91–127.

Gunnerus, J.E. 1772. *Flora Norvegica: Pars posterior*. Copenhagen: Hafniæ.

Gunther, E. 1973. *Ethnobotany of Western Washington*. Revised edn. Seattle, Wash.: University of Washington Press.

Harmaja, H. 1990. 'New Names and Nomenclatural Combinations in *Rhododendron* (Ericaceae)', *Annales Botanici Fennici* 27: 203–204.

——— 1991. 'Taxonomic Notes on *Rhododendron* Subsection *Ledum* (*Ledum*, Ericaceae), with a Key to its Species', *Annales Botanici Fennici* 28: 171–173.

Høeg, O.A. 1974. *Planter og tradisjon: Floraen i levende tale og tradisjon i Norge 1925–1973*. Oslo, Bergen & Tromsø: Universitetsforlaget.

Hofsten, N. von. 1960. 'Pors och andra humleersättningar och ölkryddor i äldre tider', *Acta Academiæ Regiæ Gustavi Adolphi* 36: 1–248.

Holmboe, J. 1911. 'Linné's botaniske "Prælectiones privatissimæ" paa Hammarby 1770: Utgit efter Martin Vahl's referat', *Bergen museums aarbog* 1910(1): 1–69.

Holmes, E.M. 1884. 'Medicinal Plants Used by Cree Indians, Hudson's Bay Territory', *Pharmaceutical Journal and Transactions* 15: 302–304.

Hultén, E. 1971. *Atlas over växternas utbredning i Norden*. 2nd edn. Stockholm: Generalstabens litografiska anstalts förlag.

——— and M. Fries. 1986. *Atlas of North European Vascular Plants North of the Tropic of Cancer*. Koenigstein: Koeltz Scientific Books.

Kron, K.A. 1997. 'Phylogenetic Relationships of Rhododendroideae (Ericaceae)', *American Journal of Botany* 84: 973–980.

——— and W.S. Judd. 1990. 'Phylogenetic Relationships within the Rhodoraea (Ericaceae) with Specific Comments on the Placement of *Ledum*', *Systematic Botany* 15: 57–68.

Lagerberg, T., J. Holmboe and R. Nordhagen. 1956. *Våre ville planter*. Vol. 5. Oslo: Johan Grundt Tanum.

Leighton, A.L. 1985. 'Wild Plant Use by the Woods Cree (Nihithawak) of East-central Saskatchewan', *National Museums of Canada, Mercury Series* 101: 1–136.

Linnaeus, C. 1737. *Flora Lapponica*. Amsterdam: Salomo Schouten.

——— 1745. *Carl Linnæi Ölandska och Gotländska Resa på Riksens Högloflige Ständers befallning förrättad År 1741*. Stockholm and Uppsala.

Locke, J. 1859. 'On the "Heath-Beer" of the Ancient Scandinavians', *Ulster Journal of Archaeology*, Third series 7: 219–226.

Moerman, D.E. 1998. *Native American Ethnobotany*. Portland, Oreg.: Timber Press.

Nielsen, K. 1926. *Lærebok i lappisk*. Vol. 2. Oslo: Brøgger.

——— 1934. 'Lappisk ordbok: Grunnet på dialektene i Polmak, Karasjok og Kautokeino', 2. G-M, *Instituttet for sammenlignende kulturforskning, serie B, skrifter* 17: 1–718.

Nordhagen, R. 1946. 'Studier over gamle plantenavn. I. Motiver i nordiske navn på skinntryter og blåbær (*Vaccinium uliginosum* og *V. myrtillus*)', *Bergens museums årbok, naturvitenskapelig rekke* 1945(10): 1–144.

——— 1948. 'Kveldkippa og skvattram: Dynamiske motiv i nordiske plantenavn', *Nysvenska studier* 27: 1–26.

Novik, E.S. 1989. 'Ritual and Folklore in Siberian Shamanism: Experiment in a Comparison of Structures, the Archaic Epic and its Relationship to Ritual', *Soviet Anthropology & Archeology* 28: 20–99.

Porsild, A.E. 1953. 'Edible Plants of the Arctic', *Arctic* 6: 15–34.

Qvigstad, J. 1901. 'Lappiske plantenavne', *Nyt magazin for naturvidenskaberne* 39: 303–326.

——— 1932. 'Lappische Heilkunde', *Instituttet for sammenlignende kulturforskning, serie B, skrifter* 20: 1–270.

Rasmussen, S. 2006. 'Lisahkkus hus', *Varanger årbok* 2006: 223–235.

Rätsch, C. 1998. *Enzyklopädie der Psychoaktiven Pflanzen*. Aarau: AT Verlag.

Rautavaara, T. 1980. *Miten luonto parantaa, kansanparannuskeinoja ja luontaislääketiedettä*. Helsinki: WSOY.

Rousseau, J. 1948. 'Ethnobotanique Abénakise', *Archive de Folklore* 11: 145–182.

Rydén, M. 2001. 'Växternas namn: knärot, skvattram och tranbär', *Svensk Botanisk Tidskrift* 95: 24–27.

——— 2003. 'Botaniska strövtåg: Svenska och engelska', *Acta Academiae Regiae Gustavi Adolphi* 82: 1–182.

Rynning, L. 2001. *Bidrag til norsk almenningsrett*. Volume 5. Oslo: Det Norske videnskaps-akademi i Oslo.

Sandermann, W. 1980. 'Berserkwut durch Sumpfporst-Bier', *Brautwelt* 120 (50): 1870–1872.

Schübeler, F.C. 1888. *Viridarium norvegicum. Norges Væxtrige*. Vol. 2. Christiania: Universitets-program.

Seidemann, J. 1993. 'Sumpfporstkraut als Hopfenersatz', *Naturwissenschaftliche Rundschau* 46(11): 448–449.

Small, E. and P.M. Catling. 2000. 'Poorly Known Economic Plants of Canada. 26: Labrador Tea, *Ledum palustre sensu lato* (*Rhododendron tomentosum*)', *Canadian Botanical Association Bulletin* 33(3): 31–36.

Smith, H.I. 1929. 'Materia Medica of the Bella Coola and Neighboring Tribes of British Columbia', *National Museum of Canada Bulletin* 56: 47–68.

Solberg, O. 1942–1943. 'Finnmark omkring 1700: Lilienskiolds speculum boreale', *Nordnorske samlinger utgitt av Etnografisk museum* 4: 49–327.

Sommerfelt, S.C. 1799. 'Kort Beskrivelse over Finmarken', *Topographisk Journal for Norge* 7(24): 101–179.

Sørensen, O.J. 1990. 'Bjørnen', in A. Semb-Johansen (ed.), *Pattedyrene 1*. Oslo: J.W. Cappelens Forlag, pp. 64–89.

Steen, A. 1961. 'Samenes folkemedisin', *Samiske Samlinger* 5(2): 1–62.

Svanberg, I. 2004. 'Samiska växtnamn och folkbotaniska uppgifter hos Johan Turi', *Svenska landsmål och svensk folkliv* 330: 43–50.

Tonning, H. 1773. *Norsk Medicinsk og Oeconomisk Flora: Første Deel*. Kiøbenhavn: L.N. Stare.

Tunón, H. 2005. 'Skvattram *Rhododendron tomentosum*', in H. Tunón (ed.), *Människan och floran: Etnobiologi i Sverige*. Volume 2. Stockholm: Wahlström & Widstrand, p. 430.

Tuovinen, J. 1984. *Tietäjistä kuppareihin, kansanparannuksesta ja parantajista Suomessa*. Helsinki: WSOY.

Turner, N.J. and M.A.M. Bell. 1973. 'The Ethnobotany of the Southern Kwakiutl Indians of British Columbia', *Economic Botany* 27: 257–310.

———, L.C. Thompson, M.T. Thompson and A.Z. York. 1990. 'Thompson Ethnobotany: Knowledge and Usage of Plants by the Thompson Indians of British Columbia', *Royal British Columbia Museum, Memoir* 3: 1–335.

Tyler, V.E. 1985. *Hoosier Home Remedies*. West Lafayette, Ind.: Purdue University Press.

——— 1987. 'Some Potentially Useful Drugs Identified in a Study of Indiana Folk Medicine', in J. Scarborough (ed.), *Folklore and Folk Medicines*. Madison, Wis.: American Institute for the History of Pharmacy, pp. 98–109.

Vars, L.S. 2000. 'Sámi šattut – Vajálduvvon Dálkasat', *Š* 15: 24–26.

Vorren, K.-D. 1971. 'Finnmarkspors', *Ottar* 67: 15–17.

——— 1978. 'Pors', *Polarflokken* 2: 95–102.

Ylipahkala, T.M. and J.E. Jalonen. 1992. 'Isolation of Very Volatile Compounds from the Leaves of *Ledum palustre* Using the Purge and Trap Technique', *Chromatographia* 34: 159–162.

CHAPTER 14

Chamomiles in Spain

The Dynamics of Plant Nomenclature

MANUEL PARDO-DE-SANTAYANA AND RAMÓN MORALES

'En summa, es la mançanilla excellente y muy familiar remedio, contra infinitas enfermedades, que affligen el cuerpo humano'
(In sum, chamomile is an excellent and very familiar remedy against infinite illnesses that afflict the human body)
(Laguna 1555: 361)

Introduction

Historically, the folk botanical category known as *manzanilla* or *camomila* in the languages spoken in Spain referred only to *Matricaria recutita* and a few very similar species. In a rather exceptional example of dynamic consolidation in European ethnobotanical knowledge, this category has grown to include more than sixty similar species (Casermeiro et al. 1995; Álvarez 2006). These include some of the most popular digestive Spanish beverages (e.g., *Matricaria recutita, Matricaria aurea* and *Chamaemelum nobile*), their substitutes (e.g., *Helichrysum stoechas, Santolina chamaecyparissus*) and adulterants (e.g., *Tanacetum parthenium, Anacyclus clavatus*). They are all used in a similar way, as digestive herbal teas, and most of them are members of the daisy family, Asteraceae, with their flowers arranged in flower heads; that is, densely packed clusters of many small flowers, commonly surrounded by long, strap-like flowers termed *ligules*.

How has this happened? To what extent has a common morphology interacting with a common utility among so many species contributed to such a large folk-generic complex? What other factors might be at work? Are ecological, economic or cultural changes in the use of chamomiles driving the extension of the category to other species? In conjunction with evidence from botany and economic history, the names themselves may provide valuable evidence to answer these questions. The existence of binomials, with translatable descriptive terms, may give us clues as to why certain species have been classified together. For instance, appreciative or pejorative epithets can denote whether the plants are valued or not, and may be one of the underlying criteria that people use to group a variety of species together under one label. While the meaning of names shared by different species often reflects underlying cultural criteria of local importance, such polysemy can also derive from misidentifications or intentional substitutions, as when commercial species are adulterated by less valuable but more accessible species (Pardo-de-Santayana, Blanco and Morales 2005; Akerreta et al. 2007).

In order to better understand the meaning and historical evolution of this folk category, we reviewed and analysed more than forty-eight studies (see Table 14.1), for information on the historical and popular uses and names of plants called *manzanilla/camomila* in Spain (e.g., Pardo-de-Santayana, Blanco and Morales 2005; Álvarez 2006; Pardo-de-Santayana, San Miguel and Morales 2006). Our findings are detailed in the sections to follow, but to preview our conclusions, we argue that *manzanilla/camomila* is a complex folk-generic composed of:

1. A small number of highly valued and widespread prototypical species that have little variation in vernacular names.
2. A few highly valued, but ecologically restricted species that tend to have local vernacular names.
3. A large number of chamomile substitutes with a great variety of local names.

There is evidence from old botanical texts, herbals and other literature from across Europe that this folk generic has always been variable and expanding. In Spain and elsewhere the expansion of the category appears to have been driven by the discovery of morphologically similar species, due to increased travel, commerce and scientific study, and the inclusion of morphologically similar, but functionally dubious, substitute species, due to the expanding recognition of chamomile as a cure-all medicinal plant.

Chamomile and Chamomiles

The word *chamomile,* like *camomila* in Spain and Portugal, *camomilla* in Italy, *camomille* in France, and *kamille* in Germany, comes from the Latin *chamaemelum,* which in turn comes from the Greek *chamaimelon,* which means 'ground-apple' (*chamai*=on the ground; *melon*= apple). The name may well have originated from the herb's strongly aromatic and distinct scent of apples. Additionally, *chamomile* is called in Spanish *manzanilla*, that is, little apple (*manzana*=apple, *illa*=diminutive suffix). This term seems to be a translation of the original Greek name, *chamaimelon* (Covarrubias 1611). The Portuguese word *macela* is likely to have the same origin (*maçã*=apple, *ela*= diminutive suffix) (Feijao 1960–1963). In Spanish, *manzanilla* and *camomila* are synonyms, labelling the same category of botanical species.

While chamomile is a generic name that can refer to many herbs, it mainly denotes the so called 'true chamomile', common chamomile or German chamomile, *Matricaria recutita*, although due to a number of inaccuracies concerning its nomenclature, it is known by many other synonyms, such as *Matricaria chamomilla* and *Chamomilla recutita. Matricaria recutita*, a member of the daisy family (Asteraceae), is native to southern Europe, north Africa, and west, southwest and central Asia, but after centuries of cultivation and breeding on a wide scale, it has been naturalized in many other regions.

Matricaria recutita is one of the most widely used medicinal herbs in the world. It is not only highly popular as a home remedy, but is also recommended in allopathic medicine and of great interest for the pharmaceutical industry. Not surprisingly, it is one of the species of which a higher number of pharmacological, experimental and clinical studies are available (Franke and Schilcher 2005).

Its use dates back at least to ancient Greece and Rome. Hippocrates (460–377 BC) and Dioscorides (AD first century), for instance, described the plant and some of its uses. During the Renaissance it was so popular that it was stated that 'there is no herb in medicine for people being more usual than chamomile flowers because they are used against nearly all kinds of ailments' (Bock 1539). Modern scientific experiments have corroborated most of the virtues that people attributed to them. In fact, chamomile appears in a significant number of internationally known pharmacopoeias (Schilcher 2005). Chamomile, used internally or externally, contains volatile oils, flavonoids and other therapeutic substances (Font Quer 1962; Schilcher, Imming and Goetersal 2005). These produce anti-inflammatory, anti-spasmodic, choleretic and cholagogic activity, all of which help to improve digestive functions as well as having sedative and relaxing effects. Nowadays it is sold both as a foodstuff (for preparing domestic herbal teas) and as a drug, and is mainly consumed as a digestive and relaxant infusion.

However, the term *chamomile* is ambiguous since it refers to many other species (see Appendix 14.1). *Chamaemelum nobile* (syn.: *Anthemis nobilis*), similar in properties and applications to *Matricaria recutita*, is also highly appreciated and therefore also included in many pharmacopoeias (Schilcher 2005). It is only native to western Europe (British Isles, France, Portugal and Spain), northwestern Africa, Madeira and Azores, although cultivated and naturalized in many other regions. Usually known as Roman chamomile, although unknown to the Romans and Greeks, the name seems to have originated during the sixteenth century, perhaps because the plant was already cultivated around Rome at this time. It is also known as English chamomile, since it is a native English species.

The idea of chamomiles as a group of species or a generic class of plants appeared as early as the ancient Greek and Roman periods. Dioscorides (AD 65: III, 148) stated in his *De Materia Medica* that there are three species of *Anthemis* (chamomile): *Leucanthemum*, a name that indicates its white ligules, *Chrysanthemum*, a term that points out its golden yellow ligules, and *Eranthemon*, a Greek name that means the flower that blooms in spring. All of them are daisy-like plants, that is, ligulated species of the Asteraceae family.

Over centuries and across Europe the generic category appears to have expanded to include more botanical species; thus the German physician Tabernaemontanus (1522–1590) and the Spanish Botanist Quer (1695–1764) distinguished six and five different taxa of chamomile respectively (Tabernaemontanus 1664; Quer 1762–1764). Both Tabernaemontanus and Quer relate the 'common chamomile' (*gemeine Chamillenblum* and *manzanilla común*), to what Dioscorides called *Leucanthemum*. Based on their botanical descriptions it seems obvious they are referring to *Matricaria recutita*. To this they contrast the 'Roman chamomile' (*Römisch Chamillen* and *manzanilla romana*), which they clearly identify as *Chamaemelum nobile*. A third species mentioned by Tabernaemontanus was the 'full Roman chamomile' (*Gefüllt Römisch Chamillen*), a variety of *Chamaemelum nobile* with more ligulated flowers, developed by medieval herbalists (Franke 2005).

The *Chrysanthemum* of Dioscorides is likely to be the 'yellow chamomile' (*Geel Chamillen*) of Tabernaemontanus, probably referring to *Anthemis tinctoria* (Franke 2005). According to Manniche (1989), this is the chamomile that Egyptians knew and greatly esteemed. Dioscorides wrote that the third chamomile, *Eranthemon*, has red ligules, which seems to refer to the 'red chamomile' (*Rothe Chamillen*) of Tabernaemontanus. It has been suggested that this name refers to *Adonis* (*A. aestivalis* L. or *A. flammea* Jacq.) (Franke 2005), a genus in the Ranunculaceae. However, this seems unlikely since Dioscorides says that all chamomiles have the typical yellow button of the flower heads of daisies, not found in Ranunculaceae.

The other species mentioned by Tabernaemontanus is the 'full Roman chamomile of another genus' (*Gefüllt Römisch Chamillen anderer Gattung*), whose botanical identity remains a mystery. Three other types of medicinal chamomiles were mentioned by Quer: 'Delicate chamomile' (*manzanilla fina*) clearly refers to *Matricaria aurea*, 'stinking chamomile' (*manzanilla hedionda*) is probably *Anthemis arvensis* or *Anthemis cotula*, and 'scentless chamomile' (*manzanilla inodora*) could refer to *Anacyclus clavatus* or *Tripleurospermum inodorum* (syn. *Matricaria inodora, Matricaria maritima*).

The similarities between the botanical species of *Matricaria*, *Chamaemelum*, *Anthemis* and several other related genera may have led to the extension of the label 'chamomile' in many European languages to many other species, such as *Matricaria dioscoidea, Chamaemelum mixtum* and *Chamaemelum fuscatum*. Not all of these related species share the same medicinal qualities, so this expansion of the class of 'chamomiles' could have resulted in misidentifications, substitutions and adulterations of the collected plant and its medicinal products. We also speculate, below, that perhaps this semantic expansion was in part driven by the marketing of chamomiles as an important, 'cure-all' medicinal product. Once a demand was created for something called *chamomile,* in an area where no one was sure what this referred to, then it would have been easy for plant collectors to make mistakes in identification or herbalists or vendors to unintentionally or even intentionally mislead consumers.

Spanish Chamomiles

Our survey of the literature showed that the Spanish terms *manzanilla* and *camomila* refer to sixty-two species across Spain, but only thirty have been considered in modern ethnobotanical studies (Álvarez 2006). These popular generics include mainly ligulate *Anthemidae* (Asteraceae) of the genera *Matricaria, Chamaemelum* and *Anthemis.* They include also other Asteraceae without ligules, such as *Achillea, Artemisia, Helichrysum* and *Santolina* (see Table 14.1), and members of other families, including Caryophyllaceae (*Herniaria glabra, Paronychia argentea*), Convolvulaceae (*Convolvulus boissieri*), Dipsacaceae (*Pterocephalus spathulatus*), Geraniaceae (*Erodium foetidum*) and Apiaceae (*Bupleurum falcatum*) (Villar et al. 1987; González Tejero 1990; Mesa 1996; Guzmán Tirado 1997; Fajardo et al. 2000; Fernández Ocaña 2000).

The term *manzanilla* has been used in Spain for a very long time. As early as the twelfth century, the Jewish rabbi, physician and philosopher Maimonides (1135–1204) mentioned that chamomile was called *masanilah* or *masanalah* in Al-Andalus (Navarro and Hernández Bermejo 1994). Some decades later, Ibn Al-Baytar (c. 1197–1218), an Andalusian Arab who was one of the most influential Medieval writers on botany and pharmaceuticals, also refers to *massanallah,* a synonym of *jamamilun* (Greek) and *babunay* (Arab) (Navarro and Hernández Bermejo 1994).

In Spanish, *manzanilla* prevails over the term *camomila,* which according to Covarrubias (1539–1613), the author of one of the first and most influential Spanish dictionaries (Covarrubias 1611), was the name used by the 'barbarians' (a pejorative term for referring to foreign people). At this time, *manzanilla* was not only used to refer to *Matricaria recutita* or *Chamaemelum nobile,* but also to their substitutes. L'Écluse (1526–1609), a Flemish botanist who travelled around Spain collecting and describing plants, stated that people from Murcia used *manzanilla* for the plant that

Table 14.1. Most important species called *manzanilla, camomila* or derived names in Spain

Asteraceae	Popular names	Popular uses
Achillea millefolium	Manzanilla (4, 24, 47), manzanilla romana (12, 23, 38), manzanillón (23, 42)	Digestive (17, 33, 37, 38, 42, 47), analgesic, vulnerary (4, 38, 47), enhance blood circulation (38, 42, 47), bronchial disorders, emmenagogue (47, 38), haemorrhoids, diuretic, headache, fever, laxative (47), diarrhoea (38)
Achillea odorata	Manzanilla de la sierra (30), manzanilla real (41)	Digestive (30), vulnerary (41)
Achillea ptarmica	Camamilla, camamilla de muntanya, camamilla de Rojà (40)	Digestive, intestinal ache, diarrhoea, laxative, anti-catarrhal, sedative, heart disorders, vulnerary, eye infections (40)
Anthemis arvensis	Manzanilla (10, 12, 18, 38, 43, 47), manzanilla basta (1), manzanilla bastarda (10, 12, 17, 18, 35, 48), manzanilla silvestre (10, 12, 18), manzanilla borde (10, 17, 18, 45)	Digestive and stomach ache (1, 3, 10, 14, 18, 43, 45, 47), laxative (9)
Artemisia granatensis	Manzanilla de la sierra (17, 20), manzanilla de Sierra Nevada (11, 17)	Digestive (17, 20)
Chamaemelum nobile	Camamilla romana (9, 17), camomila (27), camomila romana (17, 19), kamamila (1), manzanilla (1, 2, 3, 4, 7, 11, 19, 22, 23, 36, 41, 38, 42, 43), manzanilla amarga (4, 43, 46), manzanilla de campo (38, 43), manzanilla de monte (1), manzanilla de Urbasa (1), manzanilla de Aralar (1), manzanilla fina (1), manzanilla romana (11, 12, 17, 19, 25, 43)	Digestive, carminative (1, 2, 3, 4, 7, 9, 19, 23, 41, 36, 38, 41, 42), eye infections (2, 4, 19, 36; 38, 7, 42), laxative (4, 19, 38, 42), relaxant (4, 36, 42), sore throat (4, 36), hepatoprotector (38), aperitif (2) headache (23), emmenagogue, earache (38)
Helichrysum italicum	Manzanilla basta (20), manzanilla borde (13, 19), manzanilla silvestre (44), mançanella borda, (32), mançanilla del bosc (37)	Digestive (19, 32, 37, 44), toothache (20)

Table 14.1 *continued*

Helichrysum stoechas	Camamilla (9), mançanilla (17, 29, 37), manzanilla (1, 5, 6, 11, 32, 38, 46), manzanilla basta (4, 11, 12, 18, 46), manzanilla bastarda (11, 12, 17), manzanilla dulce (1), manzanilla fina (1), manzanilla de monte (41, 46), manzanilla de pastor (6, 11, 12, 17), manzanillón (12, 41)	Digestive, intestinal ache (1, 20, 22, 32, 37, 38), anti-catarrhal (6, 38, 41), wounds (32, 41), tooth ache (41), anti-helmintic (38)
Matricaria aurea	Manzanilla (22, 43), manzanilla fina (11, 17, 43)	Digestive (22, 43, 47), eye infections, laxative (2)
Matricaria dioscoidea	Manzanilla falsa (42), manzanilla silvestre (38), manzanillón (12), manzanilla dulce (43), mencenilla (3)	Digestive (3, 38, 43)
Matricaria recutita	Camamilla (1, 8, 9, 16, 17, 32, 33, 37), camamilla dolça (40), camomila (12, 17, 43), manzanilla (3, 4, 5, 6, 7, 11, 12, 14, 15, 16, 17, 20, 21, 22, 23, 28, 30, 31, 32, 34, 36, 39, 43, 45, 46, 47), manzanilla buena (15, 45), manzanilla dulce (4, 5, 14, 21, 32, 39, 43, 46)	Digestive, intestinal ache (1, 3, 4, 5, 6, 9, 14, 31, 33, 40, 41, 45), eye infections (3, 4, 7, 9, 14, 33, 39, 41), anti-cattarrhal (9, 22, 33), laxative (9, 33, 39), sedative (9, 14, 33), antacid (9, 39), emmenagogue (14, 33), hepatoprotector, emetic (9, 33), earache (33, 40), depurative, bad breath (41), headache, haemorrhoids, vulnerary (33)
Santolina chamaecyparissus	Camamilla de botó (40), camamilla de botó groc (40), camamilla de l'hort (40), camamilla de muntanya (11, 29, 37), camamilla de parets (40), mançanilla (29, 32), manzanilla (1, 11, 16, 17, 32, 47), manzanilla amarga (13, 14, 34, 41), manzanilla basta (1), manzanilla de burro (1), manzanilla de monte (1), manzanilla del campo (14, 41)	Digestive, intestinal ache (1, 14, 16, 19, 20, 32, 34, 37, 40, 41, 47), hepatoprotector (16, 33), sedative (1, 40), anti-catarrhal, anti-rheumatic, vulnerary, anti-helmintic (41), headache, depurative, women's hygiene (1), eye infections (40)
Santolina oblongifolia	Manzanilla de Gredos (17, 26), manzanilla dulce (26)	Digestive (26, 17)
Santolina rosmarinifolia	Manzanilla fina (11), manzanilla del campo (22)	Digestive (22)

Table 14.1 *continued*

Tanacetum parthenium	Camamila de los huertos (11, 17), camamilla (40), camamilla amargant (40), camamilla borda (9, 11, 17, 40), camamilla de jardí (8, 29), manzanilla (11), manzanilla amarga (1), manzanilla brava (12), manzanilla de huerta (1), manzanillón (12), manzanillota (24)	Digestive, stomach and intestinal ache (1, 40, 9, 47), purgant (9, 47), anti-catarrhal, depurative, hepatoprotector (9), anti-helmintic, relaxant (47), diarrhoea (40)

1 Akerreta et al. 2007; 2 Barandiarán and Manterola 1990; 3 Blanco 1996; 4 Blanco 1998; 5 Blanco 2002; 6 Blanco and Cuadrado 2000; 7 Blanco and Diez 2005; 8 Bonet 1991; 9 Bonet 2001; 10 Casana 1993; 11 Colmeiro 1885–1895; 12 Esgueva 1999; 13 Esteso 1992; 14 Fajardo et al. 2000; 15 Fernández Ocaña 2000; 16 Ferrández and Sanz 1993; 17 Font Quer 1962; 18 Galán 1993; 19 Gil Pinilla 1995; 20 González Tejero 1990; 21 Granzow de la Cerda 1993; 22 Guzmán Tirado 1997; 23 Lastra 2003; 24 Lastra et al. 2000; 25 Lastra and Bachiller 1997; 26 López Sáez 2002; 27 Losada, Castro and Niño 1992; 28 Martínez Lirola, González Tejero and Molero 1997; 29 Masclans; 30 Mesa 1996; 31 Molina Mahedero 2001; 32 Mulet 1991; 33 Muntané 1994; 34 Obón and Rivera 1991; 35 Oria de Rueda, Diez and Rodríguez 1996; 36 Panero and Sánchez 2000; 37 Parada et al. 2002; 38 Pardo-de-Santayana 2003; 39 Rabal 2000; 40 Rigat, Garnatje and Vallès 2006; 41 Rivera et al. 1994; 42 San Miguel 2004; 43 Tardío, Pascual and Morales 2002; 44 Triano 1998; 45 Verde, Rivera and Obón 1998; 46 Verde et al. 2000; 47 Villar et al. 1987; 48 Villar, Sessé and Ferrández 2001

he called *Chrysocome altera* (L'Écluse 1576). According to Colmeiro (1885–1895), *manzanilla* also referred during the Renaissance to *Gnaphalium luteo-album*, but it is likely that this plant was *Helichrysum stoechas* or another species of the same genus, since Laguna (1499–1559), the Spanish translator of Dioscorides, also mentioned the name *manzanilla bastarda* (false manzanilla) for *Helichrysum stoechas* (Laguna 1555). Laguna also reported the use of *manzanilla loca* ('crazy manzanilla') for *Tanacetum parthenium.*

Up until the twentieth century, all of the plants known by *manzanilla, camomila* or some cognate term were members of the Asteraceae (see Colmeiro 1885–1895). Importantly, the Spanish pharmacopoeias from 1739 to 1884 included more than thirty species called *manzanilla,* but the pharmocopoeias of the twentieth century have restricted the category of *manzanilla* to *Chamaemelum nobile, Matricaria recutita* and *M. aurea.* This represents a logical and somewhat understandable switch in the criteria used to classify and label plants in pharmacopoeias, from those with similar morphological characteristics, daisy-like species, to those species of the same family with similar medicinal properties, mainly used for digestive disorders (Casermeiro et al. 1995), including even those that did not have ligulated flower heads.

While this contraction of the folk generic is not surprising given the function of pharmacopoeias, in common usage the category has widened its meaning, to cover also other highly valued species with digestive functions from other families, and of course, with very different appearances. Our surveys of the literature showed 215 vernacular names assigned to 62 species, including many synonyms, which yields a total of 369 vernacular denominations, or combinations of vernacular and botanical names.

This richness of names applied to *manzanilla/camomila* provides interesting information about the underlying criteria being used in plant classification. The epithets, secondary lexemes used to label the different species, refer to places of origin or use, habitat, morphology, function and if they are appreciated or not (Table 14.2). Common epithets that appear across the country include terms for 'bitter', 'true', 'sweet' and 'fragrant'. There are also restricted names, found only in certain parts of the country that generally refer to rare or local species, such as *Sierra Nevada*, *Gredos* or *Urbasa*. Other epithets indicate if they are valued (good, genuine, noble) or not (coarse, rude, crazy).

Table 14.2. Epithets that are used to label plants known as *manzanilla/camomila* and derived names in Spanish

Meaning of epithets	Spanish epithets (English translation)
Morphological, organoleptical and phenological characteristics	*Amarillo* (yellow), *blanca* (white), *dorada* (golden); *invierno* (winter), *San Juan* (Saint John); *espatulada* (spatulated), *estrellada* (star), *flor* (flowered), *margarita* (daisy), *rastrera* (creeping); *olorosa* (fragrant), *sin olor* (scentless), *hedionda* (stinky); *amarga* (bitter), *dulce* (sweeet), *fina* (delicate, mild), *fuerte* (strong)
Habitat (cultivated, noncultivated)	*Corral* (courtyard), *huerto* (kitchen garden), *jardín* (flower garden); *silvestre* (wild); *campo* (country), *alpina* (Alpine), *bosque* (forest), *lastra* (rocky), *marina* (sea), *montaña* (mountain), *puerto* (mountain pass), *sierra* (high country)
Region of origin or region of use	*Alemana* (German), *aragonesa* (Aragonese), *americana* (American), *francesa* (French), *gallega* (Gallician), *portuguesa* (Portuguese), *romana* (Roman), *valenciana* (Valencian), and from many other regions and localities (Aralar, Granada, Gredos, Mágina, Maó, Moncayo, Navarra, Nuri, Pirineo, Sierra Nevada, Soria, Urbasa , Urgel)
Uses	*Purgante* (purgant), *tinte* (dyeing), *yesquera* (kindling)
Pejorative epithets	*Basta* (coarse), *bastarda* (illegitimate), *borde* (rude), *borriquera* (donkey), *falsa* (false), *gorda* (fat), *loca* (crazy), *mala* (bad), *morisca* (moorish), *manzanillón* (big)
Appreciative epithets	*Buena* (good), *común* (common), *legítima* (genuine), *noble* (noble), *real* (true), *vera* (true)

At first glance, the correspondence between vernacular and scientific names seems to be rather chaotic. The same species is called by different and contradictory names: *Anthemis cotula*, for example, is called *sin olor* (scentless), *fina* (delicate, mild) or *hedionda* (stinking). Moreover, the same label can be used for species with different characteristics. *Manzanilla loca* (crazy), for instance, is used for the disgusting species *Anacyclus clavatus*, but also for the pleasant *Matricaria recutita*. On the other hand, epithets that indicate appreciation are used both for the best chamomile substitutes and for other, less valued species; so *vera* (true) refers to the most esteemed *Chamaemelum nobile* and *Matricaria recutita*, as well as their substitutes *Phagnalon saxatile*, *Achillea millefolium*, *Helichrysum stoechas*. However, these examples are not the rule but the exception. Typically, appreciative names refer to delicious species and pejorative names to those species of a worse quality.

In Navarra, for instance, *Santolina chamaecyparissus* is widely called *manzanilla*. There are people who prefer it to *Chamaemelum nobile* and they call *Santolina chamaecyparissus* just *manzanilla*, while those who prefer *Chamaemelum nobile* call *Santolina chamaecyparissus* by pejorative names such as *manzanilla basta* or *manzanilla de burro* (rough or donkey chamomile) (Akerreta et al. 2007).

The process of extending the meaning of the term *manzanilla* from *Matricaria recutita* and *Chamaemelum nobile* to many other plants exemplifies the links between oral and written traditions in so-called 'popular knowledge'. Names used by botanists and herbalists and 'popular names' have been interchanged, producing a rich and varied, but confusing lexicon.

The compilation of Colmeiro (1885–1889) is particularly useful, since it includes 'popular' and 'botanical' names in use at the end of the nineteenth century, which can be compared with those used in the twentieth and twenty-first centuries. According to Colmeiro, 32 species, 81 names and 110 vernacular denominations were known as *manzanilla*, *camomila* or some derived names. It is difficult to know which of these names were in fact known only to botanists and people familiar with botanical and herbal books, and which of them were 'popular names' used by common people. In fact, less than 25 per cent of these denominations have been discovered in use in modern ethnobotanical surveys (Álvarez 2006), likely evidence that most of them were names that never became widespread.

The binomial *manzanilla romana* is a good example of the influence of written botanical knowledge and popular knowledge. This name appears in the botanical texts compiled by Colmeiro, only for referring to *Chamaemelum nobile*, but it was later borrowed for many other species, such as the chamomile substitutes *Achillea millefolium* and *Anthemis arvensis*. It is not possible to reconstruct how this process of transferring names actually happened, but, to speculate for a moment, one possibility is that herbal sellers could have begun to use and sell these species as

Chamaemelum nobile substitutes, at first telling people that they have similar properties to *manzanilla romana*, and then eventually just calling them by that name. Associating new products with older and successful ones seems a logical, if deceptive, marketing strategy, and if there were no noticeable differences in the functions of these new plants, then perhaps consumers would have been willing to go along with the new names. The lack of availability of the original species, for whatever reason, would also allow for its name to be transferred to other species.

From the consumer's point of view, there may have been good social reasons to adopt the names of what are considered 'exotic' or high status kinds of chamomile. As with other luxury goods such as wine, coffee and even chocolate, chamomile types would have been ranked on both an economic and a prestige scale. Serving an expensive or high status tea would have indicated a higher social standing. Similarly, local or familiar types of chamomile might be renamed with more exotic names in order to associate oneself with a wider, more 'developed' and therefore prestigious social context.

Recent survey work shows that rural people in many parts of Europe often lack confidence in their own culture and language, and think that their local name is not the 'correct' or 'proper' name. This devaluation is said to arise from the adoption by recent rural migrants to the cities of the prevailing urban stereotype of rural life as simple and backward (Gómez Pellón 2004; A. Pieroni personal communication).

On the other hand, it is possible that some of the names used by botanists have been taken from the pool of popular epithets and used to label species that did not have any Spanish, Catalan or Galician names. This seems a likely explanation for the names *manzanilla bastarda* for *Anthemis tuberculata* (Blanca and Morales 1991) and *camamilla borda* for *Chrysanthemum coronarium* (Masclans 1981), which have never been recorded in ethnobotanical studies.

Other tendencies can be outlined if the popular names compiled in modern ethnobotanical surveys are compared with those that appear in botanical books. Pejorative names and those inspired by morphology, uses or habitat are more frequently found in popular nomenclature, while names that indicate a region of use or origin are less common (see Table 14.1). The use of geographical epithets in scientific nomenclature is very frequent, since it is easier for a botanist to have access to knowledge about the global distribution of the species and of their uses. Therefore, it is possible that some of the popular names that include a geographical epithet derive from botanists' names.

Unfortunately, there is just not enough information to come to any firm conclusions about whether such shifts between vernacular and scientific plant nomenclature have occurred, but the subject is both fascinating and necessary and deserves more attention in future research.

Highly Valued, Widely Used Chamomiles

In Spain, as in many other European countries, the two chamomiles that have been most widely and regularly used are *Chamaemelum nobile* and *Matricaria recutita* (San Miguel 2004; Pardo-de-Santayana 2005). Both herbs are used for the same purposes: they are prepared as a tea and drunk mainly for aiding the digestive process, after the main meals, or as a relaxant before going to bed. However, they serve social functions as well, being an important component of the socializing and conversation that occurs after meals (Pardo-de-Santayana, San Miguel and Morales 2006). Another highly valued species but less commonly used nowadays is *Matricaria aurea*. All three species were included in previous editions of the Spanish Pharmacopoeia (Farmacopea Oficial Española IX 1954) and have very similar pharmacological actions, including tonic, carminative, antispasmodic, emmenagogue, choleretic and cholagogic, anti-inflammatory, analgesic, antiseptic, antifungic and antiparasitic properties. Used internally, they fight colic and improve the digestion, stimulate the secretion and production of bile, and act as a sedative. Applied externally, they treat eye infections and dye hair blond. They are also used to aromatize many traditional liqueurs (Tardío, Pardo-de-Santayana and Morales 2006).

As might be expected of such a valued and widespread species, there is less variation in the local vernacular, so *Matricaria recutita* (Figure 14.1A) is usually known as *manzanilla*, without epithets, but it can also be labelled by such binomials as *manzanilla dulce* (sweet). This annual blossoms in spring,

Figure 14.1. Highly valued, widely used chamomiles in Spain. A. *Matricaria recutita* B. *Matricaria aurea* C. *Chamaemelum nobile*

grows wild around cultivated fields and on fallow land, and is cultivated in homegardens or fields in the south and east of Spain. The most common of chamomiles in Spain, *M. recutita* is marketed and sold in tea bags in supermarkets or served in bars or restaurants and its use is not restricted to rural areas. Besides chamomile being a lucrative crop domestically, Spain produces approximately fifty tonnes of chamomile's flowering tops for export to the U.S.A., Germany and other countries. However, this is just half of the production of ten years ago, as countries from the former Eastern Bloc are now the most important exporters of medicinal plants (J.L. López Larramendi, personal communication; Lange 1998).

Chamaemelum nobile (Figure 14.1C) is also known as *manzanilla*, but is usually distinguished from *M. recutita* with secondary lexemes such as *manzanilla amarga* (bitter) or *manzanilla romana* (Roman). This perennial herb blossoms in summer and grows in grazed grasslands, though many report its decline there due to changes in agropastoral systems that have led to less intensive grazing (Barandiarán and Manterola 1990). Although it is more common in the northern half of Spain, *C. nobile* can be found in the cold and humid mountainous regions of the south. It is considered a pleasant and aromatic herb with a characteristic bitter flavour, which many people claim to prefer over the sweet varieties 'served in bars'. Although it is possible to purchase it in herbal shops, wild gathering is preferred and considered one of the central activities of the summer (San Miguel 2004); not to collect is considered by some to be a sign of slovenliness and lack of foresight! Many people collect *C. nobile* to send to their relatives who have moved to the cities (Pardo-de-Santayana 2003).

The third widespread species is *Matricaria aurea* (Figure 14.1B), known as *manzanilla fina* (delicate), a name that derives from its smooth, mild and delicate flavour. This annual blossoms at the end of the winter or at the beginning of spring, and grows in trampled sites such as paths, roads or streets that have not been surfaced, though it is less common these days because many of these locations have now been paved. So people who like this chamomile now have to protect it (Tardío, Pascual and Morales 2002). It is highly esteemed in many rural areas, and some people prefer it to the more well known *Matricaria recutita* and *Chamaemelum nobile* because its scent is so aromatic and delicate (Laguna 2006). Once listed in the Spanish pharmacopoeia (Farmacopea Oficial Española IX 1954), it has been left out of a recent edition because of the decline in its use (Real Farmacopea Española 2005).

Highly Appreciated, but Locally Used Chamomiles

Artemisia granatensis, known as *manzanilla de Sierra Nevada* or *manzanilla de la sierra*, grows only in rocky places above 2,000 m in the Sierra Nevada (south of Spain). Boissier, the botanist who described the species in the nineteenth century, commented that shepherds gathered it in great amounts and sold it in the city of Granada. These shepherd-gatherers were called *manzanilleros* ('chamomile gatherers'), and they went to great efforts to collect the herb because it was considered a panacea or a miracle cure-all, though only its digestive properties have been validated (Calle and Gómez 2009). The high demand eventually led to increased scarcity, higher prices and the threat of local extinction, so the species was officially protected in 1982 (Blanca 2003). The plant achieved national notoriety when a local shepherd was charged with illegal gathering, and threatened with a two-year prison sentence and a €1,500 fine. Most people thought the punishment far outweighed the crime and were happy when he was finally acquitted. The case brought attention to the sustainability of wild-harvested plants in Spain, and the need for environmental education and alternatives so that shepherds and other gatherers and consumers could still enjoy chamomiles and other herbs without endangering their existence. The fact that this species has already been cultivated in the Andalusian Botanical Gardens (Clemente et al. 1991) suggests that it might be possible to substitute cultivated for wild varieties.

The Sierra de Gredos, in the Centre-west of Spain, also has its own *manzanilla, manzanilla de Gredos (Santolina oblongifolia).* It grows at 1,000 m elevation in the rocky mountains of Ávila, Salamanca and Cáceres provinces, and is considered sensitive to habitat alterations and thus is also listed as endangered (Regional Catalogue of Threatened Species of Extremadura 2001 DOE – decree 37/2001, 6 March). Highly valued for its sweet, mild flavour, it is used as a digestive tonic, as a sedative, and for treating menstrual disorders and rheumatism (Silván et al. 1996; López Sáez 2002).

Chamomile Substitutes

It is common that medicinal plants are substituted or replaced by others with similar morphology or properties. This can be due to the scarcity or high price of the former and the easier accessibility of the replacement species. Vernacular plant names also reflect their function as substitutes. All of the substitutes described here are morphologically quite different from *Matricaria recutita* or *Chamaemelum nobile* and thus people do not confuse them.

The most popular chamomile substitute is *Helichrysum stoechas* (Figure 14.2B), known as *manzanilla* at least since the sixteenth century (Laguna 1555). It grows throughout the country, typically in rocky, stony and other poorly developed soils. *Helichrysum stoechas* has a distinct aroma, similar to cognac or curry, but is usually considered too strong for use by humans. Most people use it only for healing animals or when better chamomiles are not available. However, in some regions it is called *manzanilla real* (royal), *fina* (delicate) or *vera* (true) and some people even prefer it to *Matricaria recutita* or *Chamaemelum nobile* (Mulet 1991; Pardo-de-Santayana 2003). It has been reported as being sold in the local markets of Cantabria and Palencia (J. Tardío personal communication), and is typically used against toothache, respiratory disorders and intestinal parasites. It is also known as *siempreviva* (everlasting) because its flowers will last indefinitely when dried and therefore are used to adorn the house (e.g., Verde, Rivera and Obón 1998). A closely related species, *Helichrysum italicum*, is similarly used.

Achillea millefolium is also often known as *manzanilla/camomila*. It grows in grassy areas, in old fields, along roadsides and other edges, and in forest clearings throughout Spain. It is called *manzanilla* in the northern half of Spain (Esgueva 1999; San Miguel 2004), where it is valued as an herbal digestive tea. Interestingly, it is beginning to be called *manzanilla* in Segovia, a province in the centre of Spain, due to the influence of outsiders (Blanco 1998). People now use it as a *digestif* and for other complaints, and not only for treating wounds as it was traditionally used there. *Achillea millefolium* has been also employed to treat rheumatic and menstrual pain, diarrhoea, intestinal parasites, haemorrhoids, headache, fever and hypertension, and to enhance diuresis (Villar et al. 1987; Blanco 1998; Pardo-de-Santayana 2003). Other species of the genus *Achillea* such as *A. ageratum*, *A. odorata*, *A. ptarmica* and *A. pyrenaica* are also known as *manzanilla* or *camomila* (Font Quer 1962; Masclans 1981; Villar et al. 1987; Mesa 1996).

Figure 14.2. Chamomile substitutes in Spain. A. *Santolina chamaecyparissus* B. *Helichrysum stoechas*

Another common chamomile substitute is *Santolina chamaecyparissus* (Figure 14.2A), which grows in open scrubland on calcareous soils mainly in the east of the Iberian Peninsula. Known commonly as *manzanilla amarga* because of its strong bitter flavour, it is typically used where *Matricaria recutita* or *Chamaemelum nobile* are not easily found (Ferrández and Sanz 1993; Gil Pinilla 1995). However, it is highly esteemed in some regions of Catalonia (Bonet et al. 1999), Navarra (Akerreta et al. 2007) and Menorca (Font Quer 1962), and in Sierra Nevada it is consumed as a substitute for *Artemisia granatensis* (Barrero et al. 1998). Among other uses, it is appreciated as a vulnerary, emmenagogue or vermifuge (Font Quer 1962). Another species in this genus, *Santolina rosmarinifolia*, is also known as a *manzanilla* and is reportedly used in Jaén to improve the digestion (Guzmán Tirado 1997).

Many of these herbs are known to have active ingredients that influence the efficiency of the digestive process. They are rich in:

1. Anti-inflammatory agents such as chamazulenes (*Matricaria recutita, Santolina chamaecyparissus*), coumarins (*Matricaria recutita, Santolina oblongifolia*) and flavonoids (*Helichrysum stoechas*).
2. Antispasmodics such as apigenol and bisabolenes (*Matricaria recutita, Santolina chamaecyparissus*), and caffeic acid (*Helichrysum italicum*).
3. Antiseptics such as α-pinene, (*Santolina chamaecyparissus, Chamaemelum nobile*).
4. Eupeptics such as achillicin (*Achillea millefolium*).

Other species, such as *Artemisia granatensis*, also have anti-inflammatory and eupeptic properties but their phytochemistry has been little studied.

Other species that have been used as chamomile substitutes or adulterants are *Anthemis arvensis, Matricaria dioscoidea, Tanacetum parthenium* (see Table 14.1), *Herniaria glabra* (Guzmán Tirado 1997; Fajardo et al. 2000; Fernández Ocaña 2000), *Leucanthemum vulgare, Phagnalon saxatile* (Mulet 1991), *Convolvulus boissieri, Erodium foetidum* (Mesa 1996), *Anthemis cotula, Bellis perennis* (Colmeiro 1885–1895), *Artemisia barrelieri* (Obón and Rivera 1991), *Artemisia herba-alba, Bupleurum falcatum* (Villar et al. 1987), *Staehelina dubia* (Molina Mahedero 2001), *Pterocephalus spathulatus* (González Tejero 1990), *Paronychia argentea* (Guzmán Tirado 1997), *Anacyclus clavatus* (Fernández Ocaña 2000) and *Chamaemelum fuscatum* (Triano 1998).

Conclusion

In this chapter, we have reviewed historical accounts and recent ethnobotanical surveys of Spain to understand which botanical species are referred to by locals as *camomila* or *manzanilla*, and why they are grouped in this large folk-generic category. We found a total of 215 different names assigned to 62 species, including many synonyms, which yield a total of 369 vernacular denominations; that is, the combination of vernacular and botanical names. Thus, *manzanilla/camomila* is a complex folk-generic composed of:

1. A small number of highly valued and wide spread prototypical species that have little variation in vernacular names.
2. A few highly valued, but ecologically restricted species that tend to have local vernacular names.
3. A large number of chamomile substitutes with a great variety of local names.

There is evidence from old botanical texts, herbals and other literature from across Europe that this folk generic has always been variable and expanding. In Spain, *manzanilla/camomila* was first a category of plants with similar flowers, then one with similar medicinal properties, and now a broad category with a great number of species, most belonging to the Asteraceae family, all of them used as herbal teas with similar properties and uses. Thus morphological, phytotherapeutic and phytochemical characteristics, and even economic and social features, now bind this great variety of species together in peoples' minds.

The old fraud of adulterating medicinal plants and the confusions between similar species explain why some of them were included. For example, the genera *Matricaria*, *Chamaemelum* and *Anthemis* are very similar and can be easily confused and thus have been historically used as chamomile adulterants. Substitutions for unavailable or rare medicinal botanicals have also resulted in a transfer of plant names from the replaced to the substitute species (Akerreta et al. 2007). This appears to be the case for highly valuable species such as *Santolina chamaecyparissus* or the endemic *Santolina oblongifolia*, restricted to the Sierra de Gredos. It seems that once people had a concept *manzanilla/camomila*, then other valuable species with similar properties were labelled with the same name, giving them a higher market value and an overall higher cultural salience.

In general, our survey suggests that the most prototypical species of this broad and inclusive folk category are the most commonly used in Spain, and also across Europe, *Matricaria recutita* and *Chamaemelum nobile*. Both also have a long history of use in the Mediterranean region. At least one of these two species can be found in most Spanish households since chamomile belongs to the basic first-aid kit. They are either drunk daily, for their general positive impact on health and well-being, as a preventive beverage, or when needed as a digestive and sedative infusion (Raja, Blanché and Vallès 1997; Pardo-de-Santayana 2005). Both species are used in a great number of areas having the highest frequency of citation;

that is, the highest number of people who used them in each area (e.g., Raja, Blanché and Vallès 1997; Pardo-de-Santayana 2005). This strong agreement can be regarded as indicative of the efficacy of their medicinal properties and efficacy (Heinrich 2000).

While *Matricaria recutita* is the only species widely marketed and served in bars or restaurants, many people in rural areas prefer to collect their local type of chamomile since they consider it more healthy and tasty. Some chamomiles are highly valued, while others are only used when a better species is not available. However, most species have not been as widely consumed as they are now, and there is no evidence of their widespread use in the historical record (Font Quer 1962).

Herbal infusions are drunk as herbal remedies, as preventive beverages, but also for pleasure. They often originate as medicines, but after people get used to them they acquire additional uses in food contexts. Infusions begin to be consumed at breakfast, while enjoying conversation in a bar or café, or following a family meal or social occasion, and therefore come to be known as 'food medicines'. This is especially common for those herbal teas that improve digestion (Pardo-de-Santayana, Blanco and Morales 2005).

There is no evidence that the current consumption of herbal teas in food contexts, such as mint tea, have a long history of use in Europe. Before tea was introduced in Europe in the sixteenth century, mint tea was only widely used in Muslim countries, where alcohol was not permissible, or by Portuguese people who had acquired the habit from Arabs (Hobhouse 1986). Water was not really fit to drink in most European towns and villages and to avoid the risk of waterborne disease, people had to drink boiled water or alcohol strong enough to kill germs. However, other infusions such as coffee or tea contain socially tolerated stimulating drugs, and were therefore soon accepted. Chamomiles and other infusions are not stimulant beverages, but have other desirable medicinal properties, many being *digestifs*. It seems that after tea was introduced into Europe, many other hot beverages became popular as well.

Across Europe, then, wild herbs are under severe threat from expanding markets driven in part by changing perceptions of health and wellbeing and in part by an increasing association of the wild with higher social status. The rising demand for chamomile led to its cultivation many centuries ago. Today's wild populations cannot meet its huge and ever increasing demand, although it is still commonly gathered in rural areas, albeit for local use. The richness of morphological varieties of *Matricaria recutita* and *Chamaemelum nobile* in Spain offers many possibilities for developing new cultivars, but study of their agronomical and chemical characters is needed. Other species that should be moved into cultivation include those narrow endemic species such as *Artemisia granatensis* that have suffered from overexploitation.

Thus, it's likely that the dynamic category of *manzanilla / camomila* will continue to expand to include these new cultivated varieties and species. On the other hand, with growing competition, regional varieties may come to be specially demarcated, and as we have seen with wines, liquors and other botanical products, a great proliferation of names may appear to distinguish and identify them.

Acknowledgements

We are grateful to our colleagues and friends José Luis López Larramendi, Andrea Pieroni, Rajindra Puri, Javier Tardío, Laura Aceituno, Susana González and Ana Carvalho for their help and suggestions.

References

Akerreta, S., R. Cavero, V. López and M. Calvo. 2007. 'Analyzing Factors that Influence the Folk Use and Phytonomy of 18 Medicinal Plants in Navarra', *Journal of Ethnobiology and Ethnomedicine* 3: 16.

Álvarez, B. 2006. *Nombres vulgares de las plantas en la Península Ibérica e Islas Baleares*, Ph.D. dissertation. Madrid: Facultad de Ciencias, Universidad Autónoma de Madrid.

Barandiarán, J.M. and A. Manterola. 1990. *La alimentación doméstica en Vasconia*. Bilbao: Eusko Jaurlaritza, Etniker Euskalerria.

Barrero, A.F., R. Álvarez-Manzaneda, J. Quilez and M.M. Herrador. 1998. 'Sesquiterpenes from *Santolina chamaecyparissus* subsp. *squarrosa*', *Phytochemistry* 48(5): 807–813.

Blanca, G. 2003. 'El ruedo ibérico y la manzanilla de Sierra Nevada', *Conservación Vegetal* 8: 10–11.

——— and C. Morales. 1991. *Flora del Parque Natural de la Sierra de Baza*. Granada: Universidad de Granada.

Blanco, E. 1996. *El Caurel, las plantas y sus habitantes*. Santiago de Compostela: Fundación Caixa Galicia.

——— 1998. *Diccionario Etnobotánico de Segovia*. Segovia: Ayuntamiento de Segovia, Caja Segovia.

——— 2002. *Etnobotánica en los Montes de Toledo*. Toledo: Asociación Cultural Montes de Toledo.

——— and C. Cuadrado. 2000. *Etnobotánica en Extremadura: Estudio de La Calabria y La Siberia extremeñas*. Madrid: CEP Alcoba de los Montes.

——— and J. Diez. 2005. *Guía de Flora de Sanabria, Carballeda y Los Valles: Catálogo de Etnoflora selecta*. Zamora: Adisac-La Voz.

Bock, H. 1539. *New Kreuterbuch*. Strassburg.

Bonet, M.A. 1991. *Estudis Etnobotànics a la Vall del Tenes (Vallès Oriental)*. Barcelona: L'Abadia de Montserrat, Ajuntament de Bellpuig.

——— 2001. *Estudi Etnobotànic del Montseny*, Ph.D. dissertation. Barcelona: Facultat de Farmàcia, Universitat de Barcelona.

———, M. Parada, A. Selga and J. Vallès. 1999. 'Studies on Pharmaceutical Ethnobotany in the Regions of L'Alt Empordà and Les Guilleries (Catalonia, Iberian Peninsula)', *Journal of Ethnopharmacology* 68: 154–168.

Calle, M. and J.E. Gómez. 2009. *Especial Sierra Nevada: Temáticos Waste*. Retrieved 15 May 2009 from http://waste.ideal.es/manzanillareal.htm

Casana, E. 1993. *Patrimonio Etnobotánico de la provincia de Córdoba: Subbética, Campiña y Vega del Guadalquivir*, Ph.D. dissertation. Córdoba: Escuela de Agronomía, Universidad de Córdoba.

Casermeiro, M.A., F. Navarro García, M. Parramón and J.J. Rodríguez Concejo. 1995. 'Evolución histórica del uso de las manzanillas en España', in L. Villar (ed.), *Historia Natural 93: (Actas de la XI Reunión Bienal de la Real Sociedad Española de Historia Natural, Jaca, 13–18 de septiembre de 1993)*. Huesca: Instituto de Estudios Altoaragoneses, Instituto Pirenaico de Ecología, pp. 105–111.

Clemente, M., P. Contreras, J. Susin and F. Pliego. 1991. 'Micropropagation of *Artemisia granatensis*', *HortScience* 26: 240.

Colmeiro, M. 1885–1895. *Enumeración y revisión de las plantas de la Península Hispano-Lusitánica e Islas Baleares, con la distribución geográfica de las especies y sus nombres vulgares, tanto nacionales como provinciales*. Madrid.

Covarrubias, S. 1611. *Tesoro de la lengua castellana o española*. Madrid: Luis Sánchez.

Dioscorides, P. 2005[AD 65]. *De Materia Medica*. Translated by L.Y. Beck. New York: Olms-Weidmann.

Esgueva, M.A. 1999. *Las plantas silvestres en León: Estudio de Dialectología Lingüística*. Madrid: Universidad Nacional de Educación a Distancia.

Esteso, F. 1992. *Vegetación y Flora del Campo de Montiel: Interés Farmacéutico*. Albacete: Instituto de Estudios Albacetenses, Diputación de Albacete.

Fajardo, J., A. Verde, D. Rivera and C. Obón. 2000. *Las plantas en la cultura popular de la provincia de Albacete*. Albacete: Instituto de Estudios Albacetenses, Diputación de Albacete.

Feijao, R.O. 1960–1963. *Elucidário fitológico: Plantas vulgares de Portugal continental, insular e ultramarinao (classificaçao, nommes vernaculos e aplicaçoes)*. 3 vols. Lisbon: Instituto Botânico de Lisboa.

Fernández Ocaña, A.M. 2000. *Estudio Etnobotánico en el Parque Natural de las Sierras de Cazorla, Segura y las Villas: Investigación química de un grupo de especies interesantes*, Ph.D. dissertation. Jaén: Facultad de Ciencias Experimentales, Universidad de Jaén.

Ferrández, J.V. and J.M. Sanz. 1993. *Las plantas en la Medicina Popular de la Comarca de Monzón (Huesca)*. Huesca: Instituto de Estudios Altoaragoneses, Diputación de Huesca.

Font Quer, P. 1962. *Plantas medicinales: El Dioscórides renovado*. Barcelona: Labor.

Franke, R. 2005. 'Plant sources', in R. Franke and H. Schilcher (eds), *Chamomile: Industrial Profiles*. Boca Raton, FL: Taylor & Francis, pp. 39–54.

——— and H. Schilcher. 2005. 'Preface', in R. Franke and H. Schilcher (eds), *Chamomile: Industrial Profiles*. Boca Raton, FL: Taylor & Francis.

Galán, R. 1993. *Patrimonio Etnobotánico de la provincia de Córdoba: Pedroches, Sierra Norte y Vega del Guadalquivir*, Ph.D. dissertation. Córdoba: Escuela de Agronomía, Universidad de Córdoba.

Gil Pinilla, M. 1995. *Estudio Etnobotánico de la Flora aromática y medicinal del término municipal de Cantalojas (Guadalajara)*, Ph.D. dissertation. Madrid: Facultad de Ciencias Biológicas, Universidad Complutense de Madrid.

Gómez Pellón, E. 2004. 'A Rural World in Change: on Cultural Modernisation and New Colonisation', in S. Nogués (ed.), *The Future of Rural Areas*. Santander: Universidad de Cantabria, pp. 301–326.

González Tejero, M.R. 1990. *Investigaciones Etnobotánicas en la provincia de Granada*, Ph.D. dissertation. Granada: Facultad de Farmacia, Universidad de Granada.

Granzow de la Cerda, I. (ed.). 1993. *Etnobotánica (el mundo vegetal en la tradición)*. Salamanca: Archivo de Tradiciones Salmantinas, Centro de Cultura Tradicional, Diputación de Salamanca.

Guzmán Tirado, M.A. 1997. *Aproximación a la Etnobotánica de la provincia de Jaén*, Ph.D. dissertation. Granada: Departamento de Biología Vegetal, Universidad de Granada.

Heinrich, M. 2000. 'Ethnobotany and its Role in Drug Development', *Phytotherapy Research* 14: 479–488.

Hobhouse, H. 1986. *Seeds of Change: Five Plants that Transformed Mankind*. New York: Harper & Row.

Laguna, A. 1555[1991]. *Pedacio Dioscorides Anazarbeo, Acerca de la materia medicinal y de los venenos mortíferos*. Translated and illustrated by Dr. Andrés de Laguna. Facsimile edn. Madrid: Consejería de Agricultura y Cooperación de la Comunidad de Madrid.

Laguna, E. 2006. *Flora de Belalcázar*. Retrieved 22 June 2006 from http://www.belalcazar.org

Lange, D. 1998. *Europe's Medicinal and Aromatic Plants: their Use, Trade and Conservation*. Cambridge: Traffic International.

Lastra, J.J. 2003. *Etnobotánica en el Parque Nacional de Picos de Europa*. Oviedo: Ministerio de Medio Ambiente, Parques Nacionales.

——— and L.I. Bachiller. 1997. *Plantas Medicinales en Asturias y en la Cornisa Cantábrica*. Gijón: Trea.

———, X. Porta, V. Ortiz and H. Gómez. 2000. 'Fitonimia en el oriente de Asturias', *Boletín de Ciencias de la Naturaleza, Real Instituto de Estudios Asturianos* 46: 185–217.

l'Écluse, C. de (Clusius). 1576. *Rariorum aliquot stirpium per Hispanias observatarum historia*. Antwerp: C. Plantin.

López Sáez, J.A. 2002. 'Notas etnobotánicas del Valle del Tiétar, Ávila (II)', *Trasierra* 5: 141–148.

Losada, E., J. Castro and E. Niño. 1992. *Nomenclatura vernácula da flora vascular galega.* La Coruña: Xunta de Galicia.

Manniche, L. 1989. *An Ancient Egyptian Herbal.* London: British Museum.

Martínez Lirola, M.J., M.R. González Tejero and J. Molero. 1997. *Investigaciones etnobotánicas en el Parque Natural de Cabo de Gata-Níjar (Almería).* Almería: Sociedad Almeriense de Historia Natural, Consejería de Medio Ambiente, Junta de Andalucía.

Masclans, F. 1981. *Els noms de les plantes als països catalans.* Granollers, Barcelona: Centre Excursionista de Catalunya, Montblanc-Martín.

Mesa, S. 1996. *Estudio Etnobotánico y Agroecológico de la comarca de la Sierra de Mágina (Jaén),* Ph.D. dissertation. Madrid: Universidad Complutense de Madrid.

Molina Mahedero, N. (2001). *Estudio de la Flora de interés etnobotánico en el municipio de Carcabuey (Córdoba),* Graduate dissertation. Córdoba: Escuela Técnica Superior de Ingenieros Agrónomos y Montes, Universidad de Córdoba.

Mulet, L. 1991. *Estudio Etnobotánico de la provincia de Castellón.* Castellón: Diputación de Castellón.

Muntané, J. 1994. *Tresor de la saviesa popular de les herbes, remeis i creences de Cerdanya del temps antic.* Puigcerdà, Girona: Institut d'Estudis Ceretans.

Navarro, M.A. and J.E. Hernández Bermejo. 1994. 'Las manzanillas en los autores andalusíes: algunos apuntes para la interpretación de los textos', in E. García Sánchez (ed.), *Ciencias de la Naturaleza en Al-Andalus, 3: Textos y Estudios.* Granada: Centro de Estudios Árabes, pp. 143–157.

Obón, C. and D. Rivera. 1991. *Las plantas medicinales de nuestra región.* Murcia: Editora regional, Agencia Regional del Medio Ambiente.

Oria de Rueda, J.A., J. Diez and M. Rodríguez. 1996. *Guía de las plantas silvestres de Palencia.* Palencia: Cálamo.

Panero, J.A. and C. Sánchez. 2000. *Sayago: Costumbres, creencias y tradiciones.* Sayago, Zamora: PRODER, Unión Europea, Junta de Castilla y León, Diputación de Zamora.

Parada, M., A. Selga, M.A. Bonet and J. Vallès. 2002. *Etnobotànica de les terres gironines: natura i cultura popular a la Plana Interior de l´Alt Empordà i a les Guilleries.* Girona: Diputació de Girona.

Pardo-de-Santayana, M. 2003. *Las plantas en la cultura tradicional de la Antigua Merindad de Campoo,* Ph.D. dissertation. Madrid: Facultad de Ciencias, Universidad Autónoma de Madrid.

——— 2005. 'Terapeútica popular en Campoo (Cantabria, España)', *Revista de Fitoterapia* 5(Suplemento 1): 222–226.

———, E. Blanco and Morales. 2005. 'Plants Known as "*Té*" (Tea) in Spain: an Ethno-pharmaco-botanical Review', *Journal of Ethnopharmacology* 98: 1–19.

———, E. San Miguel and R. Morales. 2006. 'Digestive Beverages as a Medicinal Food in a Cattle-farming Community in Northern Spain (Campoo, Cantabria)', in A. Pieroni and L.L. Price (eds), *Eating & Healing: Traditional Food as Medicine*. London: Haworth Press, pp. 131–151.

Quer, J. 1762–1764. *Flora española o historia de las plantas que se crían en España*. Vols 1–4. Continued by C. Gómez-Ortega, 1784, vols 5–6. Madrid: Ibarra.

Rabal, G. 2000. '"Cuando la chicoria echa flor..." Etnobotánica en Torre Pacheco', *Revista Murciana de Antropología* 6: 1–240.

Raja, D., C. Blanché and J. Vallès. 1997. 'Contribution to the Knowledge of the Pharmaceutical Ethnobotany of La Segarra Region (Catalonia, Iberian Peninsula)', *Journal of Ethnopharmacology* 57: 149–160.

Rigat, M., T. Garnatje and J. Vallès. 2006. *Plantes i gent: Estudi etnobotànic de l'Alta Vall del Ter*. Ripoll, Girona: Centre d'Estudis Comarcals del Ripollès.

Rivera, D., C. Obón, F. Cano and A. Robledo. 1994. *Introducción al mundo de las plantas medicinales en Murcia*. Murcia: Ayuntamiento de Murcia.

San Miguel, E. 2004. *Etnobotánica de Piloña (Asturias): Cultura y saber popular sobre las plantas en un concejo del Centro-Oriente Asturiano*, Ph.D. dissertation. Madrid: Facultad de Ciencias, Universidad Autónoma de Madrid.

Schilcher, H. 2005. 'Legal Situation of German Chamomile: Monographs', in R. Franke and H. Schilcher (eds), *Chamomile: Industrial Profiles*. Boca Raton, FL: Taylor & Francis, pp. 7–38.

———, P. Imming and S. Goeters. 2005. 'Pharmacology and Toxicology', in R. Franke and H. Schilcher (eds), *Chamomile: Industrial Profiles*. Boca Raton, FL: Taylor & Francis, pp. 245–263.

Silván, A.M, M.J. Abad, P. Bermejo, M. Sollhuber and A. Villar. 1996. 'Antiinflammatory Activity of Coumarins from *Santolina oblongifolia*', *Journal of Natural Products* 59: 1183–1185.

Tabernaemontanus, J.T. 1664. *New vollkommenlich Kräuter-Buch*. Basel: Jacob Werenfels.

Tardío, J., M. Pardo-de-Santayana and R. Morales. 2006. 'Ethnobotanical Review of Wild Edible Plants in Spain', *Botanical Journal of the Linnean Society* 152(1): 27–72.

———, H. Pascual and R. Morales. 2002. *Alimentos silvestres de Madrid*. Madrid: La Librería.

Triano, M.C. (coord.) 1998. *Recupera tus tradiciones: Etnobotánica del Subbético Cordobés*. Carcabuey, Córdoba: Ayuntamiento de Carcabuey. Centro Botánico del Subbético cordobés.

Verde, A., J. Fajardo, D. Rivera and C. Obón. 2000. *Etnobotánica en el entorno del Parque Nacional de Cabañeros*. Madrid: Parques Nacionales, Secretaría General de Medio Ambiente, Ministerio de Medio Ambiente.

———, D. Rivera and C. Obón. 1998. *Etnobotánica en las Sierras de Segura y Alcaraz: Las plantas y el hombre*. Albacete: Instituto de Estudios Albacetenses, Diputación de Albacete.

Villar, L., J.M. Palacín, C. Calvo, D. Gómez and G. Monserrat. 1987. *Plantas medicinales del Pirineo Aragonés y demás tierras Oscenses*. Huesca: CSIC, Diputación de Huesca.

———, J.A. Sessé and J.V. Ferrández. 2001. *Atlas de la Flora del Pirineo Aragonés. Vol. II (Pyrolaceae-Orchidaceae. Síntesis)*. Huesca: Consejo de Protección de la Naturaleza de Aragón, Instituto de Estudios Altoaragoneses, Departamento de Agricultura y Medio Ambiente del Gobierno de Aragón, Instituto Pirenaico de Ecología (CSIC).

Appendix 14.1. List of species known as *manzanilla/camomila* and derived names mentioned in the text

Achillea ageratum L.
Achillea millefolium L.
Achillea odorata L.
Achillea ptarmica L.
Achillea pyrenaica L.
Anacyclus clavatus (Desf.) Pers.
Anthemis arvensis L.
Anthemis cotula L.
Anthemis tinctoria L.
Anthemis tuberculata Boiss.
Artemisia barrelieri Besser
Artemisia granatensis Boiss.
Artemisia herba-alba Asso
Bellis perennis L.
Bupleurum falcatum L.
Chamaemelum fuscatum (Brot.) Vasc.
Chamaemelum mixtum (L.) All.
Chamaemelum nobile (L.) All. (= *Anthemis nobilis* L.)
Chrysanthemum coronarium L.
Convolvulus boissieri Steud.
Erodium foetidum (L.) L´Hér.
Gnaphalium luteo-album L.
Helichrysum italicum (Roth) G. Don
Helichrysum stoechas (L.) Moench
Herniaria glabra L.
Leucanthemum vulgare Lam.
Matricaria aurea (Loefl.) Sch.Bip.
Matricaria dioscoidea DC.
Matricaria recutita L. [=*Matricaria chamomilla* L., *Chamomilla recutita* (L). Rauschert]
Paronychia argentea Lam.
Phagnalon saxatile (L.) Cass.
Pterocephalus spathulatus (Lag.) Cout.
Santolina chamaecyparissus L.
Santolina oblongifolia Boiss.
Santolina rosmarinifolia L.
Staehelina dubia L.
Tanacetum parthenium (L.) Sch.Bip.
Tripleurospermum inodorum (L.) Sch.Bip. (= *Matricaria inodora* L., *Matricaria maritima* L.).

CHAPTER 15

A Preliminary Study of the Plant Knowledge and Grassland Management Practices of English Livestock Farmers, with Implications for Grassland Conservation

JENNY L. MCCUNE

Introduction

The importance of recognizing local knowledge (LK) in conservation and development projects in less-developed countries is widely acknowledged. LK is considered distinct from scientific knowledge (SK). It is gained by the long experience of a group of people living in and interacting with a specific local environment, and is embedded within a cultural context (McClure 1989; van Dusseldorp and Box 1993; Clark and Murdoch 1997; Purcell 1998; Ellen and Harris 2000). Techniques like 'participatory rural appraisal' (PRA) have been developed in order to collect and apply LK (Chambers 1994), but these efforts are confined almost exclusively to less-developed countries. There is often an assumption that LK no longer exists in the more developed countries of Europe, having been replaced by science and technology (Ellen and Harris 2000: 6). However, there are ecosystems in Europe that continue to be managed by local, long-term residents – usually farmers. Therefore it is likely that these people possess LK regarding these ecosystems, which may be useful in conservation efforts.

England's grassland is an example of such a situation. English pastures and hay meadows were created thousands of years ago by early agropastoralists, and continue to be managed by livestock farmers today.

These grasslands are an important habitat for much of England's flora and fauna (Hopkins and Hopkins 1993; Pain, Hill and McCracken 1997; Vickery et al. 2001). But grassland management has changed dramatically in the past century. Technological advances, coupled with the drive for self-sufficiency following food shortages during the First World War, led to intensification of agricultural practices. There have been enormous increases in the application of synthetic fertilizers and pesticides, more frequent ploughing-up and resowing of grassland with nonnative species, higher stocking rates and harvest frequencies, and overall a reduction both in the area of grassland and its biodiversity (Hopkins, 1979; Fuller, 1987; Green, 1990; Hopkins and Hopkins 1993; Smith 1993; Frame, Baker and Henderson 1995; Vickery et al. 2001). Conservationists are calling for a return to more traditional grassland management, and the British Government has created voluntary 'agri-environment schemes' which provide monetary incentives for farmers to de-intensify management. While there has been extensive research on farmers' attitudes towards the natural environment and their receptiveness to the schemes (e.g., Gasson and Potter 1988; Carr and Tait 1991; McEachern 1992; Young, Morris and Andrews 1995; Beedell and Rehman 1996; Wilson 1996; Macdonald and Johnson 2000), relatively little research has considered the LK these farmers may possess. In this chapter I report on a study where methods commonly used by ethnobotanists and anthropologists to study the ethnobiological knowledge of indigenous peoples (Martin 1995; Alexiades 1996) were used to explore the plant knowledge and grassland management practices of a small group of English livestock farmers from two different regions. The goals of the study were to find out:

1. If LK exists in this European context.
2. How it may differ between farmers from different regions.
3. What its significance might be for grassland conservation.

Defining Knowledge

The distinction between LK and SK, and the way in which the two interact, is crucial to issues of agricultural sustainability everywhere (Clark and Murdoch 1997; Winter 1997; Cleveland and Soleri 2002). SK is said to be standardized, dispassionate and universally applicable, while LK is characterized as more descriptive, is embedded in a local context, and may be subjective. However, both LK and SK 'may consist of a complex combination of intuition, empiricism and theory, and of verifiable objective observation and social construction, and may, therefore, be similar as well as different' (Soleri et al. 2002: 20). Any distinction between the two may therefore be artificial (e.g., Agrawal 1995). The industrialization and intensification of agriculture in Europe and North America is thought by some to have displaced LK there (Morgan and Murdoch 2000), but this displacement may be incomplete (Winter 1997: 376).

The purpose of this research was to explore English farmers' environmental knowledge and to determine if this knowledge includes information from outside the realm of what we call science. Since it is difficult to define LK based on its characteristics, I define it according to its origin. LK is specific knowledge about a local environment that has been gained with experience in the local environment, or from the experience of previous generations. SK is knowledge gained from 'scientific experts' – from industry representatives, the press, books, conservation advisors or government agents. Like Soleri et al. (2002) in their 'holistic model' of farmers' knowledge, I hypothesised that English farmers' knowledge consists of knowledge from both sources. Farmers have undoubtedly adopted much SK in their understanding of the environment, but they also make use of LK in applying technology to their particular needs.

Studying English Livestock Farmers

The Study Sites and Interview Methods

Farms and farmers in England are not homogeneous. One major source of variation is geography. Following the Second World War, due to increasing industrialization and changing government policy, mixed farming gave way to a polarization of agriculture in England. The drier, flatter south and east became largely devoted to arable crops, while livestock rearing was concentrated in the hillier, wetter north and west (Newby 1979: 77; Mabey 1980; Stoate 1996; Robinson and Sutherland 2002). In order to explore potential regional differences caused by this polarization, the study focused on two counties: Lancashire in the northwest, and Kent in the southeast (see Figure 15.1). Table 15.1 outlines some of the differences in agricultural statistics between the two regions. Kent is much more highly populated than Lancashire, and although it has a comparable amount of agricultural land, only about 32 per cent is grassland, as compared to 82 per cent in Lancashire. Cash-cropping has been important in Kent for centuries due to its proximity to the London markets, and this has contributed to the loss of many permanent grasslands to make way for grain crops (e.g., Green 1990). In Lancashire, with its wetter climate and extensive upland areas, there are fewer arable farms. The uplands include many designated 'Less Favoured Areas' (LFAs), where livestock farmers receive subsidies to help compensate for the difficulty of farming.

Names and contact information for approximately twenty livestock farmers (sheep, beef, dairy or mixed) in each county were obtained

Figure 15.1. Map highlighting the two counties where farmers were interviewed

from county farm advisors of the Farming and Wildlife Advisory Group (FWAG). FWAG is a nongovernmental, independent agency that offers conservation advice to farmers who request it, and assists them in applying for entrance into agri-environment schemes. It is important to note that, as a result of the source of the contacts, all of the farmers had at some point consulted their county FWAG, and most, but not all, had applied for or were already participating in an agri-environment scheme. This suggests that the sample was biased towards farmers who had some interest in conservation schemes. The limit of twenty farmers per county was necessary due to a timeframe of only a few months to complete fieldwork. Fieldwork took place in April (Lancashire) and June (Kent), and therefore many farmers were busy with lambing, in the former instance, and shearing and hay-making in the latter. In the end, fifteen Lancashire farmers and thirteen Kent farmers were interviewed. Interviews were conducted in farmers' residences or outdoors on their farms, and lasted between thirty minutes and two hours. Before interviews began, farmers

Table 15.1. Summary of agricultural and environmental statistics for Kent and Lancashire counties

Statistic	Kent	Lancashire
Population	1 329 718	108 378
Full-time farmers	2 225 (0.2% total pop.)	4 133 (3.8% total pop.)
Total area (ha)	373 000	307 000
Average precipitation (mm/year)	776.5	1 290.6
Total agricultural area (ha)	223 028	212 905
Area of rough grazing* (ha)	5 874	39 501
Area of permanent grass$^{\Psi}$ (ha)	53 982	117 885
Area of temporary grass+ (ha)	11 613	16 716
Total area grass (ha)	71 469 (32% ag. area)	174 102 (82% ag. area)
Total number of farms	5 323	6 420
Dairy farms	76	913
Lowland cattle/sheep farms	1 023	791
Upland (LFA) cattle/sheep farms	0	973
Mixed farms	258	150
Total livestock farms	1 357 (25% total farms)	2 827(44% total farms)

Sources: DEFRA 2002, 2003; Lancashire Rural Partnership 2002; National Statistics 2003

* Rough grazing is typically in large blocks, with limited control of grazing, often on hill or mountain land, and usually on poor soil with an unfavourable climate
$^{\Psi}$ Grassland that is more than five years old, not part of an arable rotation
+ Grassland that is less than five years old

were given a letter of introduction and were asked for permission to record the interview.

Interviews were semistructured and designed to allow as much freedom as possible for the respondent to elaborate on areas he or she considered most important. Each interview consisted of three broad areas concerning livestock farming and grasslands:

1. The first set of questions focused on basic farm and biographical information, for example, the farmer's age and number of years farming, the size of the farm, etc.
2. The second set of questions sought information on the grassland management regime, including fertilizer application rates and timing, stocking rates, hay-cutting dates and herbicide use.

3. The third part of the interview requested a free list of grassland plants from the respondent – that is, a list from memory of all the plants the farmer could recall seeing in his or her pastures and/or meadows at any time. The plants listed were then discussed with the farmers to gain additional information about their knowledge of grasslands and their management.

Data Analysis

Ethnobiological knowledge can be divided into two kinds. The first is formal knowledge, which is encoded in language. Ellen (2003: 48) calls this 'lexical' knowledge, and distinguishes it from 'substantive' knowledge, which is applied in everyday work but is not easily describable in words. Collecting and analysing free lists is a method used by anthropologists to visualize the lexical knowledge of a particular group of people (Martin 1995: 213–215; Borgatti 1996a). Free lists are a typical way to explore a particular cultural domain, such as 'plants', 'livestock', or 'diseases', with a group of respondents; they help define the contents of the domain and thus its boundaries. They also allow the interviewer to get an idea of who knows and who does not know the domain, and thus can help identify local experts or key informants. Free lists of the domain 'grassland plants' were analysed using the computer programme ANTHROPAC (Borgatti 1996b). ANTHROPAC calculates the frequency with which each plant is listed and its average rank in the free list of each respondent, then combines these to produce a measure of the cultural importance, or salience, of each plant.

A problem arose during the analysis of free lists due to the fact that some farmers gave specific names (e.g., 'white clover'), while others used more generic terms (e.g., 'clover'). In order to account for this, those who listed a specific term were assumed to know the corresponding generic term, and the generic term was added to their list for the purpose of analysis. Thus, those who listed 'white clover' were assumed also to know the term 'clover', and therefore the frequency of generic terms was not underestimated by the analysis.

After the average salience of each plant was calculated, a consensus analysis (*sensu* Romney, Weller and Batchelder 1986) was performed in order to:

1. Determine the extent of agreement, or shared knowledge, of the domain of grassland species among the farmers of each region and among the group as a whole.
2. Explore cross-group and within-group patterns of agreement and disagreement (Borgatti 1996a).

ANTHROPAC's consensus analysis programme is essentially a factor analysis over a matrix of the agreement of respondents with regard to the inclusion or exclusion of species in their free lists (Borgatti 1996c). A consensus is found to the extent that the data conform to a single-factor solution. The programme constructs a hypothetical model of the consensus answer; that is, the shared knowledge of the group. If the reliability of the model is significant and the population of informants does not vary greatly, then the model can be said to represent the typical answer of a member of that population. Each respondent's first factor score (labelled as 'competence' in ANTHROPAC) represents the proportion of agreement to this consensus model. Those respondents with a low competence score disagree with the consensus, and may be experts with specialist knowledge or novices with little knowledge of the domain. It is also possible that low scores occur because the respondent has misunderstood the question being asked, or may even be from another location, or of a different linguistic or cultural group.

The agreement among farmers was then plotted using ANTHROPAC's nonmetric multidimensional scaling (MDS) program to give a visual representation of the relationship between farmers as regards their given free lists. The closer the points are to the origin, the more similar they are to the consensus, and the larger the distance between two points, the more dissimilar are the free lists given by those two respondents.

Finally, a property fitting (PROFIT) analysis was run in order to determine which attributes of farmers might be responsible for the distribution produced by the MDS program (Borgatti 1996a, 1996c). Essentially, PROFIT analysis is a multiple regression of attribute variables on the agreement matrix. PROFIT analysis works only for continuous variables, so a quadratic assignment procedure (QAP) was used to regress nonmetric, categorical data against the agreement matrix (Borgatti 1996c). The attributes used in the PROFIT analysis and QAP regression were those hypothesised to influence farmers' knowledge of plants and included five factors describing the farmer and the farm, and seven describing the intensity of management of grassland. Those factors describing the farmer and the farm were:

1. Farmer age.
2. Years of farming experience.
3. Farmer education.
4. The history of family farming on the same farm.
5. Area of grassland on the farm.

For example, it might be expected that an older farmer whose family had farmed the same land for generations would have greater plant knowledge than a younger farmer who had recently moved to a new area.

The first two factors were entered directly, as reported by the farmer. Education level was coded by assigning a value of *1* to farmers with primary education only (not beyond A-levels), *2* to farmers with a college certificate or diploma, and *3* to those with higher degrees – bachelor's or above. Factor 4 was coded according to the number of generations the farmer's family had farmed and on whose land. A first generation farmer was coded as *1*, a second generation farmer on a different farm than his or her predecessor was coded as *2*, a second generation farmer on the same farm was coded as *3*, a third generation or higher farmer on a different farm than his or her predecessors was coded as *4*, and a third generation or higher farmer farming on the same land as his or her predecessors was coded *5*. The area of grassland was entered as reported by the farmer in hectares. Factors describing the intensity of grassland management were:

1. The average stocking rate.
2. The degree of herbicide use.
3. The amount of synthetic nitrogen applied per hectare per year to pastures.
4. The amount of synthetic nitrogen applied per hectare per year to meadows.
5. The average length of time between ploughing-up and resowing grasslands.
6. The date of the earliest hay or silage cut.
7. The number of hay or silage cuts taken per year.
   A factor analysis of all twelve variables revealed that management factors 5 and 6 could be eliminated as they explained little of the variation between free lists, while the remaining variables explained approximately 48 per cent of the variation.

A relative scale of overall management intensity was created by using the remaining five management variables in ANTHROPAC to generate a Likert Scale (Bernard 2002; Borgatti 1996c). This overall scale of intensity was then used in the PROFIT analysis for explaining variation in the free list results in order to test overall management intensity as well as individual measures of intensity. Clearly the way a farmer manages grassland affects its floristic diversity. Since intensification of grassland management has been shown to decrease grassland diversity (e.g., Fuller 1987; Green 1990), I expected farmers who managed their grasslands more intensively to have less diverse grasslands, and hence shorter free lists.

Of course, a farmer's ability to name grassland plants is only one small part of his or her knowledge, which includes much substantive knowledge. To complement the free list analysis, all interviews were carefully analysed to ascertain the source of the knowledge farmers use in their day-to-day grassland management, and the basis of their understanding of grassland flora. Many respondents discussed these subjects in the course of responding to the broader questions mentioned above. For example, if a farmer related making management decisions based on the recommendations of industry, textbooks or scientists, this was considered evidence of SK. On the other hand, if a farmer spoke of basing grassland management on particular local conditions and/or the past experience of the farmer himself or his predecessor, it was considered evidence of LK. Statements made during interviews which unambiguously indicated either a scientific or a local source of knowledge were tabulated for each farmer, which proved useful in interpreting the statistical data (Bernard 2002).

As this was a preliminary study of a cultural domain – that is, what farmers know, or say they know, rather than what they actually do – there are obvious limits to the use of the results. For example, since plant surveys were not carried out in each of the farmers' fields, it is impossible to say how closely farmers' free lists matched the actual species composition in the fields, or how faithfully they actually followed the management regimes they described. No botanical identifications were made or voucher specimens collected, and farmers were not asked to prove that the plants they listed in free lists actually grew in their grasslands. Thus, the purpose of the study was not to compare farmers' stated knowledge of plant species with a biologist's perception of the species composition of pastures and meadows, but to explore the extent and sources of farmers' overall knowledge of grassland plants and their management.

What Farmers Know

In Lancashire 106 plant names were listed altogether, with free lists ranging from five to thirty plants. Kentish farmers gave a cumulative list of ninety-three plant names, individual free lists ranging from four to twenty-seven plants. The average number for both counties pooled was fifteen plants per free list. Altogether, the respondents listed 153 different plant names.

Consensus analysis of each county separately, and then all farmers together, gave the consensus lists of plants shown in Table 15.2. These lists represent a hypothetical model, the cultural consensus model, of shared knowledge about the domain (Romney, Weller and Batchelder 1986). For each group of farmers and all farmers pooled, there was a very high level of consensus in their free lists of grassland species (Lancashire: First factor = 9.37 (95.8 per cent of variation explained), average competence score = 78 per cent; Kent: First factor = 6.68 (94 per cent of variation explained), average competence score = 70 per cent; All farmers: First factor = 20.08 (95.6 per cent of variation explained), average competence score = 84.3 per cent). Thus we can say that these farmers share a common model on what constitutes the core of the domain of 'grassland plants'. The consensus lists indicate that the most commonly mentioned plants were the economically important sown species (perennial ryegrass, clover and timothy), the most pernicious weeds (thistles, docks and nettles), and other common or conspicuous species (e.g., dandelions, buttercups and daisies). Most free lists were lacking in rare or threatened species. In Lancashire, more than eighty grassland species are listed in the Guidelines for Site Selection of Biological Heritage Sites (Lancashire County Heritage Sites Scheme 1998) and/or in Lancashire's Biodiversity Action Plan as 'associated species' of threatened grassland habitats (Lancashire's Biodiversity Partnership 2001). Only nineteen of these were mentioned by farmers. Similarly, in Kent more than fifty grassland species appear in the Kent Red Data Book (Kent Wildlife Trust 2000) and/or as 'notables' or 'quality indicator species' for grassland habitats (Kent Biodiversity Action Plan Steering Group 1997), but only nine of these appeared on free lists.

The MDS plots (Figures 15.2, 15.3 and 15.4) show that despite a high consensus there were differences between and within the two groups of farmers. In Figure 15.2, for Lancashire, farmers 9, 6, 3 and 13 are outliers in the four corners of the graph. All four have fairly long free lists, but share only 13–17 per cent of the plant names on their lists. Many species listed by these informants were plants that were named only once in the entire study. PROFIT analysis (Table 15.3, first column) shows that grassland

Table 15.2. Consensus lists of grassland plants in each county, and for both counties pooled

Lancashire	Kent	Pooled
Ryegrass (*Lolium perenne* L.)	Ryegrass	Ryegrass
Clover (*Trifolium* spp.)	Clover	Clover
White clover (*Trifolium repens* L.)	Cocksfoot (*Dactylis glomerata* L.)	Timothy
Timothy (*Phleum pratense* L.)	Thistle	Buttercup
Buttercup (*Ranunculus* spp.)	Nettle	Dandelion
Dandelion (*Taraxacum officinale* Weber)	Timothy	Thistle
Thistle (*Cirsium* spp.)	Fescue (*Festuca* spp.)	Nettle
Nettle (*Urtica* spp.)	Kent wild white clover	Dock
Dock (*Rumex* spp.)	Dock	
Daisy (*Bellis perennis* L. or *Leucanthemum vulgare* Lam.)	Perennial ryegrass	

area and farmer age and experience were the only variables to explain a significant ($p<0.05$) amount of the variation in farmers' free lists, although the amount of variation that was explained was quite small. The vectors on the graph represent the direction of increasing age and experience of the farmer and increasing grassland area on the farm. Thus, part of the reason farmers 3 and 6 are outliers can probably be attributed in part to their age and experience, and indeed they are 67 and 68 years old respectively, each with over fifty years of experience. Farmer 13, as the grassland area vector suggests, farms a very large moor of close to 3,000 acres. However, he has relatively little experience; his plant list was about average in length, but unique in its inclusion of a few rare moor species. Similarly, farmer 3 manages over 1400 acres of land including a high moor. Farmer 9, also an outlier, farms a relatively small acreage (less than 300 acres), and is relatively young, with twenty-four years of experience, which is about average. In this case, the explanation for the inclusion in his free list of many plants not mentioned by others is probably due to a factor not tested by the PROFIT analysis: both farmer 9 and his wife are amateur naturalists who keep their own database of daily sightings of flora and fauna on the farm. This undoubtedly contributed to the uniqueness of his free list.

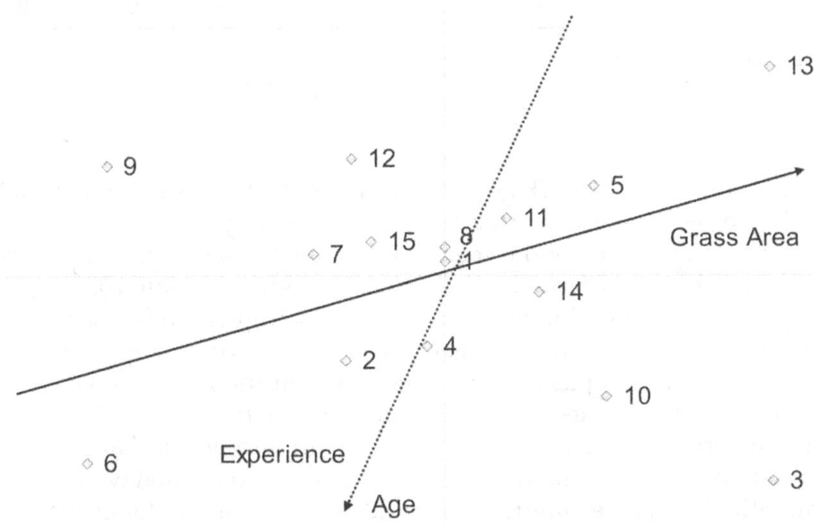

Figure 15.2. MDS plot showing the relationships between Lancashire farmers as regards their given free lists of plants. Numbers were assigned to farmers based on the order of interviews. Each point represents a particular farmer. The two vectors indicate the highly correlated dimensions of age and experience, increasing from top to bottom, and the dimension of grassland area increasing from lower left to upper right, as determined by the PROFIT analysis. See Table 15.3 for R-Squared values

Table 15.3. Summary statistics from PROFIT and QAP analysis of independent variables regressed on freelist agreement matrices

County	Lancashire		Kent		All	
Variable	**R-Squared**	**Probability**	**R-Squared**	**Probability**	**R-Squared**	**Probability**
Farmer age	0.394	0.047*	0.047	0.808	0.010	0.883
Farmer experience	0.551	0.009**	0.004	0.977	0.013	0.870
Farmer education	0.004	0.696	0.005	0.540	0.001	0.668
Family heritage	0.007	0.624	0.001	0.898	0.010	0.268
Grassland area	0.457	0.047*	0.387	0.092	0.271	0.046*
Stocking rate	0.035	0.835	0.313	0.150	0.243	0.037*
N inputs to meadow	0.091	0.582	0.176	0.400	0.106	0.259
N inputs to pasture	0.102	0.523	0.293	0.157	0.127	0.179
Herbicide inputs	0	0.950	0.018	0.296	0	0.668
Age of grass	0.115	0.447	0.813	0.001**	0.147	0.144
Cuts per year	0.038	0.799	0.062	0.697	0.027	0.707
Date of first cut	0.002	0.452	0.016	0.440	0.001	0.676
Intensity index	0.030	0.850	0.276	0.193	0.048	0.554
County	-	-	-	-	0.056	0.000**

* p<.05, **p<.01

For farmers in Kent (Figure 15.3), among the factors tested by PROFIT, only the age of grassland (the average number of years grassland was left before ploughing-up and reseeding) explained a significant amount of the variation between farmers (see Table 15.3, second column). In general, then, farmers who plough up and reseed their grasslands less frequently name different grassland species. Farmer 27, for example, had never known any of his pasture to be ploughed in the last sixty years. None of the tested variables help to explain the variation among Kent farmers in the vertical dimension. I suspect, as with farmer 9 in Lancashire, this variation has to do with unique aspects of the farmers that were not easily quantifiable. For example, farmer 22 had one of the longest free lists given in Kent (twenty-seven plants), and four of the plants mentioned were rare or 'quality indicator' species. I believe this is due to his very keen involvement in conservation on his farm, which includes a Site of Special Scientific Interest, and his being enrolled in more than one agri-environment scheme.

The two counties were very similar in the average length of free lists, the most salient plants, and the consensus lists. The main differences

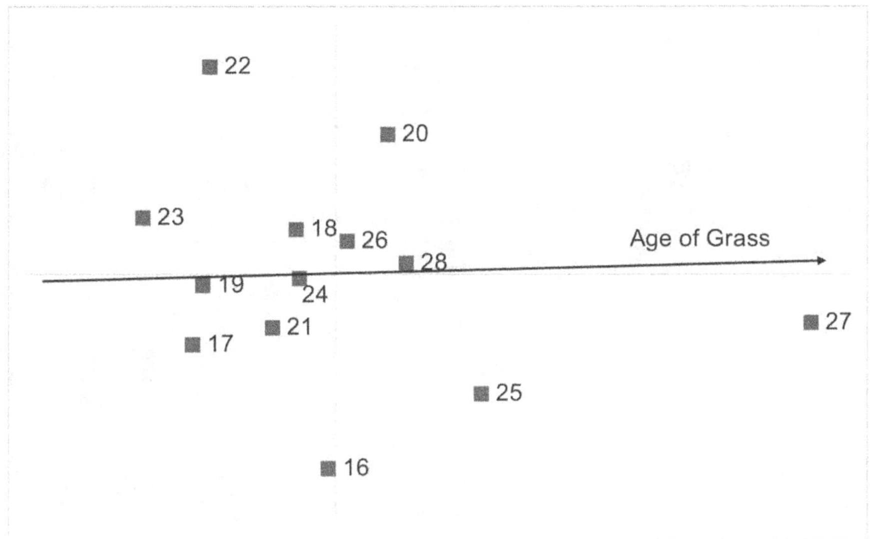

Figure 15.3. MDS plot showing the relationships between Kentish farmers as regards their given free lists of plants. Numbers were assigned to farmers based on the order of interviews. Each point represents a particular farmer. The vector indicates the dimension of age of grassland (or average number of years between ploughing up and resowing grasslands) increasing from left to right, as determined by the PROFIT analysis. See Table 15.3 for R-Squared values

were in the less-common, regional species usually noted further down in free lists. In the MDS plot for all farmers (Figure 15.4), farmers from each region cluster together, the Kentish farmers are found mainly in the upper half, while the Lancashire respondents are mainly on the lower half. It is likely that the different regional species are responsible for the vertical separation. For example, many Kentish farmers listed the regional variety of clover, Kent wild white clover (Table 15.2). The PROFIT analysis of both counties pooled (Table 15.3, third column) suggests that the amount of grassland managed by a farmer is an important factor, though this clearly co-varies with county, as Kent farms had on average less grassland acreage than Lancashire farms. Stocking rate appears to vary independently of size, though two Kent farms had the highest rates observed and two Lancashire farms had the lowest. Factors measuring overall intensity of grassland management and characterizing the farmers showed no significant influence.

Table 15.4 shows a tabulation of times in each farmer's interview where the source of the farmer's knowledge could be pinpointed as either local or scientific. A simple t-test shows that for Lancashire farmers, but not Kent farmers, the proportion of responses indicating sources of knowledge as

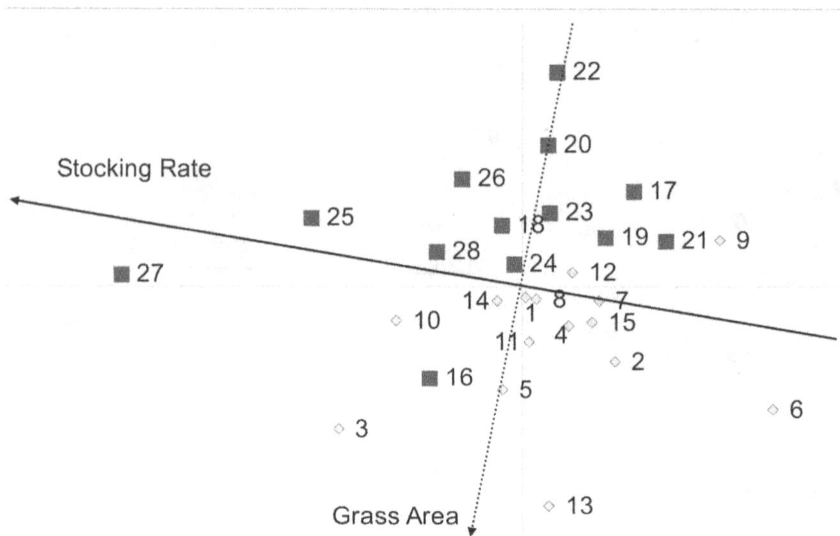

Figure 15.4. MDS plot showing the relationship between all farmers pooled as regards their given free lists of plants. Each point represents a particular farmer. Small diamonds = Lancashire farmers, Large squares = Kent farmers. The vectors indicate the dimension of grassland area increasing from top to bottom, and stocking rate increasing from right to left, as determined by the PROFIT analysis. QAP determined that the counties were significantly different. See Table 15.3 for R-Squared values

local was higher ($p<0.002$), and that for all farmers there was a significant inverse relationship between the intensity of management (as measured by the calculated Likert Scale) and the proportion of responses indicating a source as local knowledge ($R^2= 0.26$, $t=-3.02$, $df=26$, $p<0.01$). Thus, the more intensive a farmer's grassland management regime, the less likely he or she was to rely on LK, at least in explaining his or her management regime during the interview.

Evidence of local knowledge was identifiable by its intuitive, site-specific nature, and by its obvious basis in personal experience, or that of previous generations. The site-specific knowledge of farmers first became apparent during the freelisting exercise. Many plants were characterized according to whether or not grazing animals liked them, or where they grew on the farm. Almost all farmers described species they had seen growing 'down by the river', or 'only in the gate openings', or in specific farm fields at specific times of year or under certain grazing regimes. At times this site-specific knowledge calls into question official management plans for land under an environmental designation. One farmer spoke

Table 15.4. Tabulation of farmer statements that showed either a distinctly scientific source of knowledge, or a distinctly local source of knowledge. Farmers 1 to 15 are from Lancashire; farmers 16 to 28 are from Kent

Farmer	Intensity*	# Statements based on SK	per cent	# Statements based on LK	per cent
1	1.76	1	0.17	5	0.83
2	2.71	5	0.63	3	0.38
3	1.02	1	0.20	4	0.80
4	2.20	1	0.33	2	0.67
5	1.02	1	0.50	1	0.50
6	0.34	2	0.20	8	0.80
7	1.64	2	0.50	2	0.50
8	2.46	1	0.50	1	0.50
9	2.69	2	0.50	2	0.50
10	2.84	1	0.50	1	0.50
11	1.19	0	0.00	3	1.00
12	1.02	1	0.17	5	0.83
13	0.60	3	0.60	2	0.40
14	2.36	1	0.50	1	0.50
15	0.80	2	0.33	4	0.67
16	1.64	1	0.50	1	0.50
17	1.75	4	0.67	2	0.33
18	2.33	2	1.00	0	0.00
19	0.43	0	0.00	1	1.00
20	1.31	3	0.60	2	0.40
21	1.84	7	0.58	5	0.42
22	1.13	1	0.33	2	0.67
23	3.15	1	1.00	0	0.00
24	1.35	1	0.33	2	0.67
25	0.86	1	0.17	5	0.83
26	0.80	3	0.43	4	0.57
27	0.57	4	0.50	4	0.50
28	0.69	3	0.75	1	0.25

* Index of intensity of grassland management (see text). This correlates inversely with the proportion of farmers' statements based on local knowledge (R^2=0.26, t= -3.02, $p<0.01$)

of part of his farm that had been designated a Site of Special Scientific Interest (SSSI) by English Nature, the main body in charge of conservation in England. The management regime on SSSIs is arranged by English Nature. The farmer explained:

> The lower moss, that side, they got me to graze it off a number of years ago ... and the year after it was just awash with orchids. They were gorgeous, they really were. I did that for two years, but they've stopped me now ... and it's just gone back again. The old coarse grass has started coming again.

Even the most intensive farmers make use of experiential knowledge. One of the three most intensive farmers interviewed – a dairy farmer in western Lancashire – gave an interesting account when asked if he ever included timothy (*Phleum pratense*) in seed mixtures:

> I find my dry cows like timothy grasses. They seem to calve so much easier on that field that's got timothy in it. I have only one field with timothy grass in it .... [W]e put dry cows in there for the six weeks they're dry, and we never have any calving problems in there .... [N]ow, just why, I don't know!

That this kind of experiential knowledge is still very much in use also became evident when farmers were asked about specifics of their management timetable: for example, when they put the animals to grass, how they know it is time to rotate animals to a new field, or when they cut hay. Most farmers were hard pressed to explain exactly how they made such decisions, and invariably answered with statements like: 'when the ground's about right'. Although they found it difficult to explain, it was clear that most farmers could tell intuitively when to do things by the appearance of the grassland and the condition and behaviour of their animals, rather than by any 'scientific' advice. Farmers may depend on outside SK for advice on seed mixtures or fertilizers, but many important management decisions are still based on past experience of farming the land. Both their management strategies and plant knowledge are a hybrid mixture of SK and intuitive, experience-gained LK.

The interviews suggested that SK, as defined above, also plays a role in grassland management for most of the farmers. For example, one farmer explained exactly how much nitrogen is required per day by grassland during the growing season. He applies synthetic nitrogen on his meadow ground as soon as a certain ground temperature is reached. This is monitored by an agent of his fertilizer supplier, who tells him when to begin. Other instances of SK included use of soil tests to measure phosphorus and potassium to determine when to apply lime or potash. Most farmers, even those who ranked lowest in management intensity, have such soil tests done periodically. When reseeding, some farmers trust outside experts to determine what to sow. As one farmer said when asked what species he prefers in a grassland mixture: 'I just ask [the seed supplier] for a good long-term ley .... I don't take much notice what's in it'. SK also figured prominently in farmers' plant knowledge. Almost every farmer noted the nitrogen-fixing capacity of clover. Some quoted the exact percentage of digestible protein found in particular plants. The influence of conservation science and the environmental schemes themselves was evident in statements like: '25 per cent ground cover of rushes is the optimum level for ground-nesting birds' or 'our stocking rate is determined by our agreement with English Nature'. These statements all indicate an external, 'scientific' source of farming knowledge.

Conclusion

The contention that farmers need to be 'environmentally re-skilled' in order to learn how to farm in an ecologically sympathetic way (see e.g., Curry and Winter 2000: 108) is at least partly related to the difference in lexical knowledge between farmers and scientists. Clark and Murdoch (1997: 53) note that the species which conservationists consider valuable are often invisible to farmers. As consensus analysis confirms here, farmers' knowledge of plant names concentrates mainly on sown varieties and weeds. Consequently, some scientists assume that farmers are 'unaware of "wild" nature and ignorant of how to look after it properly' (Clark and Murdoch 1997: 51). They have failed to take farmers' substantive, site-specific knowledge into account. I found that most farmers do not know the names of very rare plants. However, in the course of their work, they may gain substantive knowledge about species and their habits, even those that are rare.

Qualitative analysis of interviews made it clear that farmers employ a mixture of LK and SK in order to manage their grasslands. The more intensively a farmer managed his or her grasslands, the less likely they were to make statements that gave evidence of LK. However, even the most intensive invoked site-specific, intuitive knowledge when describing management practices and even the least intensive made use of SK in the form of advice from outside, 'scientific' sources. Also, there is some variation in the knowledge held by individual farmers. As the MDS plots show (Figures 15.2, 15.3, 15.4), most farmers are clustered in a group showing high consensus, but a few are 'outliers', whose free lists differ substantially from the consensus. The distribution of farmers in relation to the consensus was in part explainable by attributes of the farmers, their farms, and their management intensity. For example, in Lancashire the age and experience of the farmer, as well as the area of grassland on the farm, contributed significantly to the differences among farmers' free lists. However, other factors that were not tested in the PROFIT analysis were highlighted by the particular identities of the outlier farmers. A farmer's interest in nature as a hobby, their involvement in conservation schemes, or simply the unique location of their farm can also lead to unique lexical knowledge. For these reasons, a farmer who farmed more intensively than another did not necessarily give a shorter free list of grassland plants, or match the consensus more closely.

To compensate for their perceived lack of knowledge, farmers who participate in agri-environment schemes are given very detailed management prescriptions, which include precise dates when grazing is permitted, heights to which grass swards should be grazed and restrictions on fertilizer applications. Such prescriptions contrast sharply with the way farmers described their management practices, in which decisions are often made in response to changing conditions. One farmer summed up the problem with such restrictive prescriptions, which he calls 'blueprints':

> Nothing in nature, nothing is cut and dried. Every season is different, and you cannot have a blueprint, you've got to have a sort of give and take all the time, and the only person who's ever going to be able to do that is the farmer, who's there all the time.

This statement embodies the collision between a science-based management style, which stipulates precise rules for environmentally sound management, and LK, which depends on site-specific experience and intuition in adapting to environmental conditions as they change. A few researchers have pointed out that this conflict occurs not only in less-developed countries, but in developed countries as well (Clark and Murdoch 1997; Winter 1997; Harrison, Burgess and Clark 1998; Morgan and Murdoch 2000). In many cases a farmer's dissatisfaction with environmental regulations may not be because of an unsympathetic attitude towards nature, but because the specificity of the 'blueprint' constrains the customary use of LK in managing the land.

This small, preliminary study suggests that LK is still very much in use by these English livestock farmers. Although agriculture in England has changed much with scientific and technological advances, LK has not been completely supplanted by science. The two counties show subtle differences in free lists and the proportion of farmers' statements identifiable as either LK or SK. However, in both counties, farmers' knowledge of plants and their management consists of a mixture of SK and LK, and the identity of farmers with the most (or most unique) knowledge of grassland plants cannot easily be predicted from straightforward characteristics like age, farming experience, or even the intensity of grassland management. While free lists can help identify 'experts', they are not the best indicators of knowledge, since they gather only the lexical knowledge of farmers, and not their substantive knowledge, which is less easily communicated.

Where the implementation of conservation measures occurs on a site-specific basis, LK may significantly benefit these efforts. For example, in their review of the success of agri-environment schemes in conserving lowland grasslands, Critchley, Burke and Steven (2003) found that a greater use of site-specific targets, as opposed to overgeneralized prescriptions, is needed to prevent damage caused by over- or undergrazing. Consultation with participating farmers in order to make use of their LK could help to define these site-specific targets, but the assumption that LK does not exist in modern Europe has meant that consultation with local people regarding conservation has been neglected here (e.g., Harrison, Burgess and Clark 1998: 308). This study suggests that further research – using these methods and others among a larger sample of farmers – could reveal a significant amount of LK in use by English farmers and help to clarify how and why that knowledge is being maintained or lost in the context of new policies to protect grassland environments. It also implies that the same participatory methods used to gather and understand LK in less-developed countries are just as valid in Europe, and may be just as important to the conservation of diverse ecosystems.

Acknowledgements

I would like to thank the farmers who generously gave their time for interviews. This study was undertaken in 2003 as part of the M.Sc. Ethnobotany Programme at the University of Kent, Canterbury. The British Council and the John Ray Trust provided funding. Thanks also to Dr Rajindra Puri, Prof. Roy Ellen, Prof. Christian Vogl and two anonymous reviewers for their comments.

References

Agrawal, A., 1995. 'Dismantling the Divide Between Indigenous and Scientific Knowledge', *Development and Change* 26: 413–439.

Alexiades, M.N. 1996. Selected Guidelines for Ethnobotanical Research: a Field Manual. Bronx, NY: New York Botanical Garden.

Beedell, J.D.C. and T. Rehman. 1996. 'Farmers' Perceptions of Conservation and the Provision of Advice on the Management of the Countryside', *Agricultural Progress* 71: 29–35.

Bernard, H.R. 2002. *Research Methods in Anthropology: Qualitative and Quantitative Approaches.* 3rd edn. Walnut Creek, Calif.: AltaMira Press.

Borgatti, S.P. 1996a. *ANTHROPAC 4.0 Methods Guide.* Natick, Mass.: Analytic Technologies.

——— 1996b. *ANTHROPAC 4.0.* Natick, Mass.: Analytic Technologies.

——— 1996c. *ANTHROPAC 4.0 Reference Manual.* Natick, Mass.: Analytic Technologies.

Carr, S. and J. Tait. 1991. 'Differences in the Attitudes of Farmers and Conservationists and their Implications', *Journal of Environmental Management* 32: 281–294.

Chambers, R., 1994. 'The Origins and Practice of Participatory Rural Appraisal', *World Development* 22: 953–969.

Clark, J. and J. Murdoch. 1997. 'Local Knowledge and the Precarious Extension of Scientific Networks: a Reflection on Three Case Studies', *Sociologia Ruralis* 37: 38–60.

Cleveland, D.A. and D. Soleri. 2002. 'Introduction: Farmers, Scientists and Plant Breeding: Knowledge, Practice and the Possibilities for Collaboration', in D.A. Cleveland and D. Soleri (eds), *Farmers, Scientists and Plant Breeding: Integrating Knowledge and Practice.* Wallingford: CAB International, pp. 1–18.

Critchley, C.N.R., M.J.W. Burke and D.P. Steven. 2003. 'Conservation of Lowland Semi-natural Grasslands in the UK: a Review of Botanical Monitoring Results from Agri-environment Schemes', *Biological Conservation* 115: 263–278.

Curry, N. and M. Winter. 2000. 'The Transition to Environmental Agriculture in Europe: Learning Processes and Knowledge Networks', *European Planning Studies* 8: 107–121.

DEFRA. 2002. *Agricultural and Horticultural Census: 5 June 2002, England.* Retrieved 20 August 2003 from http://statistics.defra.gov.uk

——— 2003. *Economics and Statistics.* Retrieved 20 August 2003 from http://statistics.defra.gov.uk

Ellen, R. 2003. 'Variation and Uniformity in the Construction of Biological Knowledge across Cultures', in H. Selin (ed.), *Nature across Cultures: Views of Nature and the Environment in Non-Western Cultures.* Dordrecht: Kluwer Academic Publishers, pp. 47–74.

——— and H. Harris. 2000. 'Introduction', in R. Ellen, P. Parkes and A. Bicker (eds), *Indigenous Environmental Knowledge and its Transformations: Critical Anthropological Perspectives.* Amsterdam: Harwood, pp. 1–33.

Frame, J., R.D. Baker and A.R. Henderson. 1995. 'Advances in Grassland Technology over the Past Fifty Years', in G.E. Pollot (ed.), *Grassland into the 21st Century: Challenges and Opportunities, Proceedings of the 50th Annual Meeting of the British Grassland Society.* Reading: BGS, pp. 31–63.

Fuller, R.M. 1987. 'The Changing Extent and Conservation Interest of Lowland Grasslands in England and Wales: a Review of Grassland Surveys 1930–84', *Biological Conservation* 40: 281–300.

Gasson, R. and C. Potter. 1988. 'Conservation through Land Diversion: a Survey of Farmers' Attitudes', *Journal of Agricultural Economics* 39: 340–351.

Green, B.H. 1990. 'Agricultural Intensification and the Loss of Habitat, Species and Amenity in British Grasslands: a Review of Historical Change and Assessment of Future Prospects', *Grass Forage Science* 45: 365–372.

Harrison, C.M., J. Burgess and J. Clark. 1998. 'Discounted Knowledges: Farmers' and Residents' Understandings of Nature Conservation Goals and Policies', *Journal of Environmental Management* 54: 305–320.

Hopkins, A. 1979. 'The Botanical Composition of Grasslands in England and Wales: an Appraisal of the Role of Species and Varieties', *Journal of the Royal Agricultural Society of England* 140: 140–150.

——— and J.J. Hopkins. 1993. 'UK Grasslands Now: Agricultural Production and Nature Conservation', in R.J. Haggar and S. Peel (eds), *Grassland Management and Nature Conservation.* Occasional Symposium No. 28, British Grassland Society. Reading: BGS, pp. 10–19.

Kent Biodiversity Action Plan Steering Group. 1997. *The Kent Biodiversity Action Plan.* Kent County Council. Retrieved 20 June 2003 from http://www.kent.gov.uk/sp/biodiversity

Kent Wildlife Trust. 2000. *The Kent Red Data Book.* Kent County Council. Retrieved 25 June 2003 from http://www.kent.gov.uk/sp/biodiversity

Lancashire's Biodiversity Partnership. 2001. *Lancashire's Biodiversity Action Plan.* Retrieved 1 May 2009 from http://www.lancspartners.org/lbap/

Lancashire County Heritage Sites Scheme. 1998. *BHS Guidelines for Site Selection.* Lancashire County Council. Retrieved 20 August 2003 from http://www.lancashire.gov.uk/environment/ecology/bhs

Lancashire Rural Partnership. 2002. *Lancashire Rural Recovery Action Plan.* Retrieved 15 April 2003 from http://www.lancashireruralpartnership.gov.uk/action_plan.htm

Mabey, R. 1980. *The Common Ground.* London: Hutchinson/The Nature Conservancy Council.

Macdonald, D.W. and P.J. Johnson. 2000. 'Farmers and the Custody of the Countryside: Trends in Loss and Conservation of non-Productive Habitats 1981–1998', *Biological Conservation* 94: 221–234.

Martin, G.J., 1995. *Ethnobotany, a Methods Manual.* London: Chapman & Hall.

McClure, G. 1989. 'Introduction', in M.D. Warren, L.J. Slikkerveer and S.O. Titilola (eds), *Indigenous Knowledge Systems: Implications for Agriculture and International Development.* Ames, Iowa: Iowa State University, pp. 1–2.

McEachern, C. 1992. 'Farmers and Conservation: Conflict and Accommodation in Farming Politics', *Journal of Rural Studies* 8: 159–171.

Morgan, K. and J. Murdoch. 2000. 'Organic Versus Conventional Agriculture: Knowledge, Power and Innovation in the Food Chain', *Geoforum* 31: 159–173.

National Statistics. 2003. *National Statistics Website.* Accessed 20 August 2003: http://www.statistics.gov.uk

Newby, H. 1979. *Green and Pleasant Land? Social Change in Rural England.* London: Wilwood House.

Pain, D.J., D. Hill and D.I. McCracken. 1997. 'Impact of Agricultural Intensification of Pastoral Systems on Bird Distributions in Britain 1970–1990', *Agriculture, Ecosystems and Environment* 64: 19–32.

Purcell, T.W. 1998. 'Indigenous Knowledge and Applied Anthropology: Questions of Definition and Direction', *Human Organization* 57: 258–272.

Robinson, R.A. and W.J. Sutherland. 2002. 'Post-war Changes in Arable Farming and Biodiversity in Great Britain', *Journal of Applied Ecology* 39: 157–176.

Romney, A.K., S.C. Weller and W.H. Batchelder. 1986. 'Culture as Consensus: A Theory of Culture and Informant Accuracy', *American Anthropologist* 88: 313–338.

Smith, R.S. 1993. 'Effects of Fertilisers on Plant Species Composition and Conservation Interest of UK Grassland', in R.J. Haggar and S. Peel (eds), *Grassland Management and Nature Conservation*. Occasional Symposium No. 28, British Grassland Society. Reading: BGS, pp. 64–73.

Soleri, D., D.A. Cleveland, S.E. Smith, S. Ceccarelli, S. Grando, R.B. Rana, D. Rijal and H. Ríos Labrada. 2002. 'Understanding Farmers' Knowledge as the Basis for Collaboration with Plant Breeders: Methodological Development and Examples from Ongoing Research in Mexico, Syria, Cuba and Nepal', in D.A. Cleveland and D. Soleri (eds), *Farmers, Scientists and Plant Breeding: Integrating Knowledge and Practice*. Wallingford: CAB International, pp. 19–60.

Stoate, C. 1996. 'The Changing Face of Lowland Farming and Wildlife. Part 2', *British Wildlife* 7: 162–172.

van Dusseldorp, D. and L. Box. 1993. 'Local and Scientific Knowledge: Developing a Dialogue', in W. DeBoef, K. Amanor, K. Wellard and A. Bebbington (eds), *Cultivating Knowledge: Genetic Diversity, Farmer Experimentation and Crop Research*. London: Intermediate Technology Publications, pp. 20–27.

Vickery, J.A., J.R. Tallowin, R.E. Feber, E.J. Asteraki, P.W. Atkinson, R.J. Fuller and V.K. Brown. 2001. 'The Management of Lowland Neutral Grasslands in Britain: Effects of Agricultural Practices on Birds and their Food Resources', *Journal of Applied Ecology* 38: 647–664.

Wilson, G.A. 1996. 'Farmer Environmental Attitudes and ESA Participation', *Geoforum* 27: 115–131.

Winter, M. 1997. 'New Policies and New Skills: Agricultural Change and Technology Transfer', *Sociologia Ruralis* 37: 363–381.

Young, C., C. Morris and C. Andrews. 1995. 'Agriculture and Environment in the UK: Towards an Understanding of the Role of "Farming Culture"', *Greener Management International* 12: 63–80.

CHAPTER 16

A Comparative Study of Rural and Urban Allotments in Gravesham, Kent, U.K.

CHRISTINE WILDHABER

Allotments as a Type of Homegarden

Homegardens (also called household, kitchen and 'dooryard' gardens) occur worldwide and can be defined as a 'supplementary food production system which is under the management and control of household members' (Cleveland and Soleri 1987: 259). The primary function of homegardens is generally food production (Fernandes and Nair 1986), although tropical homegardens typically include a variety of plants used for other purposes such as medicine, fodder, firewood, construction materials, market products and ornamentals (Lamont, Hardy Eshbaugh and Greenberg 1999).

Vogl, Vogl-Lukasser and Puri (2004: 287) state that the defining criterion of homegardens is that they are 'adjacent to the house where their gardener(s) live', and as such are distinguishable from other parks and gardens including allotments. However, allotments serve many of the functions of homegardens, hence I believe they can be considered in the same theoretical framework. Allotment gardens in the U.K. are defined by law as 'not exceeding forty poles in extent which is wholly or mainly cultivated by the occupier for the production of vegetable or fruit crops for consumption by himself or his family' (Clayden 2002: 11).[1] Allotment gardens differ from homegardens in two important aspects. First, they are not owned by the household or gardener; the land is comprised of a number of individual allotments (or plots), and the site is managed by an organizational body (a local authority or an association of plotholders).

Second, allotment sites are governed by legislation, which specifies that they be used primarily to grow fruit and vegetables for household consumption and prohibits the sale of produce (Buckingham 2005).

Homegardens contain high levels of species and genetic diversity (Eyzaguirre and Watson 2002) and as such can be important for in-situ conservation of agricultural biodiversity (Eyzaguirre and Linares 2001). The variation in diversity may be explained by social, cultural or geographical factors, whilst differences within a single community are often correlated with wealth, ethnicity and family organization of owners (Martin 2004). Women have been seen to play an important role as managers of biodiversity in tropical homegardens (Thrupp 2000; Greenberg 2003; Howard 2003) and in Europe (Vogl and Vogl-Lukasser 2003). In the U.K., Buckingham (2005) believes that as more women participate in allotment gardening there is likely to be a move towards greater urban biodiversity. Migrants to urban areas bring plants with them, which can result in increased biodiversity, as has been documented in the U.S.A. (Corlett, Dean and Grivetti 2003), Britain (Michaud 2006), Germany (Gladis 2002) and Mexico (Lazos Chavero and Alvarez-Buylla Roces 1988; Greenberg 2003).

Many studies have been carried out on tropical homegardens, but less work has been conducted in temperate climates (Vogl, Vogl-Lukasser and Puri 2004), and the aim of this study was to contribute to our knowledge about homegardens, in regard to rural–urban relationships, the biodiversity of homegardens and their social nature, set in the context of allotments in Britain.

Allotments in Britain

Allotments in England derive from enclosure legislation of the eighteenth and nineteenth centuries (such as The General Enclosure Act 1845) that ensured provision was made for the landless poor in the form of 'field gardens' (HOC 1998). Initially allotments were largely confined to rural areas. However, in the latter half of the nineteenth century, allotments were recognized as being important for the urban working classes, and allotments spread, particularly in areas of high-density housing, often without gardens. At this time, hunger or economic need was the primary reason for growing food on an allotment, and this was reinforced in the early twentieth century by the national demands of the First World War and interwar depression, when the number of allotments rose from 600,000 to 1,500,000 (HOC 1998). From the 1950s onwards, allotments gradually declined in the U.K. to just fewer than 300,000 in 1996 as a result of readily available cheap food and pressures on the use of land for development (HOC 1998; Clevely 2006).

In the 1960s a government inquiry was instigated to rationalize allotment land and its use (Crouch 2003). The resulting Thorpe Report found rural allotments (those in the country and under parish council authority) were profitable and had an 'inherent stability' (MHLG 1969: 290), whilst urban allotments (located in the country and under Parish Council authority) suffered from vandalism, theft, and a reduction in the amount of allotment land due to development. The main function of allotments was now as a recreational hobby, and growing fresh, high quality, cheap food was of secondary importance. Pressure from the National Society of Allotment and Leisure Gardeners concerned about the loss of allotment sites to development resulted in a survey of allotments in England in 1996, which confirmed that total plots had declined (from 532,964 in 1970 to 296,923 in 1996), and the number of vacant plots had increased (from 23,178 in 1978 to 43,586 in 1996). The reasons given for vacant plots included lack of promotion by local authorities, lack of maintenance (which discourages participation), site problems (such as vandalism), rumours of forthcoming disposal of the site, and plot sizes being too large (Crouch 1997).

This provided the impetus for an inquiry into the future of allotments (HOC 1998) and the publication of an advocacy paper: 'A New Future for Allotments' (LGA 2000). Allotments were acknowledged as having a critical role in modern (urban) life, providing benefits both to the general public and to allotment holders, including exercise, the supply of affordable fresh vegetables, increased biodiversity, 'green space' and the potential for educational and therapeutic benefits for sections of the community. Local authorities and parishes were recommended to support allotment practice by promoting them to attract new plotholders, and by modernizing allotment provision and management regimes.

Limited research has been carried out since then to determine whether these new roles are being implemented and to understand the general impact of the guidelines. In 2003, allotment holders in west London did not think sites were being effectively advertised or promoted (Buckingham 2003). There is little evidence of positive change, as a 2005 study carried out in London, and in Ashford and Wye (in Kent), identifies similar negative factors affecting allotment development, including urbanization pressure (sites being sold to developers), lack of policy schemes to promote and support allotments, lack of facilities, theft and vandalism and abandoned or poorly maintained sites (Perez-Vazquez 2002; Perez-Vazquez, Anderson and Rogers 2005), suggesting that the long-term future for allotments continues to be under threat. My study adds to the existing research by comparing allotments in rural and urban settings in terms of their botanical and social characteristics and the roles they play for their holders.

The Research Setting

This study was carried out between April and June 2006, and took place on a rural and an urban allotment in Gravesham, Kent, in the southeast of England. Gravesham covers an area of 10,117 hectares, housing a population of approximately 95,000 inhabitants (results from 2001 census, Gravesham Borough Council 2006). There are two towns, Gravesend and Northfleet (with a combined population of 75,000); and six villages, Cobham, Higham, Luddesdown, Meopham, Shorne and Vigo (with a total population of roughly 20,000). Of the rural areas, Meopham is a large parish with a population of approximately 9,500 people residing in an area of 8.5 square miles. At the time of this study, the towns of Gravesend and Northfleet had eight and seven allotment sites respectively, whilst there were two rural allotment sites, one in Higham and one in Meopham. This study focused on the Central Avenue (also known as Northridge Road) site in Gravesend, and Southdown Shaw (the Meopham allotment site).

The Central Avenue allotment site is the largest of the urban sites (forty-two plots, 0.54 ha in area), has both fencing and water, and is very popular, as evidenced by a long waiting list to join. Each plot is five rods (approximately 126.5 m^2). The site is managed by Gravesham Borough Council and has been in operation since the 1950s (Gravesham Borough Council, pers. comm.). The site is boxed in behind the back gardens

Figure 16.1. Gravesham (urban) allotments

of residential housing, and access is by small alleyways between the houses. There are three pedestrian entrances and one vehicular gate used for manure deliveries, although there is no access for other vehicles (Figure 16.1).

The Meopham allotment site is owned by the local parish council and managed by an Allotments Committee (a subcommittee of the Parish Council Environment and Amenities Standing Committee), comprised of three local councillors and six allotment holders. Sites are allocated by the committee secretary, who is a plotholder, but rents are payable to Meopham Parish Council. The allotment site has been in operation for twenty-eight years, is 4.86 hectares in size, with 103 cultivated plots as of May 2006 (Meopham Parish Council, pers. comm.). Each plot is six rods (approximately 151.8 m^2). There is a large area to one end of the site that was previously laid out to allotments (there were originally 220 plots under cultivation in 1978), and this area is mown two to three times a year. There is water on site, with water butts and taps at intervals across the site. The site has one vehicular gate, and a pedestrian gate that allows access to the public and is used by dog walkers. The site is fenced, and is bounded by a barrier of trees and shrubs on one side, a railway cutting, a barrier of shrubs adjacent to a smaller country lane, and expansive views out to large, open fields on the opposite side (Figure 16.2).

Figure 16.2. Meopham (rural) allotments

Working in Allotments

The nature of the rural and urban allotment sites was considered on different levels. Semistructured interviews were used to gather information about the plotholders on various topics, including where planting material was sourced, reasons for growing particular crops, renting an allotment and perceptions about the allotment site. The relationship between plotholders (social units) and plots produced a complex social structure, as there was often more than one person to a plot (for example, couples often gardened together), and people also rented more than one plot (see Figure 16.3). Therefore, the research analyzed and compared each social unit's plots.

Allotment plots were surveyed and maps drawn to ascertain the abundance of species. The maps were drafted during the initial phase of the study period, and were finalized in mid June, when all summer crops were in the ground, and plotholders were at a similar stage of cultivation. Harvesting of summer crops had not begun, and the harvesting of winter and late spring crops, such as purple sprouting broccoli (*Brassica oleracea*), had finished. However, this also meant the diversity of the plots reflected the summer season only, and not the whole year. Plant species present were initially identified by the author, and later verified with the plotholder when carrying out the final mapping. Voucher specimens were not taken because the majority of plants were easily identified. Cultivar names were noted where known by plotholders or where labels were present on plants. Plant nomenclature and family names were verified using The International Plant Names Index (IPNI 2006) and the Royal Horticultural Society horticultural database (RHS 2006).

The maps were analysed first by recording the presence or absence (richness) of species in all plots held by each social unit, and second by recording the relative abundance of each species, calculated as the percentage of the total plot(s) area. To prevent error, when carrying out

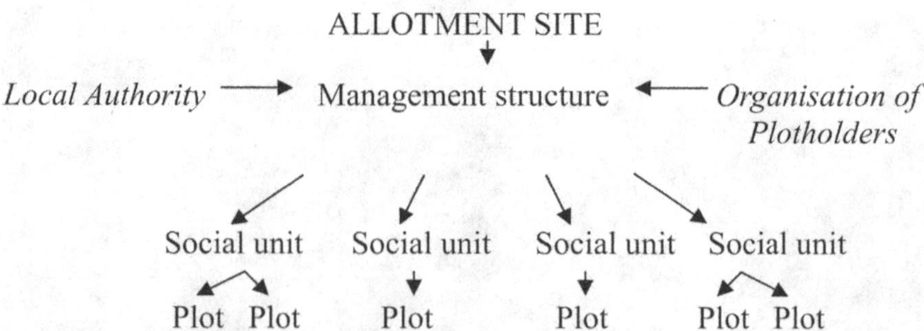

Figure 16.3. Diagram representing the social structure of the rural and urban allotments

analysis, the percentage occupancy of plant species was described in terms of proportions, and these were then transformed using the arcsine transformation (Fowler and Cohen 1991). Correspondence analysis in ANTHROPAC 4.0 (Borgatti 1996) was used to compare floristic diversity among the social units, in terms of the plant species' richness and abundance.

Details about each social unit were entered into a database of individual or joint attributes (characters that could be assigned to a social unit). Social units were compared to each other using similarities programs in ANTHROPAC (Borgatti 1996). Similarity matrices were calculated for each of the categorical variables (such as gender and allotment site) and dissimilarity matrices were calculated for all the metric variables (such as plot size, and the number of species cultivated). A similarity matrix was also created for both species abundance and richness. Quadratic Assignment Procedure (QAP) regression was then used to test how well the social attributes account for variation in species diversity. Chi-square tests and t-tests were carried out using Statistical Package for the Social Sciences (SPSS). Other data collected from the interviews were analysed qualitatively by examining the patterns and similarities of statements made by participants (Bernard 1995) on coded topics.

Who Are the Plotholders?

The sample size from the rural site (sixteen social units) was larger than that from the urban site (ten social units). A total of 28.3 plots, rented by sixteen social units (a total area of 4178.60 m^2) were mapped on the rural site. Ten of the social units were single persons, five social units were couples, and one social unit consisted of three people. On the urban site, seventeen plots rented by ten social units (a total area of 2072.90 m^2) were mapped. Seven of the social units were single persons, and three social units were couples. The following discussion refers to the sample size in this study, and may not reflect all the social units or plots of each site.

The profile of allotment holders on both the urban and the rural sites were very similar. All except two of the allotment holders are white British (Figure 16.4); the two exceptions are white European (on the rural site). No significant differences were found in terms of age (the mean age of rural participants was fifty-eight years, and urban participants sixty-five years), although only the age of people responsible for making decisions on the allotment was taken into account. One person on the rural site has a small area for his granddaughter and one woman regularly brings her nine-year-old son with her. Young children were also observed on the urban site. However, the perception that the profile of allotment holders is changing in regards to the age and gender of plotholders was evident in this study.

> I was surprised, you always associate it with old boys with flat caps on and wellies and, ah, there are quite a few young people, or younger, let's say, 30s, 40s... and you know, they're having a bash, having a go, it's not just always the old boys that are king of the compost heap (rural plotholder).

> [T]here's a lot more women come on site, and generally speaking the ones taking them on are a lot younger, in the time I've been here (13 years) (rural plotholder).

No significance differences were found between the sites either in the presence of men or women ($p<0.48$), or the 'gender' of the social units ($p<0.28$).[2] Crouch and Ward (1994: 82) write that 'women have always worked on the land out of necessity, and allotments – though seen traditionally as a male preserve – have often been, from the same necessity, a focus of activity for the whole family'. This kind of social organization was also apparent in this study; although a few 'single' plotholders garden exclusively on their own, family members might assist at specific times.

> [S]he'll come down when it's picking time, which is handy, because it gets hectic then, you know, you don't get time to do all your weeding and bits like that (rural plotholder).

The rural site also has one plot rented by a special needs school. However, while some people on the urban site are amenable to having a school on site, other plotholders are not:

Figure 16.4. Meopham plotholders

> we did suggest, but it was voted down, there's a school up the road here for... um... children with special needs and I thought it would be great if a couple of the allotments were put over to them, and that they came with their teachers and worked it, but no nobody liked that idea (urban plotholder).

The one significant difference found in regards to the nature of plotholders from urban and rural allotment sites in this study was how people travel to the allotment, and where people live relative to the site. Most of the urban plotholders walk to the site, whilst nearly all of the rural plotholders come by car. This relationship was found to be significantly different ($p<0.01$). However, no significant difference was found in the time taken to travel to the site ($p<0.37$).

All the urban plotholders live in Gravesend, whilst those at Meopham came from neighbouring villages or the town of Gravesend. This was either because of the lack of a local allotment site, or because the rural site was more pleasant and did not have the same issues with vandalism found on urban allotment sites.

> I always have fancied an allotment up here, to be truthful with you, I've always looked when I've passed the gate... but um, it never materialized until I retired... but there's a load of people from Gravesend who's got allotments here.... [He] told me he used to have a big shed on there... [Gravesend site] and um, he caught all the druggies in there and that and then they set it alight (rural plotholder).

This is a new phenomenon. In 1969, the Thorpe Report (MHLG 1969) found that most people lived within half a mile of their plot (83 per cent of allotment holders in urban areas, and 93.3 per cent of plotholders in rural parishes). Similarly, Perez-Vazquez, Anderson and Rogers (2005) found that plotholders in both urban and rural areas lived within half a kilometre of their allotment. In this study, only one rural allotment holder was within easy walking distance of their plot.

Floristic Diversity amongst Plotholders and Sites

The overall plant diversity across all plots in this study was very high, with a total of 44 plant families, 130 species and 201 cultivars documented, but even more interesting is the variation in species richness and abundance from plot to plot. A total of fifty-four species (41.5 per cent) are shared across both sites; fifty-five species are unique to the rural site and twenty-one species are unique to the urban site. The ten most common species present were rhubarb, potatoes, onions, runner beans, tomatoes, carrots, raspberries, greens, marrows and courgettes, and beetroot. Only twenty-one species had a frequency of more than 50 per cent across both sites (see Appendix 16.1). With the exception of rhubarb, which had no named

Figure 16.5. Gravesend plotholders with produce

cultivars, most of the named cultivars were found for species with high frequencies. People with more plots tend to have a greater variety of species. A significant positive correlation was found between the number of species and the area of the plots cultivated by a social unit ($p<0.01$, $r=0.51$). Species with high abundance are potatoes, raspberries, greens, leeks, onions, marrows and courgettes, broad beans, runner beans, French beans, parsnips, squash and pumpkins, lettuce and tomatoes (Figure 16.5).

The only significant factor found to determine species absence or presence was whether the site was urban or rural ($p<0.1$, see Table 16.1). A significant correlation ($p<0.05$) was also found between species abundance and the site studied (rural or urban). However, the R-square (0.02) indicates that this factor only accounts for 2 per cent of the variation in abundance found. When mapping the plots some of the urban plotholders created large areas of nettles (*Urtica dioica*), brambles (*Rubus fruticosus*) or other plants (such as *Syringa vulgaris*, lilac) towards the fence boundary of their plots to discourage intruders. This may partly explain the differences found. Thus, when the same test was carried out on the ten most abundant species found in each plot excluding the above boundary plants, no significant difference was found between the sites.

Other explanations might include site-specific conditions (such as soil type or prevalent pests or diseases), or the influence of ethnic minority cultures. Winklerprins (2002) found that the agrobiodiversity of rural and urban houselots in Brazil were similar, although flooding in the rural area limited perennial fruit-tree growth. Corlett, Dean and Grivetti (2003) found that Vietnamese migrants to the U.S.A. grew crops previously

Table 16.1. Summary of QAP regression test results

Dependant variable	Independent variable	R-square	Probability	Result
Species present	Site (rural or urban)	0.045	0.006	Significant at $p<0.01$
Species present	Garden at home	0.050	0.070	No significant difference
Dependant variable	**Independent variable**	**R-square**	**Probability**	**Result**
Species abundance	Site (rural or urban)	0.020	0.028	Significant at $p<0.05$
Species abundance	Garden at home	0.076	0.022	Significant at $p<0.05$
Dependant variable	**Independent variable**	**R-square**	**Probability**	**Result**
Top ten abundant species	Site (rural or urban)	0.008	0.140	No significant difference
Top ten abundant species	Garden at home	0.065	0.024	Significant at $p<0.05$

unknown to them, but found in local stores or introduced to them by neighbouring gardeners. Although there are no ethnic minority groups on the urban site, some people are growing peppers and okra, and this may be due to the influence of different ethnic cultures resident in Gravesend. Further research would be required to clarify this. Overall, more species were found on the rural site. However, the sample size was also larger, and the plot sizes are also larger.

Another significant factor affecting the abundance of species grown was the type of garden people had at home ($p<0.05$): this accounts for approximately 8 per cent of the variation observed (R-square=0.076), although there was no significant difference found in species presence and the type of garden people have at home ($p>0.05$).[3] This suggests that if people do grow vegetables or fruit at home, it has an effect on the abundance of these species that are grown on the allotment, but not on whether they decide to grow the same species at the allotment. The interviews revealed that in some cases the same species are grown in the allotment as at home, but not in others.

> [T]he runner beans are all ready to plant out at home: we won't put any out up here because they are the one thing that do well at home (rural plotholder).

> [W]e have got some potatoes in a pot (at home), but that's only because ... they wouldn't do too good, they get blight up here [reference to a particular blue variety of potato], so we thought we'd keep them at home, but give them a try (urban plotholder).

So what are some of the other factors that might account for the variability in species grown by different social units and the quantities in which they are grown? Perez-Vazquez, Anderson and Rogers (2005) found that women grew more herbs and flowers, but that potatoes were seldom grown by women as potato cultivation was said to be hard work, and this is echoed by the Thorpe Report (MHLG 1969), which states women prefer to cultivate flowers because vegetable growing demands more physical effort. There was a tendency for flowers to be seen as a woman's domain, although there is no significant difference between the gender of the social unit and the number of flower or herb species grown ($p<0.80$).

> [S]he's the flower queen, and I do the lawns (male plotholder).

> [A] garden to me is not flowers' (male plotholder).

> I'm interested in flowers, but she's the person that sort of does the flowers (male plotholder).

Potatoes are the most or second most abundant species, grown by twenty-four of the twenty-six social units. Although not everyone grows potatoes in large quantities, gender did not appear to be the reason.

> [W]hen I first came round here, everybody traditionally grows tons of spuds, and I would end up with these bags of spuds and I'd think, 'Why am I growing these bloody things?' We don't eat them, don't even like them much, so I don't grow many potatoes, but they're very traditional here, everybody discusses the merits of you know, their earlies and their lates, and all that (urban female plotholder).

The diversity in species richness and abundance found amongst the plots is not unlike other homegardens worldwide. Kimber (1973: 6) found a range of variation in the 'door yard' gardens of Puerto Rico and suggests that that each garden is the result of 'thousands of decisions about plants which a person makes in his own space'. In this study a number of reasons were mentioned as to why people chose to grow particular vegetables (and in particular quantities), that can account for the variability observed. These included personal health, family needs and preferences, and whether people could get certain crops from neighbours.

> I didn't grow peas this year because ... they always have the pea moth and maggots ... and I think I threw more peas away unnecessarily last year than I needed to cause I couldn't see properly, you know and I still can't so... (urban plotholder).

> I just think what we will use and what will store well over the winter and what my two girls will eat. For example, ... they used to love beetroot ... so last year, I grew a lot of beetroot and the previous year I grew [extra] peas, because they like shelling them (rural plotholder).

> I'm hoping that ... is going to grow some [leeks] and he'll give me a few (rural plotholder).

Some people prefer to grow what is considered easy or cultivars known to be reliable for the area, whilst others enjoy the challenge of growing something unusual or persisting with difficult crops.

> [T]here was a chap down there; he advised me to not to grow the potato King Edwards, although they're a very good potato, he never got a good crop here and he grew Desirée, so that's why I took his advice and I still grow them and I find them one of the best potatoes to grow (urban plotholder).

> I went to a talk by a local farmer and he said, 'Oh you can't grow cauliflowers around this area', so of course that was a red rag to a bull.... I said, 'I'm going to grow these' and I had some whoppers (rural plotholder).

> I like growing novelty crops – scorzonera, salsify, seakale – and in the greenhouse at home I've got heritage varieties of plants so they're a bit more of a challenge because they're not so disease resistant, um pest resistant like F1 varieties[4] (rural plotholder).

> [C]arrots are not very successful with me.... They might be growing, but we get carrot fly. I'm trying a different method of um, ah, trying to prevent the carrot fly disease to attack 'em, you know, it's a constant battle but it's enjoyable in one way (urban plotholder).

Other people preferred not to grow certain crops if they had problems with pests.

> [W]e did do Brussels ... They were overtaken with whitefly ... so I haven't bothered with them too much until we can get to the point where we can look after them a bit more (urban plotholder).

The diversity of plants grown will also reflect where people source their seeds. Whilst most people obtain their seeds and planting material from local garden centres (77 per cent), a significant difference was found between the sites ($p<0.01$), and this may be because all the urban plotholders buy some seeds through their contacts with an allotment association in nearby Rochester.

Tropical homegardens have been touted as important sites for *in-situ* conservation of rare crop varieties and local landraces (Eyzaguirre and Linares 2001), and species with endangered status have also been found in Austrian homegardens (Vogl-Lukasser and Vogl 2004). However, most of the species and the cultivars found in this study are commercially available and not currently under threat, except for the heritage seeds grown by one plotholder. These originate from the Heritage Seed Library established by the Organic Association (formerly HDRA, Henry Doubleday Research Association) to collect varieties of seed that are being rapidly lost. Losses have occurred mainly through changing practices, but also as a result of the Plant Varieties Act (1964), which set up the National Seed Register and made the sale of small-scale, unregistered varieties illegal (Purdue 2000). The Organic Association encourages seed saving and the exchange of these varieties and cultivars, but the presence of heritage seeds on plots in this study is not common, even though gift-giving of plants was evident (the author received plants from more than five different people). People will refuse plants if they do not have sufficient room on their plot or for other reasons and the perception that heritage seeds are expensive may be a deterrent to some. 'There's a gentleman over there that only grows old crops .... [T]he seeds cost a fortune'. Another factor could be that these seeds are less resistant to pests and diseases. Also, although a few people save their own seed – a mechanism that aids the survival of unique cultivars – these are usually readily available varieties, primarily runner beans, which have large seeds and are easy to save 'I used to save my own peas, but they were too fiddly, so I buy them now'. Purdue (2000: 164) writes that these 'backyard biologists' share a concern over the global biodiversity crisis. However, the gardener growing heritage seeds claimed to be motivated by the challenge of growing them, which suggests the cultivation of heritage seeds may be due to the character of the grower, rather than environmental concerns. Other authors have also observed that it is the personality of the gardener that can assist is maintaining species diversity. Nabhan (1989: xxii) writes that 'the only factors that have stood between some old fashioned beans and their extinction are the kindness and curiosity found amongst exceptional individuals', while Nazarea (2005: 28) states that 'the nurturance of biodiversity … has existed for as long as farmers have cultivated and tinkered with their plants, whether driven by scarcity and need, curiosity and fun'. On both sites there are some individuals who find it interesting and fun to grow unusual vegetables.

To summarize, the results indicate that the variation in floristic diversity can only be explained by taking into account a large number of variables – the origins and sociodemographic characteristics of plotholders, the ecological aspects of particular sites, the sources and availability of seeds, and the multiple functions of allotments for gardeners and their wider social units – which ultimately suggests that every garden is unique.

The Role of the Allotment

Participants cited a number of reasons as to why they chose to have an allotment and they did not always say one reason was more important than another.

> I like growing my own vegetables, I think they taste nice. It works out cheaper for me. It's an enjoyment as well: I like coming over here and watching things grow (rural plotholder).

The most frequently given reason for people having an allotment (70.5 per cent) was related to the quality of growing their own food. This included taste and freshness, growing organically, or, if any chemicals were used, knowing what they were.

> I think they are so much nicer than what you can buy in the shops. I mean, our grandchildren won't eat carrots bought from the shop, but they will eat ours (rural plotholder).

> [W]hen you have carrots from here and then you buy some, the difference, the flavour and everything, it's marvellous – even the beans (rural plotholder).

> I mean, that tastes so much better than tomatoes from the supermarket – they're pretty disgusting really – so for the taste, the fact that I know they've got no chemicals, or very little… (rural plotholder).

> I like fresh vegetables and not out of the supermarket if possible … because they're not sprayed with all these different things (urban plotholder).

Overall, eighteen people (53 per cent) mentioned that it was important to them to know that the food was organic or had fewer chemicals, or simply to 'know what's in it'. One person was also concerned about general environmental issues,

> like doing your part for the environment …, [knowing] how many litres [of water] it takes to grow a lettuce. And a lot of them are grown in semiarid, sandy[?] places … where water is at a critical point now …. [Y]ou've got to transport it all the way here as well …. Apples is a good example, isn't it? We live in the middle of what used to be England's apple-growing centre and there used be thousands of acres of apples and now a lot of them are being imported from New Zealand and, you know, it takes … lots of fuel to get it here (rural plotholder).

Overall, ten people (29 per cent, nine men, one woman) stated that their reason for having an allotment was as a retirement activity, and this was tied with doing something productive, working with the land, exercise or getting out of the house: '…partly so the wife, so I don't get under her feet' (urban plotholder). Other benefits mentioned included exercise (two people stated it had helped with back problems), fresh air, being outside, social aspects, and relaxation and enjoyment despite the hard work that vegetable growing often entails.

> Well, we enjoy it …. I had me doubts, it was a bit like yours, it looked like climbing Everest (rural plotholder).

> I like being outside; I can come down and spend hours here and it puts me in a good mood (rural plotholder).

> I think this place is an escape for a lot of people. It's a lot of hard work … It's a funny kind of escape (urban plotholder).

> I just like it here …. It's a good antidote to the office … to just come up here and you can actually concentrate on weeding and digging …. [I]t takes you away from that constantly worrying about work or anything else (urban plotholder).

> [I]t's social, it is nice …. [W]e've got a diverse amount of people which is great …, but the common bond is getting on and growing your veges (urban plotholder).

> [I]t's the fun of growing and the fun of watching things come up and knowing you've done it yourself (rural plotholder).

Similar to that of Perez-Vazquez, Anderson and Rogers (2005), this study found the main purpose for having an allotment is for growing safe, fresh food, whilst secondary purposes include having an activity, for relaxation or to be outdoors. The cost of food did not feature highly as a reason why people had an allotment. Only three people mentioned cheaper costs of growing your own vegetables (particularly if they are organic) as a reason.

> [O]ur youngest son having food intolerance …[,] I just couldn't keep affording paying supermarket prices, so …[. W]ell, this is my project, to get an allotment …. [I]t's organic, so no pesticides, and for our children having an allotment has been fantastic 'cause they don't go to Sainsbury's and see a lettuce in a packet they know exactly where it comes from and also they like to grow things (rural plotholder).

Nearly half of the plotholders commented that their reason for having an allotment was because the garden at home was too small or the soil was unsuitable for growing vegetables. 'We couldn't possibly do this in our little garden so, it's an extension of that really'. Although some vegetables or fruit might be grown at home, most of it is limited to the allotment and the garden at home is primarily lawn and flowers. Bhatti and Church (2004) state that until fairly recently many gardeners used to cultivate a mix of fruit, vegetables and flowers at home but with decreasing plot sizes since the 1980s there is less space to grow vegetables and fruit and increasingly the lawn and flowers are dominant. The findings in this study suggest that future studies of allotment biodiversity should also include what people cultivate in their garden at home (if they have one).

Management and Stability of Sites

In this study, the rural and the urban site are different in their relationship with fellow plotholders, the local authority and the problems that they encounter on site. The local authorities are also different in their approach to the management of the sites, the facilities provided, the level of promotion and their expectations of the roles of the plotholders. This can impact on the stability and future of allotment sites.

Although vandalism is more prevalent on the urban site, there is a core of people who liaise with or lobby the borough council on management issues to improve security and to evict plotholders not utilizing their plots.

> I came round here and they were all vacant [twelve years prior], hardly any of them were done so I rang the council ... and 'Oh no, they're all taken, they all pay their rent'. I said, 'Yes, but none of them are being done' (urban plotholder).

> [T]his whole area, you might not think, ah, it's not worked, but it's virtually worked one hundred per cent more than what it was then .... We couldn't get people to come in here because there was too much vandalism – we still get a little bit – but the fence went up and then that stopped a hell of a lot (urban plotholder).

In general most people on the urban site feel they have a good relationship with the council, although some feel that allotments should be given a higher priority.

> [W]e wanted to sort of really try and get Gravesham to realise what an asset allotments are, whereas they'd sidelined them enormously. They do the basics, you know, but as I say ... we met up with them and we talked very sensibly about a much more positive programme for Gravesham Borough Council. Why did they not, you know, make a huge thing of these wonderful plots, loads of them are just abandoned, they've got no water laid on, the

> council won't spend money on fences, so people say quite rightly, 'Christ, I'm not slogging my guts out there because, um, you know, some kids are going to all come in and all play about as though it's a playground' … and later… I think that the local authorities are very silly not to high-power them and put some money into them and really make them an asset that they could claim [them] as their own contribution towards greener living (urban plotholder).

On the rural site, the uptake of plots also increased from 64 in 1995 to the current 110 (Meopham Parish Council, pers. comm.) with the establishment of the allotments subcommittee and the appointment of a secretary (a plotholder) whose role includes showing prospective new people to the site and liaising with the parish council as to maintenance works required on site. The secretary's wife also has a regular column in a local journal (*The Meopham Review*), and allotments to rent are advertised. Both have worked hard to promote the allotments and organize an annual open day, judging of the allotments and publicity. However, over the past three years the parish council has decided that the allotment site should become self-funding and self-managed (Meopham Parish Council, pers. comm.) and this is now being insisted upon at a time when the current secretary is due to retire after eleven years, despite concerns and reluctance from plotholders.

> I think part of the issue of self management with this allotment is … people come from quite a wide area … so unlike um, shall we say up north, where all the people seem to be from a very close community usually, … the community spirit is not here because it's not a community in the same way …. We're spread all over the place and people have got different work patterns, different abilities to get here and things like that (rural plotholder).

> [T]hey're talking about self management …. I'm not interested, I mean so long as they just leave me my plot …[. T]hat's very selfish, but I really can't begin to be interested in all the politics of it (rural plotholder).

Although Thorpe's Report was critical of the allotment movement and the allotment holder ('he is primarily an individualist who considers his allotment to be as private as his home garden, who is seldom interested in anything beyond it's boundaries, and is blind to his further responsibilities', MHLG 1969: 167), gardens are considered as a private haven from work and politics (Bhatti and Church 2004) and for many people allotments are also seen as a private space, where issues of management should not intrude. People might not wish to get involved because they value the allotment for their own sense of space and relaxation, they might have other personal commitments, may not feel capable, or may not have the time.

> I used to do all the odd jobs here for [the secretary] … but now, I'm older than [him] … so it is a bit, I can't do it now (rural plotholder).

> [N]ormally he's on top of it but not this year or last year …[. F]amily things get in the way (urban plotholder).

> [A] lot of people see this as getting away from that kind of bureaucracy, you know, they don't want to come round here and have to be, oh, accountable too much (urban plotholder).

> [W]hen we went to see the council we wrote a kind of little report of it … but not many people want to be doing that, a lot of them work and they're not unhappy I don't think for us to do it, but I think they are wary that it might get a bit overorganized and I think they're quite right to have that thought; at least I'm not one of these people who thinks, 'Nobody gets involved', you know, the point is this: people are involved, but in different ways (urban plotholder).

Crouch and Ward (1994: 110) acknowledge that whilst self-management is fragile as it depends on the circumstances of the individual, 'it is always cheaper than professional management, which depends on current pay scales in public employment as well as the attitudes of council officers who may regard this responsibility as just one more dreary chore'. Crouch, Sempik and Wiltshire (2001: 51) state that achieving the benefits of devolved management depends upon:

> effective consultation with plotholders and associations in formulating the scheme, so that the degree of devolution matches the capacity of the association to deliver the services concerned, ongoing support from the local authority … to sustain the enthusiasm and develop the skills of devolved managers, and periodic review to ensure that all is well.

However, plotholders may be forced into becoming self-managed if they wish the site to continue operating. Shortly after these interviews were carried out, Meopham Parish Council had a meeting further debating the issues of self-management and threatened to close the site. Despite a statutory obligation to provide allotments, the council decided 'it was not the parish council's responsibility to manage the allotments … as some of the plots were taken by people outside of the parish' (*Gravesend Messenger*, 10 August 2006, p. 20). This placed the rural site at a critical turning point in regards to its future.

Conclusions

The allotment sites in this study – one rural, one urban – are different in many ways yet similar in others. The allotments are similar in their age and gender profiles, although this contrasts with the profiles found in studies of urban city areas, where allotment holders are likely to be younger and have a more equal balance of women to men (Buckingham 2005). Both exhibited a high degree of biodiversity and variation across the plots. Although other studies have found differences in the crops grown on the basis of gender, this was not found to be significant in this study. However, the type of garden that people had at home did affect the species abundance on the allotment, a factor that other studies on allotments have not considered. Although allotments can be seen as a type of homegarden, plotholders consider their allotment to be an extension of their homegarden, and future studies should consider both the garden at home (if people have one) and the allotment as one entity. Both rural and urban allotments play an important role in providing fresh food and have other benefits to plotholders, such as health and relaxation.

Two important differences were found between the sites that have an impact on the social nature and management of the sites, namely the size of the site and where plotholders live in relation to it. The urban site is smaller in size and plotholders are living in close proximity to it. Plotholders have developed an informal association that has a good relationship with the council, and the facilities provided are generally good, although some plotholders would like additional security. The rural site is larger and plotholders travel further from home to get there, with over half the plotholders living in towns and villages outside the parish boundary of the allotment site. Consequently the site does not have a well developed community spirit, but also due to the number of people living outside the parish, the parish council feels a diminished responsibility to the site, and is threatening closure of the site. This challenges the perception that rural sites are inherently stable.

Although Kumar and Nair (2004: 148) write that 'tropical gardens are an enigma and … because of that, the policy makers seldom recognize their importance', in Britain there has been an increased interest in allotments over the last decade. The changes in profiles of allotment holders (more women, younger plotholders and increasing ethnic diversity) and the changing primary purpose of allotments (from food security to a leisure pursuit or food growing for environmental reasons) have brought new roles for allotments, and the Local Government Association (LGA 2000) urges local authorities to promote allotments and work towards developing the role of allotments as an educational and community resource. Despite the recent guidelines, there was little evidence of promotion of the sites by either Meopham Parish or Gravesham Borough Council. Crouch (2003: 43) writes that promotion is crucial to advertise that plots are available, to secure the relevant image of what allotments mean for the local environment and for local people, and to secure 'wide public support, participation and value'.

Although allotments have benefits to the wider society (health, biodiversity, food security) they will continue to be at risk if they are perceived as an individual recreational pursuit that local authorities do not consider their responsibility to manage. Future studies need to examine the threats to allotment sites in rural parishes and to establish more effective means of securing the future of allotments.

Notes

1. 40 poles = 40 rods = 1012 sq.m =0.25 acres = roughly 0.1 hectares.
2. Here *gender* is defined as male plotholder, female plotholder, or male and female plotholders.
3. Homegarden types were grouped into the following categories: no garden at home; garden at home but no vegetables or fruit; garden at home with vegetables; garden at home with fruit; and garden at home with vegetables and fruit.
4. An F1 variety is a hybrid produced by crossing two carefully selected parents of the same species that have been developed through a process of self-fertilization or 'inbreeding' over a number of generations. F1 varieties are normally both more uniform and more vigorous than normal, open-pollinated varieties. Due to the complex breeding involved they are also more expensive and seed saved from them does not come true.

References

Bernard, H.R. 1995. *Research Methods in Anthropology: Qualitative and Quantitative Approaches.* 2nd edn. Walnut Creek, London andNew Delhi: AltaMira Press.

Bhatti, M. and A. Church. 2004. 'Home: the Culture of Nature and Meanings of Gardens in Late Modernity', *Housing Studies* 19(1): 37–51.

Borgatti, S.P. 1996. *ANTHROPAC 4.0.* Natick, Mass.: Analytical Technologies.

Buckingham, S. 2003. 'Allotments and Community Gardens: a DIY Approach to Environmental Sustainability', in S. Buckingham and K. Theobald (eds), *Local Environmental Sustainability.* Woodhead, Cambridge: CRC Publishers, pp. 195–212.

——— 2005. 'Women (Re)construct the Plot: the Regen(d)eration of Urban Food Growing', *Area* 37(2): 171–179.

Clayden, P. 2002. *The Law of Allotments.* 5th Edn. Kent: Shaw and Sons.

Cleveland, D.A. and D. Soleri. 1987. 'Household Gardens as a Development Strategy', *Human Organisation* 46(3): 259–270.

Clevely, A. 2006. *The Allotment Book.* London: Harper Collins Publishers Ltd.

Corlett, J.L., E.A. Dean and L.E. Grivetti. 2003. 'Hmong Gardens: Botanical Diversity in an Urban Setting', *Economic Botany* 57(3): 365–379.

Crouch, D. 1997. *English Allotments Survey.* Anglia University/National Society of Allotment and Leisure Gardeners Limited, for the Department of the Environment.

——— 2003. *The Art of Allotments: Culture and Cultivation.* Nottingham: Five Leaves Press.

——— and C. Ward. 1994. *The Allotment: Its Landscape and Culture.* Nottingham: Mushroom Bookshop.

———, J. Sempik and R. Wiltshire. 2001. *Growing in the Community: A Good Practice Guide for the Management of Allotments.* London: LGA Publications.

Eyzaguirre, P.B. and O.F. Linares. 2001. 'A New Approach to the Study and Promotion of Home Gardens', *People and Plants Handbook* 7: 30–33. WWF-UNESCO-RBG Kew.

——— and J.W. Watson. 2002. 'Homegardens and Agrobiodiversity: an Overview across Regions', in J.W. Watson and P.B. Eyzaguirre (eds), *Proceedings of the Second International Home Gardens Workshop: Contribution of Home Gardens to In Situ Conservation of Plant Genetic Resources in Farming Systems,* 17–19 July 2001, Witzenhausen, Federal Republic of Germany: International Plant Genetic Resources Institute Rome, pp. 10–13.

Fernandes, E.C.M. and P.K.R. Nair. 1986. 'An Evaluation of the Structure and Function of Home Gardens', *Agricultural Systems* 21(4): 279–310.

Fowler, J. and L. Cohen. 1991. *Practical statistics for Field Biology.* Chichester: John Wiley & Sons Ltd.

Gladis, T. 2002. 'Ethnobotany of Genetic Resources in Germany – Diversity in City Gardens', in J.W. Watson and P.B. Eyzaguirre (eds), *Proceedings of the Second International Home Gardens Workshop: Contribution of Home Gardens to In Situ Conservation of Plant Genetic Resources in Farming Systems,* 17–19 July 2001, Witzenhausen, Federal Republic of Germany. International Plant Genetic Resources Institute Rome, pp. 171–173.

Gravesham Borough Council. 2006. *About Gravesend.* Retrieved 10 April 2006 from http://www.towncentric.co.uk/Learn/Gravesend/Gravesend.htm

Greenberg, L. 2003. 'Women in the Garden and Kitchen: The Role of Cuisine in the Conservation of Traditional House Lots among Yucatec Mayan Immigrants', in P.L. Howard (ed.), *Women and Plants: Gender Relations in Biodiversity Management and Conservation.* London & New York: Zed Books, pp. 51–65.

HOC (House of Commons). 1998. *The Future for Allotments: Fifth Report of the House of Commons Environment, Transport and Regional Affairs Committee.* Retrieved 30 March 2006 from http://www.parliament.the-stationery-office.co.uk/pa/cm199798/cmselect/cmenvtra/560/56002.htm

Howard, P. L. 2003. *The Major Importance of Minor Resources: Women and Plant Biodiversity.* Gatekeeper Series 112. London: IIED.

IPNI (The International Plant Names Index). 2006. Database. Accessed 22 August 2006 at http://www.ipni.org

Kimber, C.T. 1973. 'Spacial Patterning in the Dooryard Gardens of Puerto Rico', *Geographical Review* 63(1): 6–26.

Kumar, B.M. and P.K.R. Nair. 2004. 'The Enigma of Tropical Homegardens', *Agroforestry Systems* 61–2(1): 135–152.

Lamont, S.R., W. Hardy Eshbaugh and A.M. Greenberg. 1999. 'Species Composition, Diversity, and Use of Homegardens among Three Amazonian Villages', *Economic Botany* 53(3): 312–326.

Lazos Chavero, E. and M.E. Alvarez-Buylla Roces. 1988. 'Ethnobotany in a Tropical-humid Region: the Home Gardens of Balzapote, Veracruz, Mexico', *Journal of Ethnobiology* 8(1): 45–79.

LGA (Local Government Association). 2000. *A New Future for Allotments: An Advocacy Document for Sustainable Living*. Retrieved 30 March 2006 from http://www.lga.gov.uk/lga/advocacy1.pdf

Martin, G.J. 2004. *Ethnobotany: a Methods Manual*. London and Sterling, Colo.: Earthscan.

MHLG (Ministry of Housing and Local Government). 1969. *Departmental Committee of Inquiry into Allotments: Report*. London: HSMO.

Michaud, M. 2006. 'The World in an Allotment', *The Garden* 131(4): 272–277.

Nabhan, G.P. 1989. *Enduring Seeds: Native American Agriculture and Wild Plant Conservation*. San Francisco, Calif.: North Point.

Nazarea, V.D. 2005. *Heirloom Seeds and their Keepers: Marginality and Memory in the Conservation of Biological Diversity*. Tucson, Ariz.: University of Arizona Press.

Perez-Vazquez, A. 2002. *The Future Role of Allotments in the Southeast of England as a Component of Urban Agriculture*, Ph.D. dissertation. Imperial College Wye: University of London.

———, S. Anderson and A.W. Rogers. 2005. 'Assessing Benefits from Allotments as a Component of Urban Agriculture in England', in L.J.A. Mouguet (ed.), *Agropolis: the Social, Political and Environmental Dimensions of Urban Agriculture*. Earthscan/IDRC. Retrieved 30 October 2005 from http://www.idrc.ca/en/ev-85414- 201-1-DO_TOPIC.html

Purdue, D.A. 2000. 'Backyard Biodiversity: Seed Tribes in the West of England', *Science as Culture* 9(2): 141–166.

RHS (Royal Horticultural Society). 2006. *RHS Horticultural Database*. Accessed August 2006 at http://www.rhs.org.uk/databases/summary.asp

Thrupp, L. 2000. 'Linking Agricultural Biodiversity and Food Security: the Valuable Role of Sustainable Agriculture', *International Affairs* 76(2): 265–281.

Vogl, C.R. and B. Vogl-Lukasser. 2003. 'Tradition, Dynamics and Sustainability of Plant Species Composition and Management in Homegardens on Organic and Non-organic Small Scale Farms in Alpine Eastern Tyrol, Austria', *Biological Agriculture and Horticulture* 21: 349–366.

———, B. Vogl-Lukasser and R.K. Puri. 2004. 'Tools and Methods for Data Collection in Ethnobotanical Studies of Home Gardens', *Field Methods* 16(3): 285–306.

Vogl-Lukasser, B. and C.R. Vogl. 2004. 'Ethnobotanical Research in Home Gardens of Small Farmers in the Alpine Region of Osstirol (Austria): an Example for Bridges Built and Building Bridges', *Ethnobotany Research and Applications* 2: 111–137.

Winklerprins, A.M.G.A. 2002. 'House-lot Gardens in Santarem, Para, Brazil: Linking Rural with Urban', *Urban Ecosystems* 6: 43–65.

Appendix 16.1. Floristic diversity of allotment sites

Family	Species	Common name	% frequency, rural	% frequency, urban	% frequency, both sites	Total number of named cultivars found	Cultivars shared by more than one person	Number of cultivars shared across both sites
Polygonaceae	*Rheum x cultorum* Thorsrud & Reisaeter	Rhubarb	100	100	100			
Solanaceae	*Solanum tuberosum* L.	Potato	100	90	96	29	14	9
Alliaceae	*Allium cepa* L.	Onion and shallot	100	90	96	9	5	4
Fabaceae	*Phaseolus coccineus* L.	Runner bean	88	90	88	12	3	1
Solanaceae	*Lycopersicon esculentum* Mill.	Tomato	88	80	85	16	4	3
Apiaceae	*Daucus carota* L.	Carrot	94	60	81	18	3	1
Rosaceae	*Rubus idaeus* L.	Raspberry	88	60	77	4	1	0
Brassicaceae	*Brassica oleracea* L.	Greens	81	70	77	19	3	3
Cucurbitaceae	*Cucurbita pepo* L.	Marrow and courgette	88	50	73	7	0	0
Chenopodiaceae	*Beta vulgaris* L. subsp. *vulgaris*	Beetroot	75	70	73	5	1	1
Grossulariaceae	*Ribes uva-crispa* L.	Green gooseberry	81	50	69	3	2	2
Apiaceae	*Pastinaca sativa* L.	Parsnip	81	50	69	6	2	1
Asteraceae	*Lactuca sativa* L.	Lettuce	88	40	69	4	1	1
Rosaceae	*Fragaria x ananassa* Duchesne ex Rozier.	Strawberry	81	40	65	2	0	0
Grossulariaceae	*Ribes nigrum* L.	Blackcurrant	69	60	65	2	0	0
Alliaceae	*Allium sativum* L.	Garlic	75	50	65	1	0	0
Fabaceae	*Vicia faba* L.	Broad bean	75	40	62	5	1	1
Alliaceae	*Allium porrum* L.	Leek	75	40	62	2	1	1
Rosaceae	*Rubus fruticosus* L.	Blackberry	50	60	54			

Family	Species	Common name	% frequency, rural	% frequency, urban	% frequency, both sites	Total number of named cultivars found	Cultivars shared by more than one person	Number of cultivars shared across both sites
Fabaceae	*Pisum sativum* L.	Pea	63	40	54	10	3	2
Brassicaceae	*Raphanus sativus* L.	Radish	63	30	50	2	1	1
Poaceae	*Zea mays* L.	Sweetcorn	56	30	46	5	1	0
Grossulariaceae	*Ribes rubrum* L.	Red or white currant	44	40	42			
Cucurbitaceae	*Cucurbita* sp.	Squash and pumpkin	56	20	42	5	1	1
Fabaceae	*Phaseolus vulgaris* L.	French bean	50	20	38	8	0	0
Chenopodiaceae	*Spinacia oleracea* L.	Spinach	50	20	38	3	0	0
Asparagaceae	*Asparagus officinalis* L.	Asparagus	50	10	35			
Lamiaceae	*Mentha* sp.	Mint	44	20	35			
Chenopodiaceae	*Beta vulgaris* subsp. *cicla* (L.) W.D.J. Koch	Spinach, beet or chard	44	10	31	1	0	0
Urticaceae	*Urtica dioica* L.	Nettle patch	13	60	31			
Grossulariaceae	*Ribes* sp.	Red gooseberry	38	10	27	2	0	0
Cucurbitaceae	*Cucumis sativus* L.	Cucumber	38	10	27	5	1	0
Lamiaceae	*Lavandula officinalis* Chaix.	Lavender	31	20	27			
Apiaceae	*Petroselinum crispum* (Mill.) Fuss	Parsley	44	0	27	2		0
Rosaceae	*Prunus x domestica* L.	Plum	19	30	23	2	0	0
Brassicaceae	*Brassica napus* L.	Swede	25	20	23	1	0	0
Asteraceae	*Tagetes* sp.	French marigold	19	30	23			
Rosaceae	*Malus* sp.	Apple	19	20	19	2	2	0
Apiaceae	*Apium graveolens* var. *dulce* Pers.	Celery	19	20	19	1	0	0
Iridaceae	*Gladiolus* sp.	Gladioli	25	10	19			

Family	Species	Common name	% frequency, rural	% frequency, urban	% frequency, both sites	Total number of named cultivars found	Cultivars shared by more than one person	Number of cultivars shared across both sites
Rosaceae	*Crataegus monogyna* Jacq.	Hawthorn	19	10	15			
Rosaceae	*Prunus insititia* L.	Damson plum	6	30	15			
Brassicaceae	*Eruca vesicaria* subsp. *sativa* (Mill.) Thell.	Rocket	25	0	15			
Asteraceae	*Helianthus annuus* L.	Sunflower	19	10	15			
Papaveraceae	*Papaver somniferum* L.	Opium poppy	13	20	15			
Amaryllidaceae	*Narcissus* sp.	Daffodil	19	10	15			
Lamiaceae	*Melissa officinalis* L.	Lemon balm	25	0	15			
Lamiaceae	*Salvia officinalis* L.	Sage	25	0	15			
Lamiaceae	*Thymus* sp.	Thyme	25	0	15			
Ericaceae	*Vaccinium corymbosum* L.	Blueberry	13	10	12			
Alliaceae	*Allium fistulosum* L.	Japanese salad onion, bunching onion	19	0	12	1	0	0
Brassicaceae	*Cheiranthus* sp.	Wallflower	13	10	12			
Lamiaceae	*Origanum* sp.	Origanum	19	0	12			
Lamiaceae	*Rosmarinus officinalis* L.	Rosemary	19	0	12			
Brassicaceae	*Armoracia rusticana* P. Gaertn., B. Mey. & Scherb.	Horseradish	6	20	12			
Boraginaceae	*Symphytum* sp.	Comfrey	13	10	12			
Hippocastanaceae	*Aesculus hippocastastanum* L.	Horsechestnut	13	0	8			
Rosaceae	*Prunus avium* L.	Wild cherry	13	0	8			
Rosaceae	*Rosa* sp.	Rose	13	0	8			
Vitaceae	*Vitis vinifera* L.	Grape	6	10	8	2	0	0

Family	Species	Common name	% frequency, rural	% frequency, urban	% frequency, both sites	Total number of named cultivars found	Cultivars shared by more than one person	Number of cultivars shared across both sites
Rosaceae	*Rubus* 'Tayberry' Group	Tayberry	13	0	8	2	0	0
Rosaceae	*Rubus loganobaccus* L.H. Bailey	Logan berry	13	0	8			
Asteraceae	*Helianthus tuberosus* L.	Jerusalem artichoke	0	20	8			
Apiaceae	*Apium graveolens* var. *rapaceum* DC.	Celeriac	13	0	8			
Asteraceae	*Cynara scolymus* L.	Globe artichoke	6	10	8			
Ranunculaceae	*Aquilegia* sp.	Aquilegia	6	10	8			
Asteraceae	*Calendula officinalis* L.	Pot marigold	0	20	8			
Fabaceae	*Lathyrus odoratus* L.	Annual sweet pea	6	10	8			
Hyacinthaceae	*Hyacinthus* sp.	Hyacinth	13	0	8			
Caryophyllaceae	*Dianthus barbatus L..*	Sweet william	6	10	8			
Asteraceae	*Chrysanthemum* sp.	Chrysanthemum	13	0	8			
Primulaceae	*Primula* sp.	Polyanthus	6	10	8			
Apiaceae	*Foeniculum vulgare* Mill	Fennel	0	20	8			
Alliaceae	*Allium schoenoprasum* L.	Chives	13	0	8			
Apiaceae	*Anethum graveolens* L.	Dill	13	0	8			
Moraceae	*Ficus carica* L.	Fig	6	0	4			
Fabaceae	*Ulex europaeus* L.	Gorse	6	0	4			
	Unknown tree	Tree unknown	6	0	4			
Salicaceae	*Populus* sp.	Variegated poplar	6	0	4			
Aceraceae	*Acer pseudoplatanus* L.	Sycamore	0	10	4			
Pinaceae	*Pinus* sp.	Pine tree	6	0	4			
Juglandaceae	*Juglans regia* L.	Walnut tree	6	0	4			

Family	Species	Common name	% frequency, rural	% frequency, urban	% frequency, both sites	Total number of named cultivars found	Cultivars shared by more than one person	Number of cultivars shared across both sites
Fagaceae	*Quercus rubra* L.	Oak	6	0	4			
Salicaceae	*Salix caprea* L.	Goat willow	6	0	4			
Caprifoliaceae	*Sambucus nigra* L.	Elderflower	6	0	4			
	Unknown shrub	Shrub unknown	6	0	4			
Corylaceae	*Corylus avellana* L.	Hazel	6	0	4			
Magnoliaceae	*Magnolia stellata* L.	Magnolia	6	0	4			
Hydrangeaceae	*Hydrangea macrophylla* (Thunb.) Ser.	Hydrangea	6	0	4			
Escalloniaceae	*Escallonia rubra* (Ruiz & Pav.) Pers.		6	0	4			
Oleaceae	*Syringa vulgaris* L.	Lilac tree	0	10	4			
Salicaceae	*Salix tortuosa* Host.	Contorted salix	0	10	4			
Rosaceae	*Pyrus communis* L.	Pear	6	0	4			
Rosaceae	*Prunus persica* var. *nucipersica* (Suckow) C.K. Schneid.	Nectarine	6	0	4			
Rosaceae	*Chaenomeles japonica* (Thunb.) Lindl.	Quince1	6	0	4			
Cannabaceae	*Humulus lupulus* L.	Hop	6	0	4			
Rosaceae	*Rubus* sp.	Locumberry	6	0	4			
Grossulariaceae	*Ribes divaricatum* Douglas	Worcesterberry	0	10	4			
Grossulariaceae	*Ribes x culverwellii* Macfarl.	Jostaberry	0	10	4			
Asteraceae	*Arctium lappa* L.	Japanese burdock	0	10	4			
Fabaceae	*Cicer arietinum* L.	Chickpea	6	0	4			
Cucurbitaceae	*Cucumis melo* var. *inodorus* H. Jacq.	Honeydew melon	0	10	4			
Brassicaceae	*Crambe maritima* L.	Seakale	6	0	4			

Family	Species	Common name	% frequency, rural	% frequency, urban	% frequency, both sites	Total number of named cultivars found	Cultivars shared by more than one person	Number of cultivars shared across both sites
Brassicaceae	*Brassica chinensis* L.	Pak choi	6	0	4			
Brassicaceae	*Brassica rapa* L.	Turnip	6	0	4	1	0	0
Solanaceae	*Solanum melongena* L.	Aubergine	0	10	4			
Solanaceae	*Physalis ixocarpa* Hornem.	Tomatillo	6	0	4			
Asteraceae	*Scorzonera hispanica* L.	Scorzonera	6	0	4	2	0	0
Apiaceae	*Foeniculum vulgare* var. *azoricum* (Mill.) Thell.	Bulb fennel	6	0	4			
Asteraceae	*Cynara cardunculus* L.	Cardoon	0	10	4			
Amaryllidaceae	*Amaryllis* sp.	Amaryllis	6	0	4			
Alliaceae	*Allium* sp.	Ornamental allium	6	0	4			
Fabaceae	*Lathyrus latifolius* L.	Everlasting sweet-pea	6	0	4			
Onagraceae	*Fuchsia magellanica* Lam.	Fuchsia	0	10	4			
Boraginaceae	*Myosotis* sp.	Forget-me-not	0	10	4			
Brassicaceae	*Malcolmia maritima* (L.) R. Br. in W.T. Aiton	Virginia stock	0	10	4			
Violaceae	*Viola* sp.	Viola	0	10	4			
Malvaceae	*Alcea rosea* L.	Hollyhock	6	0	4			
Asteraceae	*Dahlia* sp.	Dahlia	6	0	4			
Asteraceae	*Leucanthemum vulgare* Lam.	Ox-eye daisy	6	0	4			
Convallariaceae	*Polygonatum x hybridum* Brugger.	Solomon's seal	0	10	4			
Fabaceae	*Lupinus* sp.	Lupin	6	0	4			
Oxalidaceae	*Oxalis* sp.	Oxalis	6	0	4			
Ranunculaceae	*Nigella damascena* L.	Nigella	0	10	4			

Family	Species	Common name	% frequency, rural	% frequency, urban	% frequency, both sites	Total number of named cultivars found	Cultivars shared by more than one person	Number of cultivars shared across both sites
Scrophulariaceae	*Digitalis purpurea* L.	Foxglove	0	10	4			
Scrophulariaceae	*Hebe* sp.	Hebe	0	10	4			
Apiaceae	*Angelica archangelica* L.	Angelica	6	0	4			
Lamiaceae	*Hyssopus officinalis* L.	Hyssop	6	0	4			
Polygonaceae	*Rumex acetosa* L.	Sorrel	6	0	4			
Fabaceae	*Medicago sativa* L.	Alfalfa (green manure)	0	10	4			

Notes on Contributors

Editors

Manuel Pardo-de-Santayana is a Senior Lecturer in Botany and Ethnobotany at the Autonomous University of Madrid. He has been researching contemporary and historical uses of Spanish medicinal and wild food plants, homegardens and folk botanical taxonomies at the Royal Botanical Garden of Madrid (CSIC) and the School of Pharmacy (University of London) since 1995. He supervises students conducting ethnobotanical projects in the Autonomous University of Madrid, where he received a Ph.D. in Biology.

Andrea Pieroni is Associate Professor of Plant Biology and Ethnobotany at the University of Gastronomic Sciences of Bra, Northern Italy. He is the Editor-in-Chief of the Journal of Ethnobiology and Ethnomedicine and the President-Elect of the International Society of Ethnobiology. His research focuses on food and medical ethnobotany in the Mediterranean and the Balkan areas (with a special interest in ethnic diasporas); ethnobiology and trans-cultural health/diet studies among migrant communities in Europe.

Rajindra K. Puri is a Senior Lecturer in Environmental Anthropology and Ethnobiology at the University of Kent at Canterbury, U.K. While his primary research area is Indonesian Borneo, he also supervises M.Sc. and Ph.D. students who conduct ethnobotanical research in Europe. He co-organizes with Professor Christian Vogl a summer field school for methods in ethnobotany in the Austrian Alps and works with the Global Diversity Foundation in Morocco, Malaysia and Namibia.

Authors

Torbjørn Alm is Associate Professor of Botany and Head Curator of Phanerogams at Tromsø Museum, University of Tromsø, Norway, with ethnobotany as one of his main interests – in particular the many and varied uses of plants by the Norwegian, Sami and Finnish ethnic groups of northern Norway.

Shamila Ayub is a pharmacist based in West Yorkshire, U.K.

Maria José Barão holds a B.Sc. in Biology and is a superior lab technician of the Department of Landscape, Environment and Planning, Universidade de Évora (Portugal). She gives support to different kinds of laboratories of this department, such as, Aquatic Ecology, Terrestrial Ecology and Biogeochemistry. She is also the chairperson of a Teachers' Association of Alentejo which deals with education on artistic, cultural and environmental issues.

Hugo J. de Boer is a Ph.D. candidate and teacher of Ethnobotany at Uppsala University in Sweden, and does diverse research on ethnobotany, biocultural diversity, herbal pharmacovigilance and molecular barcoding of medicinal plants. His main geographical areas of research are Laos, Morocco and Bulgaria.

Ana Maria Carvalho is a Professor in the Department of Biology of the Escola Superior Agrária, in the Polytechnic Institute of Bragança, Portugal, and is researching agroecosystems and ethnobotany in the northeastern Portuguese region of Trás-os-Montes.

Anja Christanell received her Ph.D. in social anthropology from the University of Vienna for her work on farmers' local knowledge of weather in two Austrian regions. She has also worked on several ethnobotanical research projects in Austria, at the University of Natural Resources and Applied Life Sciences Vienna (BOKU). Since 2007 she has been working at the Austrian Institute for Sustainable Development in Vienna on research topics regarding Energy Use and Poverty, Corporate Social Responsibility and Sustainable Consumption.

Bernd Gliwa is a researcher in linguistics. Currently he is finishing his thesis on Lithuanian plant names at Cracow University, Poland.

Marianne Guetler is a MSc in Botany, University of Vienna, specialised in vegetation ecology. Her main fields of interest are in man's interaction with plants, nature and landscape. She is currently working free-lance in nature conservation.

Michael Heinrich holds Masters degrees in biology (Univ. Freiburg, Germany) and sociocultural and medical anthropology (Wayne State Univ., Michigan U.S.A.). His Ph.D. focused on the ethnopharmacology of the Mixe in Oaxaca, Mexico and on bioactive constituents from selected species. In the last twenty years he has led research at the interface of pharmaceutical biology/pharmacognosy and anthropology with a special interest in medicinal and food plants used in Mexico and in selected regions of the Mediterranean.

Marianne Iversen holds a Ph.D. at University of Tromsø, , Norway, working with plant–reindeer interactions in the summer pastures in northernmost Norway. The Ph.D. was initiated by Sámi allaskuvla/Sami College in Guovdageaidnu/Kautokeino, Norway.

Jenny L. McCune is interested in the interactions between human culture, history and ecological communities. She received a M.Sc. in Ethnobotany from the University of Kent at Canterbury in 2003, and is currently a PhD candidate in the Department of Botany at the University of British Columbia, Canada.

Ramón Morales is a researcher at the Real Jardín Botánico, Madrid (CSIC). He has been working for thirty years on the systematic botany of the Labiate family and Iberian ethnobotany. His main research focus is on aromatic and medicinal plants traditionally used in Spain. He has been a collaborator in the Flora Iberica Project as author and editor since 1993.

Sabine Nebel has a background in pharmacy (ETH Zurich), an M.Sc. in Ethnobotany (University of Kent at Canterbury) and a Ph.D. from the School of Pharmacy in London on wild food plants used in Graecanic communities in Calabria, Southern Italy. Her research interests include ethnopharmacology and European ethnobotany.

Daiva Šeškauskaitė is a musician and researcher in ethnomusicology and ethnobotany. She lectures on cultural anthropology and ethnobotany at Kaunas College of Forestry and the University of Agriculture in Lithuania.

Alexandra Soveral Dias is a researcher in chemical ecology and ethnobotany, especially wild plant foods, herbs and spices, and lectures on general botany, plant physiology and ethnobotany at Dep. de Biologia, Universidade de Évora (Portugal). She is also coordinating an interdisciplinary group working on natural dyes.

Timothy J. Tabone is a freelance researcher of botany and ethnobotany. He has been documenting the ethnobotany of the Maltese Islands since 1996, and expects to publish a monograph on the subject in the near future.

Javier Tardío is a researcher in genetic resources and food ethnobotany in IMIDRA (Madrid Institute for Agricultural and Food Research), Alcalá de Henares, Madrid, Spain. He has been doing ethnobotanical research on wild food plants in Spain, especially in central Spain, for about ten years.

Bren Torry is a Senior Lecturer at the School of Pharmacy, University of Bradford. Her current research interests include health psychology and cultural diversity within veterinary medicine, and transcultural health and the elderly.

Veerle Van den Eynden is specialized in studying how rural communities use and manage plant resources and how this impacts on plant conservation. She has worked in Namibia, Senegal, Ecuador, Trinidad and Scotland. She is currently a Visiting Lecturer in Ethnobotany at the University of the Highlands and Islands in Scotland and a data sharing specialist for the rural economy and land use programme in the UK.

Brigitte Vogl-Lukasser is an ethnobotanist with research experience on Maya homegardens, Alpine homegardens, in-situ conservation of traditional varieties of vegetable species and ethnoveterinary medicine. She is currently a Postdoctoral Research Assistant and Lecturer at the University of Natural Resources and Applied Life Sciences Vienna (BOKU).

Christian R. Vogl is Associate Professor of Organic Farming at the University of Natural Resources and Applied Life Sciences Vienna (BOKU). He has been lecturing and conducting research on local knowledge, organic farming and traditional land-use systems for about twenty years. He currently supervises M.Sc. and Ph.D. students and other research projects and is involved with several scientific communities and scientific journals to support research on local knowledge.

Christine Wildhaber completed an M.Sc. in Ethnobotany at the University of Kent at Canterbury in 2006. Particular interests include food/medicinal plants and the use of plants in material culture. She currently works for an inner-city London local authority, facilitating community involvement in parks and open spaces.

Hadar Zaman is a pharmacist based in Bradford, U.K.

Index

H

U

www.ingramcontent.com/pod-product-compliance
Lightning Source LLC
Chambersburg PA
CBHW060020030426
42334CB00019B/2112

* 9 7 8 1 7 8 2 3 8 1 2 4 2 *